AF361799

The Benefits of Plant Extracts for Human Health

The Benefits of Plant Extracts for Human Health

Editor

Charalampos Proestos

MDPI • Basel • Beijing • Wuhan • Barcelona • Belgrade • Manchester • Tokyo • Cluj • Tianjin

Editor
Charalampos Proestos
National and Kapodistrian University of Athens
Greece

Editorial Office
MDPI
St. Alban-Anlage 66
4052 Basel, Switzerland

This is a reprint of articles from the Special Issue published online in the open access journal *Foods* (ISSN 2304-8158) (available at: https://www.mdpi.com/journal/foods/special_issues/benefit_plant_extract_health).

For citation purposes, cite each article independently as indicated on the article page online and as indicated below:

LastName, A.A.; LastName, B.B.; LastName, C.C. Article Title. *Journal Name* **Year**, *Volume Number*, Page Range.

ISBN 978-3-03943-851-8 (Hbk)
ISBN 978-3-03943-852-5 (PDF)

Contents

About the Editor

Charalampos Proestos (Associate Professor in Food Chemistry) Charalampos Proestos has a BSc in Chemistry, University of Ioannina, Greece, and an MSc in Food Science at Reading University, U.K. He obtained his Ph.D. in Food Chemistry at the Agricultural University of Athens (AUA), Greece, where he continued his post-doc working on natural antioxidants on programs funded by the EU and Greece. After further training at Wageningen University (the Netherlands), he worked as a Research Associate at AUA. He worked as a Chemist for the Hellenic Food Authority (EFET as a food industry auditor and supervisor of the Chemical Laboratory in Athens accredited with ISO 17025. He is currently Associate Professor at the Department of Chemistry, National and Kapodistrian University of Athens, and director of the laboratory of Food Chemistry. He has published more than 70 papers in reputed journals and has been serving as an editorial board member of more than 10 reputed journals. He is Member of the European Committee of the Division of Food Chemistry, European Association of Chemical and Molecular Sciences (EuChemS). His research field focuses on food antioxidants, plant bioactive compounds, foodomics, and food contaminants.

Editorial

The Benefits of Plant Extracts for Human Health

Charalampos Proestos

Laboratory of Food Chemistry, Department of Chemistry, National and Kapodistrian University of Athens, 15771 Athens, Greece; harpro@chem.uoa.gr; Tel.: +30-210-727-4160

Received: 5 November 2020; Accepted: 10 November 2020; Published: 12 November 2020

Nature has always been, and still is, a source of foods and ingredients that are beneficial to human health. Nowadays, plant extracts are increasingly becoming important additives in the food industry due to their content in bioactive compounds such as polyphenols [1] and carotenoids [2], which have antimicrobial and antioxidant activity, especially against low-density lipoprotein (LDL) and deoxyribonucleic acid (DNA) oxidative changes [3]. The aforementioned compounds also delay the development of off-flavors and improve the shelf life and color stability of food products. Due to their natural origin, they are excellent candidates to replace synthetic compounds, which are generally considered to have toxicological and carcinogenic effects. The efficient extraction of these compounds from their natural sources and the determination of their activity in commercialized products have been great challenges for researchers and food chain contributors to develop products with positive effects on human health. The objective of this Special Issue is to highlight the existing evidence regarding the various potential benefits of the consumption of plant extracts and plant extract-based products, along with essential oils that are derived from plants also and emphasize in vivo works and epidemiological studies, application of plant extracts to improve shelf-life, the nutritional and health-related properties of foods, and the extraction techniques that can be used to obtain bioactive compounds from plant extracts.

In this context, Concha-Meyer et al. [4] studied the bioactive compounds of tomato pomace obtained by ultrasound assisted extraction. In this review, it was presented that the functional extract obtained by ultrasounds had antithrombotic properties, such as platelet anti-aggregant activity compared with commercial cardioprotective products. Turrini et al. [5] introduced bud-derivatives from eight different plant species as a new category of botanicals containing polyphenols and studied how different extraction processes can affect their composition. Woody vine plants from Kadsura spp. belonging to the Schisandraceae family produce edible red fruits that are rich in nutrients and antioxidant compounds such as flavonoids. Extracts from these plants had antioxidant properties and had shown also key enzyme inhibitions [6]. Hence, fruit parts other than the edible mesocarp could be utilized for future food applications using Kadsura spp. rather than these being wasted. Saji et al. [7] studied the possible use of rice bran, a by-product generated during the rice milling process, normally used in animal feed or discarded due to its rancidity, for its phenolic content. It was proved that rice bran phenolic extracts via their metal chelating properties and free radical scavenging activity, target pathways of oxidative stress and inflammation resulting in the alleviation of vascular inflammatory mediators. Villedieu-Percheron et al. [8] evaluated three natural diterpenes compounds extracted and isolated from *Andrographis paniculata* medicinal herb as possible inhibitors of NFκB (nuclear factor kappa-light-chain-enhancer of activated B cells) transcriptional activity of pure analogues. Yeon et al. [9] evaluated the antioxidant activity, the angiotensin I-converting enzyme (ACE) inhibition effect, and the α-amylase and α-glucosidase inhibition activities of hot pepper water extracts both before and after their fermentation. These water extracts were proved to have potentially inhibitory effects against both hyperglycemia and hypertension. The hydrolyzed extracts of Ziziphus jujube fruit, commonly called jujube, were examined for their protective effect against lung inflammation in mice [10]. They contained significant amounts of flavonoids which inhibited cytokine release from macrophages and promoted antioxidant defenses in vivo. Tran at al. [11] examined the antidiabetic activity of spray-dried Euphorbia hirta L. herb extracts containing high concentrations of bioactive compounds such as phenolics and flavonoids. Li et al. [12] reported that intestinal microbiota is closely associated with the initiation and progression of diabetes mellitus and reviewed bioactive components which exhibited anti-diabetic activity by modulating these intestinal microbiotas. Essential oils have promising activity against antibiotic-resistant bacteria and chemotherapeutic-resistant tumors. This was supported by the study of Viktorová et al. [13] where lemongrass essential oil and especially citral, the dominant component, proved to have potential antimicrobial and anticancer activity. Additionally, Mitropoulou et al. [14] investigated the antimicrobial potential of *Sideritis raeseri subps. raeseri* essential oil against common food spoilage and pathogenic microorganisms and evaluated its antioxidant and antiproliferative activity. Salehi et al. [15] reviewed the Berberis plants, which contain alkaloids, tannins, phenolic compounds and essential oils, and their possible use in the food and pharmaceutical industry. Last but not least, Kiokias et al. [16] reviewed the naturally occurring phenolic acids from plants and their antioxidant activities in o/w emulsions and in vitro lipid-based model systems.

Still more research is needed to explore more and in depth the health beneficial effects of plant extracts, since nature certainly has more to give to humans.

Funding: This research received no external funding.

Conflicts of Interest: The author declares no conflict of interest.

References

1. Proestos, C.; Varzakas, T. Aromatic Plants: Antioxidant Capacity and Polyphenol Characterisation. *Foods* **2017**, *6*, 28. [CrossRef] [PubMed]
2. Langi, P.; Kiokias, S.; Varzakas, T.; Proestos, C. Carotenoids: From Plants to Food and Feed Industries. In *Microbial Carotenoids, Methods and Protocols*, 1st ed.; Barreiro, C., Barredo, J.L., Eds.; Humana Press: New York, NY, USA, 2018; Volume 1852, pp. 57–71.
3. Kiokias, S.; Proestos, C.; Oreopoulou, V. Effect of Natural Food Antioxidants against LDL and DNA Oxidative Changes. *Antioxidants* **2018**, *7*, 133. [CrossRef] [PubMed]
4. Concha-Meyer, A.; Palomo, I.; Plaza, A.; Gadioli Tarone, A.; Junior, M.R.M.; Sáyago-Ayerdi, S.G.; Fuentes, E. Platelet Anti-Aggregant Activity and Bioactive Compounds of Ultrasound-Assisted Extracts from Whole and Seedless Tomato Pomace. *Foods* **2020**, *9*, 1564. [CrossRef] [PubMed]
5. Turrini, F.; Donno, D.; Beccaro, G.L.; Pittaluga, A.; Grilli, M.; Zunin, P.; Boggia, R. Bud-Derivatives, a Novel Source of Polyphenols and How Different Extraction Processes Affect Their Composition. *Foods* **2020**, *9*, 1343. [CrossRef] [PubMed]
6. Sritalahareuthai, V.; Temviriyanukul, P.; On-nom, N.; Charoenkiatkul, S.; Suttisansanee, U. Phenolic Profiles, Antioxidant, and Inhibitory Activities of *Kadsura heteroclita* (Roxb.) Craib and *Kadsura coccinea* (Lem.) A.C. Sm. *Foods* **2020**, *9*, 1222.
7. Saji, N.; Francis, N.; Schwarz, L.J.; Blanchard, C.L.; Santhakumar, A.B. The Antioxidant and Anti-Inflammatory Properties of Rice Bran Phenolic Extracts. *Foods* **2020**, *9*, 829. [CrossRef] [PubMed]
8. Villedieu-Percheron, E.; Ferreira, V.; Campos, J.F.; Destandau, E.; Pichon, C.; Berteina-Raboin, S. Quantitative Determination of Andrographolide and Related Compounds in *Andrographis paniculata* Extracts and Biological Evaluation of Their Anti-Inflammatory Activity. *Foods* **2019**, *8*, 683. [CrossRef] [PubMed]
9. Yeon, S.-J.; Kim, J.-H.; Cho, W.-Y.; Kim, S.-K.; Seo, H.G.; Lee, C.-H. In Vitro Studies of Fermented Korean Chung-Yang Hot Pepper Phenolics as Inhibitors of Key Enzymes Relevant to Hypertension and Diabetes. *Foods* **2019**, *8*, 498. [CrossRef] [PubMed]
10. Kim, Y.; Oh, J.; Jang, C.H.; Lim, J.S.; Lee, J.S.; Kim, J.-S. In Vivo Anti-Inflammatory Potential of Viscozyme®-Treated Jujube Fruit. *Foods* **2020**, *9*, 1033. [CrossRef] [PubMed]
11. Tran, N.; Tran, M.; Truong, H.; Le, L. Spray-Drying Microencapsulation of High Concentration of Bioactive Compounds Fragments from *Euphorbia hirta* L. Extract and Their Effect on Diabetes Mellitus. *Foods* **2020**, *9*, 881. [CrossRef]

12. Li, B.-Y.; Xu, X.-Y.; Gan, R.-Y.; Sun, Q.-C.; Meng, J.-M.; Shang, A.; Mao, Q.-Q.; Li, H.-B. Targeting Gut Microbiota for the Prevention and Management of Diabetes Mellitus by Dietary Natural Products. *Foods* **2019**, *8*, 440. [CrossRef]
13. Viktorová, J.; Stupák, M.; Řehořová, K.; Dobiasová, S.; Hoang, L.; Hajšlová, J.; Van Thanh, T.; Van Tri, L.; Van Tuan, N.; Ruml, T. Lemon Grass Essential Oil does not Modulate Cancer Cells Multidrug Resistance by Citral—Its Dominant and Strongly Antimicrobial Compound. *Foods* **2020**, *9*, 585. [CrossRef]
14. Mitropoulou, G.; Sidira, M.; Skitsa, M.; Tsochantaridis, I.; Pappa, A.; Dimtsoudis, C.; Proestos, C.; Kourkoutas, Y. Assessment of the Antimicrobial, Antioxidant, and Antiproliferative Potential of *Sideritis raeseri* subps. *raeseri* Essential Oil. *Foods* **2020**, *9*, 860. [CrossRef] [PubMed]
15. Salehi, B.; Selamoglu, Z.; Sener, B.; Kilic, M.; Kumar Jugran, A.; de Tommasi, N.; Sinisgalli, C.; Milella, L.; Rajkovic, J.; Flaviana, B.; et al. *Berberis* Plants—Drifting from Farm to Food Applications, Phytotherapy, and Phytopharmacology. *Foods* **2019**, *8*, 522. [CrossRef] [PubMed]
16. Kiokias, S.; Proestos, C.; Oreopoulou, V. Phenolic Acids of Plant Origin—A Review on Their Antioxidant Activity In Vitro (O/W Emulsion Systems) Along with Their in vivo Health Biochemical Properties. *Foods* **2020**, *9*, 534. [CrossRef] [PubMed]

Publisher's Note: MDPI stays neutral with regard to jurisdictional claims in published maps and institutional affiliations.

Article

Platelet Anti-Aggregant Activity and Bioactive Compounds of Ultrasound-Assisted Extracts from Whole and Seedless Tomato Pomace

Anibal Concha-Meyer [1,2], Iván Palomo [3,*], Andrea Plaza [2], Adriana Gadioli Tarone [4], Mário Roberto Maróstica Junior [4], Sonia G. Sáyago-Ayerdi [5] and Eduardo Fuentes [3,*]

1   Facultad de Ciencias Agrarias, Universidad de Talca, Talca 3460000, Chile; anibal.concha@utalca.cl
2   Centro de Estudios en Alimentos Procesados (CEAP), CONICYT-Regional, Gore Maule, R09I2001, Talca 3460000, Chile; aplaza@ceap.cl
3   Thrombosis Research Center, Medical Technology School, Department of Clinical Biochemistry and Immunohaematology, Faculty of Health Sciences, Universidad de Talca, Talca 3460000, Chile
4   LANUM (Laboratory of Nutrition and Metabolism), FEA (School of Food Engineering), UNICAMP (University of Campinas), Rua Monteiro Lobato, 80, Campinas 13083-862, Brazil; dricagt@gmail.com (A.G.T.); mmarosti@unicamp.br (M.R.M.J.)
5   Tecnologico Nacional de Mexico, Instituto Tecnologico de Tepic, Av Tecnológico 2595, Col Lagos del Country, Tepic 63175, Nayarit Mexico, Mexico; ssayago@ittepic.edu.mx
*   Correspondence: ipalomo@utalca.cl (I.P.); edfuentes@utalca.cl (E.F.)

Received: 23 September 2020; Accepted: 21 October 2020; Published: 28 October 2020

Abstract: Tomato paste production generates a residue known as tomato pomace, which corresponds to peels and seeds separated during tomato processing. Currently, there is an opportunity to use tomato pomace to obtain a functional extract with antithrombotic properties, such as platelet anti-aggregant activity. The aim of this study was to evaluate the yield and inhibitory activity of different extracts of tomato pomace on in vitro platelet aggregation, comparing this activity with commercial cardioprotective products, and quantify bioactive compounds. Aqueous or ethanolic/water (1:1) extracts of whole tomato pomace, seedless tomato pomace, tomato pomace supplemented with seeds (50% and 20%), and only seeds were obtained with different ultrasound-assisted extraction times. The inhibition of platelet aggregation was evaluated using a lumi-aggregometer. The quantification of bioactive compounds was determined by HPLC-MS. From 5 g of each type of tomato pomace sample, 0.023–0.22 g of a dry extract was obtained for the platelet aggregation assay. The time of sonication and extraction solvent had a significant role in platelet anti-aggregant activity of some extracts respect the control. Thus, the most active extracts decreased adenosine diphosphate (ADP)-induced platelet aggregation from 87 ± 6% (control) to values between 26 ± 6% and 34 ± 2% ($p < 0.05$). Furthermore, different ultrasound-assisted extraction conditions of tomato pomace fractions had varied concentration of flavonoids and nucleosides, and had an effect on extract yield.

Keywords: tomato pomace; extraction; platelet; ultrasound; functional ingredient

---

1. Introduction

Cardiovascular disease (CVD) is the number one cause of death worldwide since more people die annually ($\approx$17 million) from CVD than from any other cause [1]. In Chile, CVD causes 27% of deaths [2]. In this context, the adherence to a diet rich in fruit and vegetables was associated with a decrease of all-cause mortality among individuals with CVD [3–5]. Furthermore, tomato (*Lycopersicon esculentum*) has been one of the most powerful vegetables associated with the reduction of this type of disease [6–10]. Emerging epidemiological and interventional data support the connection between higher tomato consumption and a lower risk of CVD. Additionally, tomato has the lowest uric acid content of any fruit and vegetable [11,12]. The latter is important because a high uric acid level in the body can cause various health issues like arthritis [13].

Tomato is an important vegetable grown in many countries across the world for fresh markets and multiple processed forms. The worldwide production of processed tomato in 2017 was 38 million tons [14]. During the season 2017–2018, Chile produced 918,000 metric tons of industrial tomato and the Maule region represented about 66% of this total [15]. This crop is industrially processed to produce concentrated tomato paste, which is used to formulate products such as ketchup, sauce, puree, and juice. This processing generates a byproduct named tomato pomace, corresponding mostly to peels and seeds that account for 3–5% (wet basis (w.b.)) of fresh tomato [16]. Currently, the tomato pomace has different uses, such as a protein supplement for growing lambs [17], ruminant feeding [18], chemical and nutritional supplementation of crackers [19], and sustainable fabrication of packaging films [20]. Furthermore, if the latter is not possible, this residue is disposed of on agricultural lands causing environmental contamination [21,22]. The results of the proximate composition of tomato pomace, on a dry weight basis, showed a relevant nutritional content of total protein (19.40%, N × 6.25), total fat (12.33%), available carbohydrates (17.15%), total dietary fiber (47.80%), ash (3.32%), sodium (13.20 mg/100 g), and sugars (9.54%) [23]. Meanwhile, tomato seed oil contains high levels of linoleic (54%) and oleic (22%) acids [24]. On the other hand, the main polysaccharides identified in tomato pomace correspond to glucose (40.5 ± 1.2%) and fructose (22.6 ± 2.1%) from total carbohydrates of 382.9 ± 10.3 mg/g dry weight [25]. Tomato pomace is also known as a source of bioactive compounds, such as carotenoids, phenolic compounds, and dietary fiber [26]. Considering the above, there is an opportunity to use tomato pomace to obtain a functional product with antithrombotic properties, such as platelet anti-aggregant activity, that could be useful as an ingredient in healthy foods for CVD prevention [5].

Different extraction procedures have been evaluated to increase the efficiency of tomato pomace compound extraction for pectin, dietary fiber, and carotenoids such as lycopene, including varied solvent extraction by mixing and heating, microwave-assisted extraction, enzymatic extraction, ultrasound-assisted extraction, and ultrasound-assisted treatment combined with subcritical water [27–30]. Ultrasound is a non-thermal technology that has shown to be particularly effective for improving the extraction of heat-labile compounds [31]. Ultrasound equipment is also commonly available in most analytic laboratories and is used to improve the efficiency of any extraction solvent, thus reducing the extraction times [32]. Furthermore, there is a need to improve the extraction of bioactive compounds, such as flavonoids and nucleosides (adenosine and guanosine), that are present in tomato (*Solanum lycopersicum*) and tomato pomace, which have significant platelet anti-aggregant activity [5,6,33,34]. This work aimed to evaluate the chemical profile and platelet anti-aggregant activity of ultrasound-assisted extracts of different tomato pomace fractions obtained using water or ethanol/water (1:1) as the solvent.

2. Materials and Methods

2.1. Chemicals

The solvents used were ethanol (Sigma-Aldrich, St. Louis, MO, USA) and type I ultrapure water (18.2 MΩ-cm), which was supplied by the Purelab Classic Elga water system (Labwater/VWS Ltd., London, UK). Adenosine diphosphate (ADP) was used as an agonist in the platelet aggregation assay (Chrono-log, Havertown, PA, USA).

2.2. Plant Material

Tomato pomace (mainly peels and seeds), which is a byproduct of the industrial production of tomato paste, was obtained from the company Sugal Chile (Talca, Chile) during the 2016 season. Given the production line of the plant, it was not possible to define specific fruit hybrids that corresponded to the tomato pomace; and it was only possible to identify middle (Sun6366, AB3, and HMX7883) and late (H9665, H7709, and H9997) hybrid tomatoes [35]. Tomato pomace was placed in trays and dried for 48 h at 60 °C in a convection oven (VHC-1A, Ventus Corp., Santiago, Chile); after this, a portion of tomato pomace (22% of seeds) was manually separated into seeds and peels for the preparation of the different extracts; and finally, all parts from tomato pomace were milled and sieved through a 425 μm mesh (No. 40) (TRF 300, TRAPP, Jaraguá do Sul, Brazil).

2.3. Preparation of Extracts

Aqueous or ethanolic/water (1:1) extracts of whole tomato pomace, seedless tomato pomace, tomato pomace supplemented with seeds (50% and 20%), and only seeds were obtained. Briefly, 5 g of each fraction was mixed with 50 mL of water for aqueous extract, and with 25 mL of water and 25 mL of ethanol for ethanolic extracts. An ultrasonic bath of 2.8 L, 220 V, amplitude 100%, and power 90 W (VWR 97043-962, Leicestershire, UK), was used with different sonication time cycles were applied as the extraction process: cycle 1 with one cycle of 20 min, cycle 3 with three cycles of 20 min each with a 10-min pause in between, and cycle 6 with six cycles of 20 min each with a 10-min pause in between. All cycles were controlled each 1 min at a frequency of 35 kHz and a temperature of 45 °C. Then, the extracts were centrifuged at 725× g (International Equipment Company IEC, Centra MP4R, Boston, MA, USA) for 5 min to obtain the supernatant, which was frozen at −86 °C for 48 h, then freeze-dried (Operon, FDU 7024, Gimpo, Korea) for 20 h with a cold trap (−70 °C), and stored at 20 °C in vacuum airtight packaging in a dark environment. Freeze-dried extracts were resuspended in physiological saline solution at a concentration of 1 mg/mL and filtered (pore filter 0.22 μm) for analysis and platelet aggregation assay.

2.4. Platelet Aggregation Assay

The inhibition of platelet aggregation was evaluated in a lumi-aggregometer (Chrono-Log) [5]. Thus, 480 μL of plasma rich in human platelets (2×10^5 platelets/μL) was added to the reaction cuvette that was pre-incubated with 20 μL of extract (all extracts at a concentration of 1 mg/mL) or control for maximum platelet aggregation (0.9% saline). After 5 min of incubation, 20 μL of agonist (ADP, 4 μM) was added to initiate platelet aggregation, which was measured for 6 min. The results were expressed as a percentage of platelet aggregation. Platelets were obtained from 20 different donors and were used during a period of 2 h after blood extraction. All volunteers signed informed consent. The maximum platelet aggregation of the controls (without extracts) was 87 ± 6%. This study was conducted following the Declaration of Helsinki and informed consent was obtained for experimentation with human subjects.

2.5. Identification and Quantification of Bioactive Compounds by HPLC-MS

2.5.1. Phenolic Compounds

Phenolic compounds were analyzed following the method of Torres et al., [36] with modifications. Briefly, 350 mg samples of each freeze-dried extract were weighed, ground, and mixed with 5 mL of 75% methanol and 100 μL of internal standard (naringenin, Sigma-Aldrich). Samples were centrifuged at 11,180× g at 10 °C for 10 min (International Equipment Company IEC, Centra MP4R, Boston, MA, USA). The supernatant was removed and diluted with 20 mL HPLC-grade water (Merck, Darmstadt, Germany). Simultaneously, extraction column C18 (No. 12102052, Agilent Technologies, Palo Alto, CA, USA) was activated with 1 mL of 75% methanol. Briefly, 5 mL HPLC-grade water was added to the C18 column and allowed to drain until dry to remove solvent residues. Samples were eluted using a vacuum pump (Welch 2545C/02, Mount Prospect, IL, USA) and a vacuum trap (20–40 kPa). Then, samples inside C18 were allowed to elute until completely drained using 2 mL of pure propanol to remove concentrated sample-specific phenolic compounds and collected in 4 mL test tubes. N_2 gas was used to dry samples that were reconstituted in 200 μL of pure methanol, mixed, and sonicated (Bandelin Sonorex TK52H, Berlin, Germany) for 1 min. Samples were analyzed using the ultra-high performance liquid chromatography (UHPLC) Dionex UltiMate 3000 chromatography system (Thermo Scientific,

Waltham, MA, USA) equipped with a refrigerated autosampler. The samples were injected into the Hypersil Gold C18 column (Thermo Fisher Scientific, Bremen, Germany) using 1 μL of samples and a gradient sample with 75% (*v/v*) acetonitrile, 24.5% (*v/v*) water, and 0.5% (*v/v*) formic acid (solution A) and 5% (*v/v*) acetonitrile, 94.5% (*v/v*) water, and 0.5% (*v/v*) formic acid (solution B) with a flow rate of 300 μL/min. Gradient elution conditions were set as follows: initial 0–1 min (10% B), 1–5 min (10% B), 5–10 min (30% B), 10–18 min (100% B), and 18–24 min (0% B), with final cleaning and reconditioning of the column. Mass quantification was carried out with an Exactive Plus Orbitrap spectrometer (Thermo Scientific) equipped with an electrospray interface operating with negative ionization mode, and data were processed using the Xcalibur 2.1 software (Thermo Scientific). Mass spectrometry conditions were 2500 spray volts, vaporization temperature of 350 °C, sheath gas pressure of 40 arbitrary units (a.u.), auxiliary gas pressure of 10 a.u., and capillary temperature of 35 °C. N_2 was used as the collision gas and all values were normalized to 350 mg/dry weight. Phenolic compounds were identified and quantified using suitable standards (Extrasynthese, Lyon, France), which were prepared as 1 mg/mL stock solutions in methanol and stored at −80 °C for up to 1 month in dark conditions. All results were expressed in mg/100 g dry weight.

2.5.2. Carotenoids Compounds

Fuentes et al. (2013) procedures were followed with modifications [37]. Briefly, 0.71 mL of 2:1 solution (pure acetone and 0.2 M HEPES buffer (pH 7.7)) was mixed with 500 mg of the freeze-dried extract sample and then agitated and centrifuged at 11,180× *g* for 5 min. The supernatant was separated; and the pellet was mixed with 0.71 mL of 2:1 solution (pure acetone and 0.2 M HEPES buffer (pH 7.7)), then stirred, and centrifuged to obtain the supernatant, which was then deposited together with the previous one. After that, 1 mL of pure acetone was added again to the pellet and centrifuged, the supernatant was deposited in the same tube, 1 mL of pure acetone was added to the pellet and it was stirred and centrifuged, the supernatant was deposited in the same tube, and 0.75 mL of pure hexane was added to the pellet, which was stirred and centrifuged. The supernatant was transferred to the same tube and 0.75 mL of hexane was added and centrifuged. The sample was recovered and evaporated with N_2 gas; and the obtained sample was reconstituted, microfiltered, and stored in vials. UHPLC-MS was used for quantification with lycopene and β-carotene standards. Results were expressed in mg/100 g dry weight.

2.5.3. Nucleosides Compounds

The method used was performed considering Dudley and Bond recommendations [38]. Briefly, 1 g of the freeze-dried extract was mixed with 20 mL of ultra-pure water and then homogenized in a mortar. The extract was sonicated with ultrasound (VWR 97043-962, Leicestershire, UK) at 35 kHz for 30 min, then centrifuged at 11,180× *g* for 10 min at 10 °C, and filtered at 0.22 μL with a microfilter disc (Millex-GN PTFE, Merck Millipore, Darmstadt, Germany). Quantification was performed by UHPLC-MS. An ESI injector with a positive charge at 35 kV, a capillary temperature of 350 °C, a flow of 250 μL were used. Mobile phase A was a solution of 10 mM ammonium acetate with 0.8% acetic acid and mobile phase B was a solution of acetonitrile with 0.1% acetic acid. A gradient of 10% at the first 6 min, then 6 min at 50% A, and then return in 6 min at 10% A was used. For quantification, adenosine, guanosine, and inosine standards were used in a curve of 0.05–1.5 μg/μL. Results were expressed in μg/100 g dry weight.

3. Competitive Study

The platelet anti-aggregant activity was compared with commercial cardioprotective products sold in the Chilean market (M1: CardioSmile, Nutrartis S.A., Providencia, Chile; M2: UltraPure Omega 3, Unicaps, Brea. CA, USA; M3: Eykosacol, Procaps S.A., Barranquilla, Colombia; M4: Benexia, Functional Products Trading S.A., Vitacura, Chile; and M5: Maqui Berry, Nativ for Life, Santiago, Chile). Liquid products (M2, M3, and M4) were evaluated directly on platelet aggregation at 1 mg/mL, while the solid product (M5) was dissolved in physiological serum in a final study concentration of 1 mg/mL. For the M1 product (given its milky consistency that affects the turbidity of the plasma in the platelet aggregation test), an aqueous extract was obtained. Thus, the product was mixed with water in a 1/8 ratio and centrifuged at 11,180× *g* for 5 min to obtain the supernatant that was freeze-dried and kept at −20 °C until use. Before the platelet aggregation assay, the aqueous extract of product M1 was resuspended in physiological saline solution at a concentration of 1 mg/mL.

4. Statistical Analysis

Data were expressed as mean ± standard deviations and analyzed by the Prism 6.0 software (GraphPad Inc., La Jolla, CA, USA). All measurements were made from six different donors. Before performing the statistical analysis, it was necessary to know if the results met with a normal distribution or not. Thus, using a significance level of 5% and according to Kolmogorov statistic with a *p*-value of 0.003, the results of platelet aggregation showed a non-normal distribution. The results of percentage of platelet aggregation were analyzed using non-parametric Kruskal–Wallis test, and subsequently analyzed by Dunn's test, used as a post-test, to establish significant differences between each extract with respect to control (p-value < 0.05).

5. Results and Discussion

5.1. Extraction Yield

Table 1 shows the extraction yield of different types of tomato pomace extracts. From 5 g of each type of tomato pomace sample, 0.023–0.22 g of a dry extract was obtained. Freeze drying allowed obtaining extracts (e.g., AWTPE3) that are considered microbiologically safe [5,23]. The process of ultrasound extraction produces a phenomenon called cavitation which generates physical, chemical, and mechanical effects responsible for the cellular wall disruption of the vegetal matrix [39,40]. According to many authors (Al-Dhabi et al., 2017; Chemat et al., 2017; Contamine et al., 1995; Delgado-Povedano and de Castro, 2017; Mason et al., 1996; Rastogi, 2011), the association of different ultrasound extraction conditions (e.g., power, temperature, time, and solvent) may change the polarity and viscosity of the system, as well as the interaction between the solute and solvent. Thus, in this study, aqueous extracts (2.8 ± 1.1% m/m) showed higher extraction yields in comparison with ethanolic extractions (1.0 ± 0.5% m/m), and this was due to the polarity and viscosity of these solvents. According to Chemat et al. (2017) and other studies, variations in viscosity, although small, may induce resistance to ultrasound waves and it may affect the extraction efficiency of the solvent system [39,41,42].

The cavitation phenomena also cause a temperature rise with the extraction time, increasing the extraction yield of phenolic and nucleosides compounds by their higher solubility and diffusion. However, when a higher extraction time exposes these compounds at high temperatures for a long time, it may promote its degradation by oxidation mechanisms, consequently decreasing the extraction yield [41–43]. Therefore, in this study, cycle 3 showed promising potential to obtain high extraction yields, since AWTPE3 preliminarily presented 3.45% (m/m). This could be explained by the higher value of soluble solids observed for AWTPE3 (6.00 °Brix) compared with whole ground dried tomato pomace (1.67 °Brix, performed in 1 g of sample powder dissolved in 10 mL ultrapure water). The latter can be explained due to mechanical effects of ultrasound that triggered the release of water-soluble compounds such as polysaccharides, polyphenols, and nucleosides from their matrices by disrupting them from cellular tissues [5,6,23,33,34,44,45]. Ultrasound is also an effective method to improve fractioning of water-soluble compounds from tomato pomace fiber, since the latter was separated as a precipitate after centrifugation [5,23].

Table 1. Extraction yield and platelet anti-aggregant activity of different types of tomato pomace extracts.

Type of Extracts	Code	Yield (%)	Platelet Aggregation (%)
Aqueous whole tomato pomace extract cycle 1	AWTPE1	3.57	46 ± 12
Aqueous whole tomato pomace extract cycle 3	AWTPE3	3.45	32 ± 9 **
Aqueous whole tomato pomace extract cycle 6	AWTPE6	2.13	48 ± 3
Ethanolic whole tomato pomace extract cycle 1	EWTPEC1	0.82	52 ± 8
Ethanolic whole tomato pomace extract cycle 3	EWTPEC3	0.68	32 ± 9 **
Ethanolic whole tomato pomace extract cycle 6	EWTPEC6	0.57	51 ± 9
Aqueous seedless tomato pomace extract cycle 1	ASTPEC1	1.31	61 ± 7
Aqueous seedless tomato pomace extract cycle 3	ASTPEC3	1.87	46 ± 9
Aqueous seedless tomato pomace extract cycle 6	ASTPEC6	0.80	26 ± 6 ***
Ethanolic seedless tomato pomace extract cycle 1	ESTPEC1	ND	ND
Ethanolic seedless tomato pomace extract cycle 3	ESTPEC3	0.59	45 ± 9
Ethanolic seedless tomato pomace extract cycle 6	ESTPEC6	0.46	29 ± 7 **
Aqueous seed extract cycle 1	ASEC1	2.51	29 ± 8 **
Aqueous seed extract cycle 3	ASEC3	1.60	34 ± 2 *
Aqueous seed extract cycle 6	ASEC6	3.45	40 ± 7
Ethanolic seed extract cycle 1	ESEC1	0.48	38 ± 4
Ethanolic seed extract cycle 3	ESEC3	1.44	47 ± 8
Ethanolic seed extract cycle 6	ESEC6	1.27	52 ± 12
Aqueous extract, 50% tomato pomace/50% seed cycle 1	AE5TPSC1	2.96	27 ± 12 ***
Aqueous extract, 50% tomato pomace/50% seed cycle 3	AE5TPSC3	4.43	36 ± 8
Aqueous extract, 50% tomato pomace/50% seed cycle 6	AE5TPSC6	2.61	54 ± 9
Ethanolic extract, 50% tomato pomace/50% seed cycle 1	EE5TPSC1	1.11	29 ± 12 **
Ethanolic extract, 50% tomato pomace/50% seed cycle 3	EE5TPSC3	ND	ND
Ethanolic extract, 50% tomato pomace/50% seed cycle 6	EE5TPSC6	ND	ND
Aqueous extract, 80% tomato pomace/20% seed cycle 1	AE8TPSC1	2.94	41 ± 7
Aqueous extract, 80% tomato pomace/20% seed cycle 3	AE8TPSC3	4.25	40 ± 11
Aqueous extract, 80% tomato pomace/20% seed cycle 6	AE8TPSC6	4.08	38 ± 11
Ethanolic extract, 80% tomato pomace/20% seed cycle 1	EE8TPSC1	2.29	39 ± 15
Ethanolic extract, 80% tomato pomace/20% seed cycle 3	EE8TPSC3	1.07	52 ± 16
Ethanolic extract, 80% tomato pomace/20% seed cycle 6	EE8TPSC6	1.28	39 ± 12
Non-ultrasound-assisted aqueous whole tomato pomace extract			64 ± 11
Control (maximum platelet aggregation)			87 ± 6

Platelet aggregation data were expressed as mean ± standard deviations (SD). Statistically significant difference in platelet aggregation (%) is considered with * $p < 0.05$, ** $p < 0.01$, and *** $p < 0.001$, analyzed with Dunn's test with respect to control. ND: not determined. Not determined (ND) extracts correspond to extracts that were generated, but obtained at a very low yield, preventing them to be weighed and evaluated. Cycle 1: 20 min, cycle 3: three cycles of 20 min each; and cycle 6: six cycles of 20 min each.

5.2. Platelet Anti-Aggregant Activity

We observed that the non-ultrasound-assisted aqueous extraction of whole tomato pomace affected the platelet anti-aggregant activity of the extract (64 ± 11% vs. control 87 ± 6%, not significant). The time of sonication and solvent used in the extraction had a significant role in platelet anti-aggregant activity [46]. As is shown in the Table 1, the extracts with significant antiplatelet activity, with respect to control (87 ± 6%), were AWTPE3 (32 ± 9%, $p < 0.01$), EWTPEC3 (32 ± 9%, $p < 0.01$), ASTPEC6 (26 ± 6%, $p < 0.001$), ESTPEC6 (29 ± 7%, $p < 0.01$), ASEC1 (29 ± 8%, $p < 0.01$), ASEC3 (34 ± 2%, $p < 0.05$), AE5TPSC1 (27 ± 12%, $p < 0.001$), and EE5TPSC1 (29 ± 12%, $p < 0.01$). These levels of inhibition of platelet aggregation induced by ADP were different, depending on ultrasound cycle times, tomato pomace composition, and solvent used. Thus, the sonication cycles only improved the platelet anti-aggregant activity in the seedless extracts. Six cycles of sonication of 20 min each significantly improved the platelet anti-aggregant activity of aqueous (ASTPEC6 26 ± 6%, $p < 0.001$) and ethanolic (ESTPEC6, 29 ± 7%, $p < 0.01$) seedless tomato pomace extracts. Tomato seeds are composed with proteins (35.5%), fiber (13.2%), fatty acids (30%), ash (3%), and also essential amino acids (except tryptophan), so they could be used in several applications, for example, as an enrichment for cereal products and foods low in lysine [22,47].

For aqueous seed extracts and aqueous 50% tomato pomace/50% seed extracts, the best activity was achieved using one ultrasound cycle (ASEC1, 29 ± 8%, $p < 0.01$ and AE5TPSC1, 27 ± 12%, $p < 0.001$), while antiplatelet activity decreased with an increase to six sonication cycles (ASEC6, 40 ± 7% and AE5TPSC6, 54 ± 9%, not significant). On the other hand, three-cycle sonication allowed the best significant activity of the extracts obtained from aqueous whole tomato pomace (AWTPE3, 32 ± 9%, $p < 0.01$) and ethanolic whole tomato pomace (EWTPEC3, 32 ± 9%, $p < 0.05$). Although, in a previous study, the ultrasound extraction time used by the authors was 5 min [10], in this study ultrasound cycle was increased to 20 min to achieve a better extraction of compounds responsible for platelet anti-aggregant activity. Furthermore, 5 min of ultrasound extraction achieved an aggregation of 55% in aqueous tomato pomace extract [10], while the 20-min cycle improved the activity of the aqueous tomato pomace extract by lowering maximum aggregation to 46 ± 12% (AWTPE1).

Based on the results of yield and platelet anti-aggregant activity, it seemed that the most suitable extract was AWTPE3 (yield, 3.45% and platelet aggregation, 32 ± 9%, $p < 0.01$). Meanwhile, ASTPEC6 (platelet aggregation, 26 ± 6%, $p < 0.001$) had a poor yield of 0.80% and showed no statistically significant differences in platelet anti-aggregant activity when compared to AWTPE3.

5.3. Identification of Bioactive Compounds

Table 2 shows the amounts of bioactive compounds by HPLC-MS on aqueous extracts (whole tomato pomace, seed, and seedless) obtained from the same original matrix. Since tomato pomace is mainly composed of peels and seeds, we compared the bioactive compounds of AWTPE3 (aqueous whole tomato pomace), previously reported [5], with two other aqueous extracts obtained with the same sonication cycles—ASTPEC3 (aqueous seedless tomato pomace) and ASEC3 (aqueous seeds).

Table 2. Identification and quantification of bioactive compounds by HPLC-MS.

	Samples		
Compounds **	**AWTPE3 ***	**ASEC3**	**ASTPEC3**
Flavonoids (mg/100 g Dry Weight)			
Gallic acid	0.83	6.94	0.53
Ferulic acid	2.44	9.08	3.68
Coumaric acid	88.56	2.58	<0.001
Phloridzin	4.71	1.35	2.62
Phloretin	97.31	26.72	1.71
Procyanidin B2	1868.49	76.62	27.95
Apigenin-7-*O*-glucoside	<0.001	0.196	<0.001
Kaempferol-3-*O*-glucoside	2032.58	415.39	<0.001
Luteolin-7-*O*-glucoside	63.34	55.77	57.63
Genistein	<0.001	0.196	<0.001
Kaempferol	77.09	<0.001	<0.001
Daidzein	<0.001	0.02	<0.001
Quercetin	408.23	<0.001	7.74
Quercitrin	1.96	0.003	<0.001
Rutin	0.262	0.065	0.012
Epicatechin	0.131	0.026	<0.001
Nucleosides (μg/100 g Dry Weight)			
Adenosine	42.90	<0.001	<0.001
Inosine	57.20	42.21	<0.001
Guanosine	20.97	7.44	7.26
Carotenoids (mg/100 g Dry Weight)			
Lycopene	<0.001	<0.001	7.74
β-Carotene	55.5	4.3	3.4

* Data reported in Palomo et al., (2019) [5]. $p < 0.001$ indicates concentrations close to the limit of detection. AWTPE3: aqueous whole tomato pomace extract cycle 3, ASEC3: aqueous seed extract cycle 3, and ASTPEC3: aqueous seedless tomato pomace extract cycle 3. ** Twenty-one flavonoid molecules were analyzed and sixteen molecules were detected by chromatography. Three nucleoside molecules were analyzed and detected by chromatography. Two carotenoid molecules were analyzed and detected by chromatography.

Several studies have observed that ultrasound is effective in assisting the extraction of bioactive compounds such as nucleosides, polyphenols, and carotenoids from different plant byproducts, including tomato pomace [5,27,48,49]. In comparison with unprocessed tomatoes and on a dry weight basis, tomato pomace contained significantly lower amounts of lycopene and increased amounts of β-carotene, tocopherols, sterols, terpenes, and flavonoids (e.g., naringenin) [50]. Thus, in this study higher concentrations of flavonoids (procyanidin B2, kaempferol-3-*O*-glucoside, and kaempferol) were observed for AWTPE3 in comparison with whole dried ground tomato pomace [5]. These results demonstrated that the ultrasound-assisted extraction and concentration using freeze drying resulted in a greater concentration of flavonoids in AWTPE3 [5]. A similar result was obtained by Navarro-González et al., where the best extraction of phenolic compounds (e.g., rutin, naringenin, and chlorogenic acid derivatives) was with ultrasonic assistance [51]. This concentration of compounds (e.g., kaempferol-3-*O*-glucoside) in AWTPE3 may be due to the fact that tomato pomace is composed of pulp residues, in addition to peels and seeds [52].

ASEC3 and ASTPEC3 had different levels of polyphenols because they were lost simply by removing their peels or seeds. Thus, the removal of the peels represented a loss of lycopene (80%), β-carotene (57%), and phenolic compounds (63%). Meanwhile, the elimination of seeds had a greater impact on polyphenols with a loss of 63% [53]. ASEC3 showed higher concentrations of flavonoid compounds than ASTPEC3, except for quercetin, which was detected in tomato peels in higher concentrations than seeds, according to literature [44]. In this context, it has been described that seed aqueous lyophilized extract from tomato contains 20,657 mg/100 g phenolic compounds, with the highest amounts of quercetin-3-*O*-sophoroside, kaempferol-3-*O*-sophoroside, and isorhamnetin-3-*O*-sophoroside [54]. However, ASEC3 did not present concentrations of kaempferol and quercetin, suggesting differences due to an extraction procedure effect. ASTPEC3 did not show concentrations of coumaric acid, apigenin-7-*O*-glucoside, kaempferol-3-*O*-glucoside, genistein, kaempferol, daidzein, quercitrin, and epicatechin, suggesting that these compounds were not effectively recovered from tomato peels by ultrasound-assisted extraction (Table 2). Chlorogenic acid, which is usually predominant in tomato byproducts, was not detected or quantified by chromatography in any of the samples. The latter could be explained due to low stability of this molecule during storage of dehydrated tomato byproducts [55].

The data available in the literature suggest a potential therapeutic effect of several flavonoids against CVD through the inhibition of platelet aggregation, such as coumaric acid [56], phloretin [57], procyanidin B2 [58], kaempferol-3-*O*- glucoside [59], and epicatechin [60]. The presence of these compounds in greater amounts in AWTPE3 and ASEC3 than in ASTPEC3 correlated with a better capacity to inhibit platelet aggregation. Several mechanisms of antiplatelet action of polyphenols have been reported, including inhibition of the arachidonic acid pathway, suppression of cytoplasmic increase in Ca^{2+}, degradation blockage, αIIbβ3 integrin-mediated signaling, inhibition of secretion of platelet granules, and inhibition of thromboxane formation [61,62]. The ultrasound was able to increase the extraction of adenosine in AWTPE3 when compared with whole dried ground tomato pomace from 6.32 μg/100 g to 42.90 μg/100 g, almost 6 times more [5]. However, the extraction of inosine and guanosine was not affected by ultrasound. The results obtained from the quantification of nucleosides (adenosine, inosine, and guanosine) demonstrated that AWTPE3 presented greater amounts of these compounds compared with ASEC3 and ASTPEC3 [5]. Authors involved in this present work have previously observed that adenosine is concentrated in higher amounts in the aqueous extract from ripe tomato pulp than whole tomato pomace and peels extracts [9]. The presence of inosine in significant concentrations in ASEC3, but not in ASTPEC3, suggests that this compound is available only in tomato seeds [6,63]. In part, this could explain why ASEC3 had better platelet anti-aggregant activity than ASTPEC3, since inosine has been reported to inhibit platelet aggregation, significantly preventing in vivo thrombus formation [34]. Both extracts (ASEC3 and ASTPEC3) presented close concentrations of guanosine and adenosine, the latter had no significant amount detected. The present research group isolated and identified adenosine and guanosine as bioactive compounds in tomato with a potent platelet anti-aggregant activity [9,33,34].

The results showed that ultrasound was not efficient to extract carotenoids in the conditions established. Lycopene was solely quantified in ASTPEC3. Lycopene is found mainly in the peels and acts as a red pigment and, although resistant to thermal processing, it is very susceptible to light degradation [64]. To determine this compound, a prior extraction process must be carried out, which requires conditions of absence of light to avoid degradation, which makes its quantification very complicated and delicate. In the case of β-carotene, a higher concentration was found in whole dried ground tomato pomace powder in comparison with the other samples. According to literature, most non-polar solvents have higher extraction yield of carotenoids than aqueous extracts, since they are lipophilic compounds that are soluble in organic solvents, such as acetone, alcohol, ethyl ether, chloroform, and ethyl acetate. Therefore, large concentrations of carotenoids in the aqueous extract of tomato pomace were not expected.

5.4. Competitive Study

In the current market, there are cardioprotective products like phytosterols, polyphenols, and omega-3 [65]. Platelet anti-aggregant activity of commercial products was compared with that of AWTPE3 (Figure 1). In the presence of M2 (platelet aggregation, 57 ± 5%), M3 (platelet aggregation, 56 ± 5%), and M4 (platelet aggregation, 54 ± 7%), platelets showed a significant decrease in aggregation compared with the control of maximum platelet aggregation (79 ± 8%). Products M1 and M5 did not show inhibition of platelet aggregation. AWTPE3 presented a significant difference compared with the control and the commercial products M2, M3, and M4.

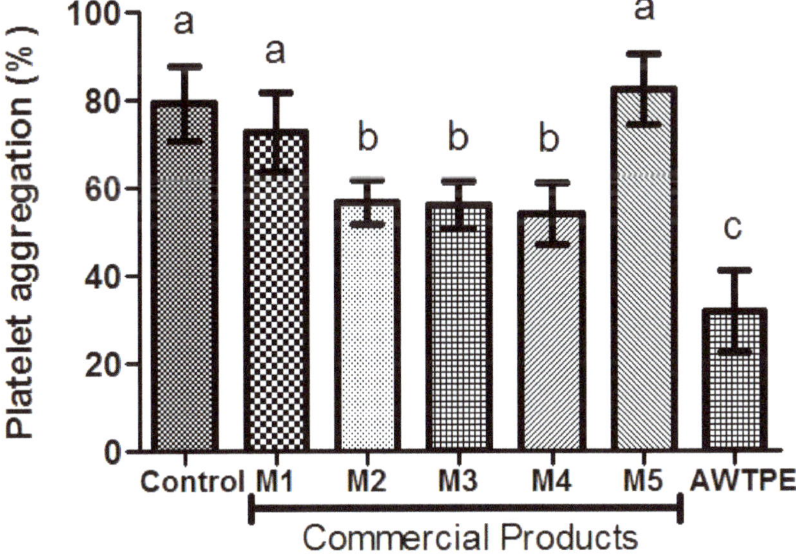

Figure 1. Platelet anti-aggregant activity of commercial products. Different letters indicate a significant difference with a *p* value < 0.05. AWTPE: aqueous whole tomato pomace extract cycle 3; M1: CardioSmile, Nutrartis S.A., Chile; M2: UltraPure Omega 3, Unicaps CA, USA; M3: Eykosacol, Laboratorio Procaps S.A., Colombia; M4: Benexia, Functional Products Trading S.A., Chile; and M5: Maqui Berry, Nativ for Life, Chile.

6. Conclusions

Currently, tomato pomace is considered as waste by the food industry; however, in this study, we showed that ultrasound-assisted extracts from this byproduct had significant platelet anti-aggregant activity. The levels of inhibition of platelet aggregation were different, depending on ultrasound cycle times, tomato pomace composition (whole, seedless, and seed), the solvent used, and the flavonoid and nucleoside content (Figure 2). In this context, the ultrasound-assisted extraction improved extraction yields, showing significant recovery of bioactive compounds from tomato pomace and increased platelet aggregation inhibition.

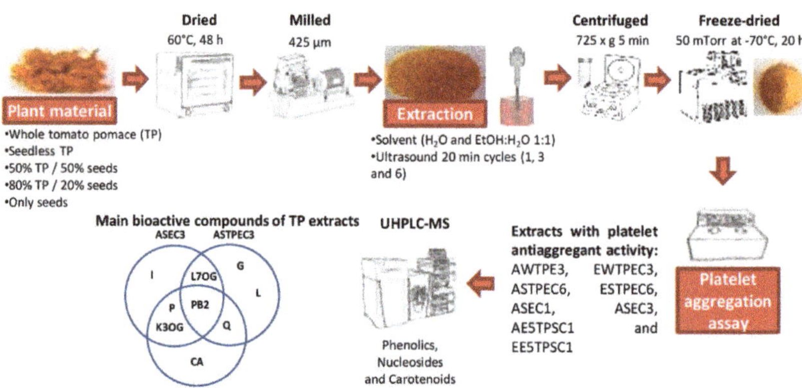

Figure 2. Platelet anti-aggregant activity and bioactive compounds of ultrasound-assisted extracts from whole and seedless tomato pomace. AE5TPSC1: aqueous extract 50% tomato pomace/50% seed cycle 1, ASEC1: aqueous seed extract cycle 1, ASEC3: aqueous seed extract cycle 3, ASTPEC3: aqueous seedless tomato pomace extract cycle 3, ASTPEC6: aqueous seedless tomato pomace extract cycle 6, AWTPE3: aqueous whole tomato pomace extract cycle 3, EE5TPSC1: ethanolic extract 50% tomato pomace/50% seed cycle 1, ESTPEC6: ethanolic seedless tomato pomace extract cycle 6, EWTPEC3: ethanolic whole tomato pomace extract cycle 3, CA: coumaric acid, G: guanosine, I: inosine, K-3-O-G: kaempferol-3-O-glucoside, L: lycopene, L7OG: luteolin-7-O- glucoside, P: phloretin, PB2: procyanidin B2, Q: quercetin.

Author Contributions: All authors have read and agreed to the published version of the manuscript. A.C.-M.: Conceptualization, Investigation and Methodology. I.P.: Investigation and Writing—Original draft preparation. A.P.: Writing—Original draft preparation. A.G.T.: Writing—Reviewing and Editing. M.R.M.J.: Writing—Reviewing and Editing. S.G.S.-A.: Writing—Reviewing and Editing. E.F.: Data curation and Writing—Original draft preparation.

Funding: This research was funded by Conicyt Regional/Gore Maule/CEAP (R0912001) and Conicyt Regional (R15F10012).

Acknowledgments: This work was supported by Proyecto Enlace Fondecyt (Dirección de Investigación/Vicerrectoría Académica/Universidad de Talca). E.F. thanks Conicyt/Fondecyt (Grant No. 1180427) and Conicyt/Redes (Grant No. 170003). I.P. thanks Redes (Grant No. 170002). This study was financed in part by the Coordenação de Aperfeiçoamento de Pessoal de Nível Superior—Brasil (CAPES) (Finance Code 001), CNPq (403328/2016-0; 301108/2016-1), and FAPESP (2015/50333-1; 2018/11069-5; 2015/13320-9)". E.F., I.P., S.G.S.-A., and M.R.M.J. acknowledges Red Iberomericana de Alimentos Autoctonos Subutilizados (ALSUB-CYTED, 118RT0543). A.G.T. thanks CNPq (140942/2016-5) for the Ph.D. scholarship.

References

1. Mensah, G.A.; Roth, G.A.; Fuster, V. The Global Burden of Cardiovascular Diseases and Risk Factors: 2020 and Beyond. *J. Am. Coll. Cardiol.* **2019**, *74*, 2529–2532. [CrossRef] [PubMed]
2. Instituto Nacional de Estadísticas. 2012. Available online: http://ine.cl/docs/default-source/demogr%C3%A1ficas-y-vitales/vitales/anuarios/anuario-2012/completa_vitales_2012.pdf?sfvrsn=6 (accessed on 6 August 2020).
3. Alissa, E.M.; Ferns, G.A. Dietary fruits and vegetables and cardiovascular diseases risk. *Crit. Rev. Food. Sci. Nutr.* **2017**, *57*, 1950–1962. [CrossRef] [PubMed]
4. Lutz, M.; Fuentes, E.; Avila, F.; Alarcon, M.; Palomo, I. Roles of Phenolic Compounds in the Reduction of Risk Factors of Cardiovascular Diseases. *Molecules* **2019**, *24*, 366. [CrossRef] [PubMed]
5. Palomo, I.; Concha-Meyer, A.; Lutz, M.; Said, M.; Saez, B.; Vasquez, A.; Fuentes, E. Chemical Characterization and Antiplatelet Potential of Bioactive Extract from Tomato Pomace (Byproduct of Tomato Paste). *Nutrients* **2019**, *11*, 456. [CrossRef] [PubMed]
6. Dutta-Roy, A.K.; Crosbie, L.; Gordon, M.J. Effects of tomato extract on human platelet aggregation in vitro. *Platelets* **2001**, *12*, 218–227. [CrossRef]
7. O'Kennedy, N.; Crosbie, L.; van Lieshout, M.; Broom, J.I.; Webb, D.J.; Duttaroy, A.K. Effects of antiplatelet components of tomato extract on platelet function in vitro and ex vivo: A time-course cannulation study in healthy humans. *Am. J. Clin. Nutr.* **2006**, *84*, 570–579. [CrossRef]
8. Fuentes, E.J.; Astudillo, L.A.; Gutierrez, M.I.; Contreras, S.O.; Bustamante, L.O.; Rubio, P.I.; Moore-Carrasco, R.; Alarcon, M.A.; Fuentes, J.A.; Gonzalez, D.E.; et al. Fractions of aqueous and methanolic extracts from tomato (Solanum lycopersicum L.) present platelet antiaggregant activity. *Blood Coagul. Fibrinolysis* **2012**, *23*, 109–117. [CrossRef]
9. Fuentes, E.; Castro, R.; Astudillo, L.; Carrasco, G.; Alarcon, M.; Gutierrez, M.; Palomo, I. Bioassay-Guided Isolation and HPLC Determination of Bioactive Compound That Relate to the Antiplatelet Activity (Adhesion, Secretion, and Aggregation) from Solanum lycopersicum. *Evid. Based Complement Alternat. Med.* **2012**, *2012*, 147031. [CrossRef]
10. Fuentes, E.; Pereira, J.; Alarcon, M.; Valenzuela, C.; Perez, P.; Astudillo, L.; Palomo, I. Protective Mechanisms of S. lycopersicum Aqueous Fraction (Nucleosides and Flavonoids) on Platelet Activation and Thrombus Formation: In Vitro, Ex Vivo and In Vivo Studies. *Evid Based Complement Alternat. Med.* **2013**, *2013*, 609714. [CrossRef]
11. Michalickova, D.; Belovic, M.; Ilic, N.; Kotur-Stevuljevic, J.; Slanar, O.; Sobajic, S. Comparison of Polyphenol-Enriched Tomato Juice and Standard Tomato Juice for Cardiovascular Benefits in Subjects with Stage 1 Hypertension: A Randomized Controlled Study. *Plant Foods Hum. Nutr.* **2019**, *74*, 122–127. [CrossRef]
12. Uddin, M.; Biswas, D.; Ghosh, A.; O'Kennedy, N.; Duttaroy, A.K. Consumption of Fruitflow((R)) lowers blood pressure in pre-hypertensive males: A randomised, placebo controlled, double blind, cross-over study. *Int. J. Food Sci. Nutr.* **2018**, *69*, 494–502. [CrossRef] [PubMed]
13. Ma, C.A.; Leung, Y.Y. Exploring the Link between Uric Acid and Osteoarthritis. *Front Med.* **2017**, *4*, 225. [CrossRef] [PubMed]
14. WPTC World Production Estimate of Tomato Processing. Available online: http://www.wptc.to/pdf/releases/WPTC%20World%20Production%20estimate%20as%20of%2031%20March%202017.pdf (accessed on 20 January 2020).
15. Food and Agriculture Organization Statistical Database FAOSTAT. Available online: http://www.fao.org/faostat/en/#data/QC (accessed on 10 July 2019).
16. Ruiz Celma, A.; Cuadros, F.; López-Rodríguez, F. Characterisation of industrial tomato by-products from infrared drying process. *Food Bioprod. Process.* **2009**, *87*, 282–291. [CrossRef]
17. Fondevila, M.; Guada, J.A.; Gasa, J.; Castrillo, C. Tomato pomace as a protein supplement for growing lambs. *Small Rumin. Res.* **1994**, *13*, 117–126. [CrossRef]
18. Marcos, C.N.; de Evan, T.; Molina-Alcaide, E.; Carro, M.D. Nutritive Value of Tomato Pomace for Ruminants and Its Influence on In Vitro Methane Production. *Animals* **2019**, *9*, 343. [CrossRef]
19. Isik, F.; Topkaya, C. Effects of tomato pomace supplementation. *Ital. J. Food Sci.* **2016**, *28*, 525–535. [CrossRef]
20. Tedeschi, G.; Benitez, J.J.; Ceseracciu, L.; Dastmalchi, K.; Itin, B.; Stark, R.E.; Heredia, A.; Athanassiou, A.; Heredia-Guerrero, J.A. Sustainable Fabrication of Plant Cuticle-Like Packaging Films from Tomato Pomace Agro-Waste, Beeswax, and Alginate. *ACS Sustain. Chem. Eng.* **2018**, *6*, 14955–14966. [CrossRef]

21. Mirzaei-Aghsaghali, A.; Maheri-sis, N.; Mansouri, H.; Razeghi, M.E.; Safaei, A.R.; Aghajanzadeh-Golshani, A.; Alipoor, K. Estimation of the nutritive value of tomato pomace for ruminant using in vitro gas production technique. *Afr. J. Biotechnol.* **2011**, *10*, 6251–6256.

22. Del Valle, M.; Cámara, M.; Torija, M.-E. Chemical characterization of tomato pomace. *J. Sci. Food Agric.* **2006**, *86*, 1232–1236. [CrossRef]

23. Concha-Meyer, A.A.; Durham, C.A.; Colonna, A.E.; Hasenbeck, A.; Sáez, B.; Adams, M.R. Consumer Response to Tomato Pomace Powder as an Ingredient in Bread: Impact of Sensory Liking and Benefit Information on Purchase Intent. *J. Food Sci.* **2019**, *84*, 3774–3783. [CrossRef]

24. Lazos, E.S.; Tsaknis, J.; Lalas, S. Characteristics and composition of tomato seed oil. *Grasas y Aceites* **1998**, *49*, 440–445. [CrossRef]

25. Lenucci, M.S.; Durante, M.; Anna, M.; Dalessandro, G.; Piro, G. Possible use of the carbohydrates present in tomato pomace and in byproducts of the supercritical carbon dioxide lycopene extraction process as biomass for bioethanol production. *J. Agric. Food Chem.* **2013**, *61*, 3683–3692. [CrossRef]

26. Cárdenas-Castro, A.P.; del Carmen Perales-Vázquez, G.; De la Rosa, L.A.; Zamora-Gasga, V.M.; Ruiz-Valdiviezo, V.M.; Alvarez-Parrilla, E.; Sáyago-Ayerdi, S.G. Sauces: An undiscovered healthy complement in Mexican cuisine. *Int. J. Gastron. Food Sci.* **2019**, *17*, 100154. [CrossRef]

27. Grassino, A.N.; Brncic, M.; Vikic-Topic, D.; Roca, S.; Dent, M.; Brncic, S.R. Ultrasound assisted extraction and characterization of pectin from tomato waste. *Food Chem.* **2016**, *198*, 93–100. [CrossRef]

28. Chen, H.M.; Fu, X.; Luo, Z.G. Properties and extraction of pectin-enriched materials from sugar beet pulp by ultrasonic-assisted treatment combined with subcritical water. *Food Chem.* **2015**, *168*, 302–310. [CrossRef]

29. Maran, J.P.; Priya, B. Ultrasound-assisted extraction of pectin from sisal waste. *Carbohydr. Polym.* **2015**, *115*, 732–738. [CrossRef] [PubMed]

30. Luengo, E.; Alvarez, I.; Raso, J. Improving carotenoid extraction from tomato waste by pulsed electric fields. *Front Nutr.* **2014**, *1*, 12. [CrossRef] [PubMed]

31. Gallo, M.; Ferrara, L.; Naviglio, D. Application of Ultrasound in Food Science and Technology: A Perspective. *Foods* **2018**, *7*, 164. [CrossRef]

32. Pacheco-Fernández, I.; Pino, V. Chapter 17—Extraction With Ionic Liquids-Organic Compounds. In *Liquid-Phase Extraction*; Poole, C.F., Ed.; Elsevier: Amsterdam, The Netherlands, 2020; pp. 499–537.

33. Fuentes, F.; Alarcon, M.; Badimon, L.; Fuentes, M.; Klotz, K.N.; Vilahur, G.; Kachler, S.; Padro, T.; Palomo, I.; Fuentes, E. Guanosine exerts antiplatelet and antithrombotic properties through an adenosine-related cAMP-PKA signaling. *Int. J. Cardiol.* **2017**, *248*, 294–300. [CrossRef]

34. Fuentes, E.; Pereira, J.; Mezzano, D.; Alarcon, M.; Caballero, J.; Palomo, I. Inhibition of platelet activation and thrombus formation by adenosine and inosine: Studies on their relative contribution and molecular modeling. *PLoS ONE* **2014**, *9*, e112741. [CrossRef]

35. Rodriguez-Azua, R.; Treuer, A.; Moore-Carrasco, R.; Cortacans, D.; Gutierrez, M.; Astudillo, L.; Fuentes, E.; Palomo, I. Effect of tomato industrial processing (different hybrids, paste, and pomace) on inhibition of platelet function in vitro, ex vivo, and in vivo. *J. Med. Food* **2014**, *17*, 505–511. [CrossRef]

36. Torres, C.A.; Sepúlveda, G.; Concha-Meyer, A.A. Effect of processing on quality attributes and phenolic profile of quince dried bar snack. *J. Sci. Food Agric.* **2019**, *99*, 2556–2564. [CrossRef] [PubMed]

37. Fuentes, E.; Carle, R.; Astudillo, L.; Guzman, L.; Gutierrez, M.; Carrasco, G.; Palomo, I. Antioxidant and Antiplatelet Activities in Extracts from Green and Fully Ripe Tomato Fruits (Solanum lycopersicum) and Pomace from Industrial Tomato Processing. *Evid Based Complement Alternat. Med.* **2013**, *2013*, 867578. [CrossRef] [PubMed]

38. Dudley, E.; Bond, L. Mass spectrometry analysis of nucleosides and nucleotides. *Mass Spectrometry Reviews* **2014**, *33*, 302–331. [CrossRef] [PubMed]

39. Chemat, F.; Rombaut, N.; Sicaire, A.-G.; Meullemiestre, A.; Fabiano-Tixier, A.-S.; Abert-Vian, M. Ultrasound assisted extraction of food and natural products. Mechanisms, techniques, combinations, protocols and applications. A review. *Ultrason. Sonochem.* **2017**, *34*, 540–560. [CrossRef]

40. Rastogi, N.K. Opportunities and Challenges in Application of Ultrasound in Food Processing. *Crit. Rev. Food Sci. Nutr.* **2011**, *51*, 705–722. [CrossRef] [PubMed]

41. Al-Dhabi, N.A.; Ponmurugan, K.; Maran Jeganathan, P. Development and validation of ultrasound-assisted solid-liquid extraction of phenolic compounds from waste spent coffee grounds. *Ultrason. Sonochem.* **2017**, *34*, 206–213. [CrossRef]

42. Delgado-Povedano, M.; Luque de Castro, M. Ultrasound-Assisted Extraction of Food Components. *Ref. Module Food Sci.* **2017**, *1*.

43. Ji, J.; Hu, J.; Chen, S.; Liu, R.; Wang, L.; Cheng, J.; Wu, H. Development and application of a method for determination of nucleosides and nucleobases in Mactra veneriformis. *Pharmacogn. Mag.* **2013**, *9*, 96–102. [CrossRef]

44. Valdez-Morales, M.; Espinosa-Alonso, L.G.; Espinoza-Torres, L.C.; Delgado-Vargas, F.; Medina-Godoy, S. Phenolic content and antioxidant and antimutagenic activities in tomato peel, seeds, and byproducts. *J. Agric. Food Chem.* **2014**, *62*, 5281–5289. [CrossRef]

45. Aguiló-Aguayo, I.; Walton, J.; Viñas, I.; Tiwari, B.K. Ultrasound assisted extraction of polysaccharides from mushroom by-products. *LWT* **2017**, *77*, 92–99. [CrossRef]

46. Azmir, J.; Zaidul, I.S.M.; Rahman, M.M.; Sharif, K.M.; Mohamed, A.; Sahena, F.; Jahurul, M.H.A.; Ghafoor, K.; Norulaini, N.A.N.; Omar, A.K.M. Techniques for extraction of bioactive compounds from plant materials: A review. *J. Food Eng.* **2013**, *117*, 426–436. [CrossRef]

47. Brodowski, D.; Geisman, J.R. Protein content and amino acid composition of protein of seeds from tomatoes at various stages of ripeness. *J. Food Sci.* **1980**, *45*, 228–229. [CrossRef]

48. Lianfu, Z.; Zelong, L. Optimization and comparison of ultrasound/microwave assisted extraction (UMAE) and ultrasonic assisted extraction (UAE) of lycopene from tomatoes. *Ultrason. Sonochem.* **2008**, *15*, 731–737. [CrossRef] [PubMed]

49. Luengo, E.; Condón-Abanto, S.; Condón, S.; Álvarez, I.; Raso, J. Improving the extraction of carotenoids from tomato waste by application of ultrasound under pressure. *Sep. Purif. Tech.* **2014**, *136*, 130–136. [CrossRef]

50. Kalogeropoulos, N.; Chiou, A.; Pyriochou, V.; Peristeraki, A.; Karathanos, V.T. Bioactive phytochemicals in industrial tomatoes and their processing byproducts. *Lebensm. Wiss. Technol.* **2012**, *49*, 213–216. [CrossRef]

51. Navarro-González, I.; García-Valverde, V.; García-Alonso, J.; Periago, M.J. Chemical profile, functional and antioxidant properties of tomato peel fiber. *Food Res. Int.* **2011**, *44*, 1528–1535. [CrossRef]

52. Zuorro, A.; Lavecchia, R.; Medici, F.; Piga, F. Use of Cell Wall Degrading Enzymes for the Production of High-Quality Functional Products from Tomato Processing Waste. *Chem. Eng. Trans.* **2014**, *38*, 355–360. [CrossRef]

53. Vinha, A.F.; Alves, R.C.; Barreira, S.V.P.; Castro, A.; Costa, A.S.G.; Oliveira, M.B.P.P. Effect of peel and seed removal on the nutritional value and antioxidant activity of tomato (Lycopersicon esculentum L.) fruits. *LWT-Food Sci. Technol.* **2014**, *55*, 197–202. [CrossRef]

54. Ferreres, F.; Taveira, M.; Pereira, D.M.; Valentao, P.; Andrade, P.B. Tomato (Lycopersicon esculentum) seeds: New flavonols and cytotoxic effect. *J. Agric. Food Chem.* **2010**, *58*, 2854–2861. [CrossRef]

55. Lavelli, V.; Torresani, M.C. Modelling the stability of lycopene-rich by-products of tomato processing. *Food Chem.* **2011**, *125*, 529–535. [CrossRef]

56. Luceri, C.; Giannini, L.; Lodovici, M.; Antonucci, E.; Abbate, R.; Masini, E.; Dolara, P. p-Coumaric acid, a common dietary phenol, inhibits platelet activity in vitro and in vivo. *Br. J. Nutr.* **2007**, *97*, 458–463. [CrossRef]

57. Stangl, V.; Lorenz, M.; Ludwig, A.; Grimbo, N.; Guether, C.; Sanad, W.; Ziemer, S.; Martus, P.; Baumann, G.; Stangl, K. The Flavonoid Phloretin Suppresses Stimulated Expression of Endothelial Adhesion Molecules and Reduces Activation of Human Platelets. *J. Nutr.* **2005**, *135*, 172–178. [CrossRef] [PubMed]

58. Murphy, K.J.; Chronopoulos, A.K.; Singh, I.; A Francis, M.; Moriarty, H.; Pike, M.J.; Turner, A.H.; Mann, N.J.; Sinclair, A.J. Dietary flavanols and procyanidin oligomers from cocoa (Theobroma cacao) inhibit platelet function. *Am. J. Clin. Nutr.* **2003**, *77*, 1466–1473. [CrossRef] [PubMed]

59. Rolnik, A.; Żuchowski, J.; Stochmal, A.; Olas, B. Quercetin and kaempferol derivatives isolated from aerial parts of Lens culinaris Medik as modulators of blood platelet functions. *Ind. Crop. Prod.* **2020**, *152*, 112536. [CrossRef]

60. Sinegre, T.; Teissandier, D.; Milenkovic, D.; Morand, C.; Lebreton, A. Epicatechin influences primary hemostasis, coagulation and fibrinolysis. *Food Funct.* **2019**, *10*, 7291–7298. [CrossRef]

61. Guerrero, J.A.; Lozano, M.L.; Castillo, J.; Benavente-García, O.; Vicente, V.; Rivera, J. Flavonoids inhibit platelet function through binding to the thromboxane A2 receptor. *J. Thromb. Haemost.* **2005**, *3*, 369–376. [CrossRef] [PubMed]

62. Khan, H.; Jawad, M.; Kamal, M.A.; Baldi, A.; Ulrih, N.P.; Nabavi, S.M.; Daglia, M. Evidence and prospective of plant derived flavonoids as antiplatelet agents: Strong candidates to be drugs of future. *Food Chem. Toxicol.* **2018**, *119*, 355–367. [CrossRef]
63. Li, Y.; Zhuang, S.; Liu, Y.; Zhang, L.; Liu, X.; Cheng, H.; Liu, J.; Shu, R.; Luo, Y. Effect of grape seed extract on quality and microbiota community of container-cultured snakehead (Channa argus) fillets during chilled storage. *Food Microbiol.* **2020**, *91*, 103492. [CrossRef]
64. Story, E.N.; Kopec, R.E.; Schwartz, S.J.; Harris, G.K. An Update on the Health Effects of Tomato Lycopene. *Annu. Rev. Food Sci. Technol.* **2010**, *1*, 189–210. [CrossRef]
65. Shah, S.M.A.; Akram, M.; Riaz, M.; Munir, N.; Rasool, G. Cardioprotective Potential of Plant-Derived Molecules: A Scientific and Medicinal Approach. *Dose-Response* **2019**, *17*, 1559325819852243. [CrossRef] [PubMed]

Publisher's Note: MDPI stays neutral with regard to jurisdictional claims in published maps and institutional affiliations.

Article

Bud-Derivatives, a Novel Source of Polyphenols and How Different Extraction Processes Affect Their Composition

Federica Turrini [1,*], Dario Donno [2], Gabriele Loris Beccaro [2], Anna Pittaluga [1], Massimo Grilli [1], Paola Zunin [1] and Raffaella Boggia [1]

[1] Department of Pharmacy, University of Genoa, Viale Cembrano 4, 16148 Genoa, Italy; pittalug@difar.unige.it (A.P.); grilli@difar.unige.it (M.G.); zunin@difar.unige.it (P.Z.); boggia@difar.unige.it (R.B.)
[2] Department of Agriculture, Forestry and Food Science, University of Torino, Largo Braccini 2, 10095 Grugliasco (TO), Italy; dario.donno@unito.it (D.D.); gabriele.beccaro@unito.it (G.L.B.)
* Correspondence: turrini@difar.unige.it

Received: 10 September 2020; Accepted: 20 September 2020; Published: 23 September 2020

Abstract: The use of herbal food supplements, as a concentrate form of vegetable extracts, increased so much over the past years to count them among the relevant sources of dietetic polyphenols. Bud-derivatives are a category of botanicals perceived as a "new entry" in this sector since they are still poorly studied. Due to the lack of a manufacturing process specification, very different products can be found on the market in terms of their polyphenolic profile depending on the experimental conditions of manufacturing. In this research two different manufacturing processes, using two different protocols, and eight species (*Carpinus betulus* L., *Cornus mas* L., *Ficus carica* L., *Fraxinus excelsior* L., *Larix decidua* Mill., *Pinus montana* Mill., *Quercus petraea* (Matt.) Liebl., *Tilia tomentosa* Moench), commonly used to produce bud-derivatives, have been considered as a case study. An untargeted spectroscopic fingerprint of the extracts, coupled to chemometrics, provide to be a useful tool to identify these botanicals. The targeted phytochemical fingerprint by HPLC provided a screening of the main bud-derivatives polyphenolic classes highlighting a high variability depending on both method and protocol used. Nevertheless, ultrasonic extraction proved to be less sensitive to the different extraction protocols than conventional maceration regarding the extract polyphenolic profile.

Keywords: bud-derivatives; *botanicals*; polyphenols; UV-Visible spectroscopic fingerprint; chemometrics; targeted chromatographic fingerprint

1. Introduction

In recent decades, food supplements have an important impact on the consumers showing a significant expectation for their health and well-being [1]. They are concentrated sources of nutrients or bioactive compounds endowed with nutritional or physiological effects and, due to their presumed health benefits, they can supplement the common diet [2,3].

In particular, the interest in herbal food supplements (botanicals) is exponentially grown and consequently the relative market has increased in all the world [4]. Botanicals are become among the most popular into the food supplements category, due to the general belief which "natural" is better, healthier and safer than synthetic drugs, although this is not always true [4]. In Italy, more than 20% of the Italian population is considered "regular" consumer of these herbal products, as highlighted from the recent European PlantLibra (Plant Food Supplements: Levels of Intake, Benefit and Risk Assessment) consumer survey [5]. The wide range of herbal food supplements on the market and the non-attendance of effective legislation to guarantee the safety and quality aspects make these products vulnerable for fraud, falsification and adulteration [6,7].

Bud-derivatives (BDs) are a relatively new category of herbal food supplements and they represent one of the supply chains investigated in the FINNOVER project (Innovative strategies for the development of cross-border green supply chains), an European Interreg Alcotra Italy/France project (2017–2020) whose aim is the green innovation of several agro-industrial chains [8]. BDs are conventionally produced, according to the European Pharmacopoeia VIII edition [9], by cold maceration of the fresh meristematic tissues of trees and herbaceous plants (i.e., buds and young sprouts) using as extraction solvent mixtures of water, ethanol and glycerol [10,11]. These natural products are already marketed, and a long history of use as dietary supplements for human well-being and health is reported in traditional medicine. No health claims are yet approved by the European Food Safety Authority (EFSA) and just for some of these botanicals pharmacognostic findings supported their use as adjuvants in several diseases. In fact, some in-vitro/in-vivo biological studies for human and veterinary use have been already reported in the literature. For examples, Allio et al. (2015) investigated whether *Tilia tomentosa* bud extracts affect hippocampal Gamma-aminobutyric acid (GABA) ergic synapses [12]. In other studies, bud extracts from *Salix caprea* L. have been demonstrated to inhibit voltage gated calcium channels and catecholamines secretion in mouse chromaffin cells [13]. Moreover, different patents have also been registered on the veterinary use of bud-extracts (e.g., Composition of *Salix caprea* bud-extract and its use in the treatment of animal endometritis, patent n. TO2015A000193) [14] and several studies are carried out on several bud-derivative biological effects [15–17].

Although gemmotherapy has been used since ancient times because of the peculiar content of buds in bio-active compounds, especially polyphenols, nowadays BDs are still a little studied "niche" production [18,19]. The lack of detailed scientific information and a clear and unique regulation, as well as for the category of herbal food supplements in overall [6,7,20], it makes these products high risk and there is an increase request for efficient quality control to ensure the proper identification of the botanical source and their content [21].

With regards to BDs, a first problem it is accidentally confusing the raw material: fresh buds must be collected, generally from spontaneous grown, in a very limited period in the late winter and/or in the early spring, corresponding to the annual germination of the plant [18]. During this period, plants may not show their distinctive characteristics and sometimes the attribution of the botanical species may be difficult for the collector. A second problem concerns the manufacturing process and the extraction protocols whose parameters are not strictly defined, and production rules are often loose and deficient [22–24].

Polyphenols play key roles in plant development processes and their synthesis increases when plants are under conditions of abiotic stress, thus helping the plant to cope with environmental constraints [25]. They form an integral part of the human diet and they are very abundant in plant-based foods, such as fruits and vegetables, tea, wine, and coffee [26]. Their chemical structure is based on at least one aromatic ring with one or more hydroxyl groups, which explains their known antioxidant and anti-inflammatory properties [27].

In recent years, many health benefits of dietetic polyphenol supplementation have been described in humans i.e., against aging and cardiovascular disease [28,29], to prevent obesity and diabetes [30,31], to modulate human gut microbiota [32] and to improve the brain cognition skills [33,34]. This knowledge guides the choice of consumers not only towards plant foods but also towards herbal food supplements, whose polyphenol content is often even more concentrated and responsible for their bioactivity [12,13]. Nevertheless, polyphenols content is strongly influenced by the manufacturing methods whose parameters are often not strictly defined (e.g., solvent ratios in the extraction mixtures, raw material/extraction mixture ratios, extraction time) and thus they could affect the final compositions [35].

In previous articles, the polyphenolic pattern of some BDs prepared starting from different botanical species have been studied [10,11,21,36].

In this research, eight species spontaneously grown and commonly used to produce BDs, i.e., *Carpinus betulus* L., *Cornus mas* L., *Ficus carica* L., *Fraxinus excelsior* L., *Larix decidua* Mill., *Pinus montana*

Mill., *Quercus petraea* (Matt.) Liebl., *Tilia tomentosa* Moench, have been taken into account as case study. Two different manufacturing methods, one conventional (maceration) and one innovative (direct sonication), as well as two different extraction protocols have been taken into account and the corresponding polyphenolic extracts' profiles have been investigated.

Pulsed Ultrasound-Assisted Extraction (PUAE), according to the six principles of the green extraction [37] and the twelve principles of green chemistry [38], has been employed as an innovative technique to quickly produce BDs comparing to the long conventional maceration [39]. Even if, the positive impacts of the ultrasound-assisted extraction, i.e., reduction of the extraction time, diminution of solvent and energy used, improvement in yield and selectivity, high reproducibility, intensification of diffusion and eliminating wastes, are known in the scientific literature [40–42], this technique is still underused in this sector.

Moreover, two different BDs manufacturing protocols, which used different extraction mixture of solvents and different solid/liquid ratio, have also been studied to evaluate both the proper identification of the botanical species and the traceability of these vegetal products regardless of extraction method and experimental conditions.

A strategy based on the untargeted UV-Visible fingerprinting coupled to chemometrics (Principal Component Analysis—PCA) has been proposed for the screening of the polyphenolic BDs profile in order to obtain a rapid control tool [43]. Finally, HPLC methods were used to obtain a targeted chromatographic profile [7,44] of the main polyphenol classes (i.e., flavonols, benzoic acids, catechins, cinnamic acids). Polyphenols are correlated with their potential health-promoting activity [45], even if they are strongly influenced both by the methods and protocols used [35].

2. Materials and Methods

2.1. Raw Samples

Buds, belonging to eight different vegetable species (*Carpinus betulus* L., *Cornus mas* L., *Ficus carica* L., *Fraxinus excelsior* L., *Larix decidua* Mill., *Pinus montana* Mill., *Quercus petraea* (Matt.) Liebl., *Tilia tomentosa* Moench) were collected from plants spontaneously grown in the Turin Province (Italy) and were immediately authenticated by an agronomist. Sampling has been performed in two years (2018 2019), from February to April, during the bud break ("balsamic period").

Table 1 reports the geo-localization coordinates of the different collection sites and the scientific naturalistic illustrations (specifically achieved during the Finnover project) of all the eight vegetable species investigated.

2.2. Chemicals

MilliQ ultrapure water, obtained by means of a Millipore equipment (Bedford, MA, USA) was used throughout. All chemicals employed for the extract preparations and for the subsequent analysis were HPLC-grade. They were supplied by VWR International S.r.l (Milan, Italy) and Sigma-Aldrich (St. Louis, MO, USA). Purity of all the used standards for HPLC analysis of BDs has been reported in the Supplementary Materials (Table S1).

Table 1. The collection sites, the corresponding geo-localization coordinates, and the scientific naturalistic illustrations of the eight different bud species.

Vegetable Species	Family (Order)	Collection Site	Geo-Localization Coordinates	Illustrations
Carpinus betulus	*Betulacee* (*Fagales*)	Bricherasio Prarostino San Germano Rostino	44.821, 7.285; 44.831, 7.272; 44.825, 7.275 44.913, 7.237 44.868, 7.253	
Cornus mas	*Cornaceae* (*Cornales*)	Bricherasio Torre Pellice Villar Pellice	44.854, 7.250; 44.855, 7.250; 44.823, 7.307 44.813, 7.181 44.804, 7.154	
Ficus carica	*Moraceae* (*Rosales*)	Brondello Pagno	44.604, 7.422; 44.603, 7.419; 44.603, 7.418 44.598, 7.424; 44.597, 7.424; 44.598, 7.425	
Fraxinus excelsior	*Oleacee* (*Lamiales*)	Angrogna Bricherasio Massello Paesana Pagno San Germano Chisone	44.869, 7.173 44.822, 7.284 44.964, 7.031 44.656, 7.261; 44.651, 7.257 44.597, 7.424; 44.598, 7.425; 44.598, 7.424 44.888, 7.261	
Larix decidua	*Pinacee* (*Pinales*)	Praly	44.902, 7.055	
Pinus montana	*Pinacee* (*Pinales*)	Masello Pramollo	44.948, 7.065; 44.948, 7.068; 44.947, 7.063 44.918, 7.193	
Quercus petraea	*Fagaceae* (*Malvales*)	Bricherasio	44.848, 7.275; 44.850, 7.274; 44.842, 7.282; 44.831, 7.270	
Tilia tomentosa	*Malvaceae* (*Malvales*)	Angrogna Bobbio Pellice Bricherasio Perrero	44.849, 7.223 44.799, 7.131 44.832, 7.265; 44.816, 7.282; 44.821, 7.273; 44.821, 7.285; 44.822, 7.283; 44.818, 7.279 44.936, 7.139	

2.3. Bud-Derivatives Manufacturing Applying Two Different Methods

Fresh buds, after their collection, were immediately processed to prepare the corresponding BDs in order to minimize any degradation preserving the peculiar phytocomplex as much undamaged as possible. The manufacturing was performed both in an Italian company (Geal Pharma Turin, Italy) and by the Authors in the analytical laboratory of the University of Genoa (Department of Pharmacy).

The following two different preparation methods of BDs were investigated: the conventional cold Maceration (M) [9], and a more rapid and innovative procedure by Ultrasounds (US) recently described by the Authors [11]. Moreover, for both preparation methods two different extraction solvents and different sample/solvent ratios were investigated too ("Protocol A" and "Protocol B", see Figure 1), in order to evaluate both the proper identification of the botanical species and the traceability of the BDs independently from their manufacturing process (Table 2). Each extraction was performed in duplicate.

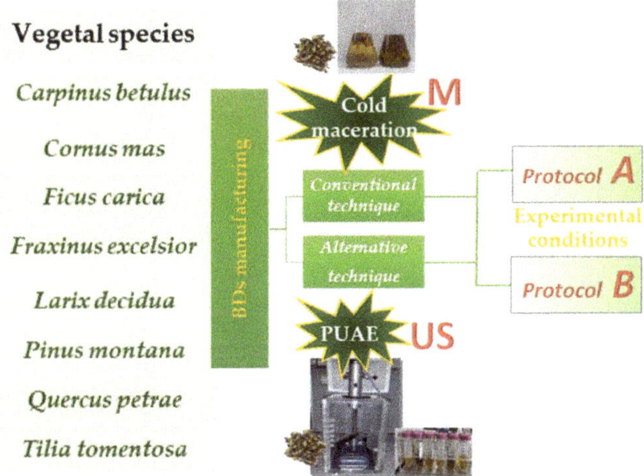

Figure 1. The global scheme of the experimental manufacturing of BDs: two methods (cold maceration, namely M, and ultrasounds, namely US) have been used. Each method has been applied following two different protocols (A and B).

2.3.1. Conventional Cold Maceration (M) as Traditional Method

BDs were prepared using a cold maceration by an Italian Company of botanicals (Geal Pharma, Bricherasio, Turin) following two different experimental manufacturing protocols, reported in Table 2:

(A) A 21 days maceration of buds in glycerol/ethanol 96% (1/1 *w/w*) with a 1:20 bud/solvent ratio (considering the dry weight) has been performed, according to the official method of glyceric macerates reported in the European Pharmacopoeia VIII edition [9] ("M_A").

(B) A 3 months maceration of buds in a mixture of water/glycerol/ethanol 96% (50/20/30 *w/w/w*) as extraction solvent with a bud/solvent ratio variable (considering the fresh weight) depending on the botanical species (see Table 2) has been used, according to the method optimized and used by the Company to produce glyceric macerates ("M_B").

In both methods, after the maceration step, the extracts, namely BDs, have been obtained by a preliminary filtration, a manual pressing and a second filtration after two days of decanting. The so obtained BDs were stored at 4 °C in the dark until their further analysis.

Table 2. BDs obtained starting from the eight vegetable species (raw materials). Two different methods (cold maceration-M and Pulsed Ultrasound-Assisted Extraction US) and two different experimental protocol (Protocol A and B) are taken into account.

	Sample Identification Code	Vegetable Species	Extraction Method	Experimental Protocol	Bud/Solvent Ratio
1	Cb_M_A	*Carpinus betulus*	M	Protocol A	1/20 DW
2	Cb_US_A	*Carpinus betulus*	US	Protocol A	1/20 DW
3	Cb_M_B	*Carpinus betulus*	M	Protocol B	1/15 FW
4	Cb_US_B	*Carpinus betulus*	US	Protocol B	1/15 FW
5	Cm_M_A	*Cornus mas*	M	Protocol A	1/20 DW
6	Cm_US_A	*Cornus mas*	US	Protocol A	1/20 DW
7	Cm_M_B	*Cornus mas*	M	Protocol B	1/20 FW
8	Cm_US_B	*Cornus mas*	US	Protocol B	1/20 FW
9	Fc_M_A	*Ficus carica*	M	Protocol A	1/20 DW
10	Fc _US_A	*Ficus carica*	US	Protocol A	1/20 DW
11	Fc _M_B	*Ficus carica*	M	Protocol B	1/10 FW
12	Fc_US_B	*Ficus carica*	US	Protocol B	1/10 FW
13	Fe_M_A	*Fraxinus excelsior*	M	Protocol A	1/20 DW
14	Fe_US_A	*Fraxinus excelsior*	US	Protocol A	1/20 DW
15	Fe_M_B	*Fraxinus excelsior*	M	Protocol B	1/10 FW
16	Fe_US_B	*Fraxinus excelsior*	US	Protocol B	1/10 FW
17	Ld_M_A	*Larix decidua*	M	Protocol A	1/20 DW
18	Ld_US_A	*Larix decidua*	US	Protocol A	1/20 DW
19	Ld_M_B	*Larix decidua*	M	Protocol B	1/20 FW
20	Ld_US_B	*Larix decidua*	US	Protocol B	1/20 FW
21	Pm_M_A	*Pinus montana*	M	Protocol A	1/20 DW
22	Pm_US_A	*Pinus montana*	US	Protocol A	1/20 DW
23	Pm_M_B	*Pinus montana*	M	Protocol B	1/10 FW
24	Pm_US_B	*Pinus montana*	US	Protocol B	1/10 FW
25	Qp_M_A	*Quercus petraea*	M	Protocol A	1/20 DW
26	Qp_US_B	*Quercus petraea*	US	Protocol A	1/20 DW
27	Qp_M_B	*Quercus petraea*	M	Protocol B	1/15 FW
28	Qp_US_B	*Quercus petraea*	US	Protocol B	1/15 FW
29	Tt_M_A	*Tilia tomentosa*	M	Protocol A	1/20 DW
30	Tt_US_A	*Tilia tomentosa*	US	Protocol A	1/20 DW
31	Tt_M_B	*Tilia tomentosa*	M	Protocol B	1/15 FW
32	Tt_US_B	*Tilia tomentosa*	US	Protocol B	1/15 FW

DW: dry weight; FW: fresh weight.

2.3.2. Green Extraction: Pulsed Ultrasound-Assisted Extraction (US) as Alternative Method

Fresh buds were finely ground by a Grindomix 200 M (Retsch, Haan, Germany) for 20 s at 5000 rpm, and then sieved by a 150 μm sieve, in order to improve the efficiency of the following extraction step [46]. PUAE was performed directly by an Hielscher UP200St sonicator (Teltow, Germany) equipped with an ultrasonic titanium sonotrode (7 mm of diameter), at a constant frequency of 26 kHz. The pulsed mode, referring to an alternation of "on" time and "off" time of the sonicator, guarantees a lowering increase in temperature, which better preserve the phytocomplex, and greater energy savings compared to continuous treatments [47]. The experimental sonication conditions (amplitude 30%, duty cycle 65%, extraction time 20 min) were previously optimized by the Authors on the same raw materials [11].

The same two experimental extraction conditions described in the paragraph 2.3.1 ("Protocol A" and "Protocol B", see Figure 1) were employed ("US_A" and "US_B", see Table 2). The extracts obtained were filtered for Buchner (Whatman n. 1 paper), centrifuged at 3000 rpm for 10 min and then stored at 4 °C in the dark until analysis.

2.4. Spectroscopic Analysis: UV-Visible Fingerprint

UV–Visible absorption spectra (200 nm–900 nm) were recorded by a spectrophotometer Agilent Cary 100 (Varian Co., Santa Clara, CO, USA) with 0.5 nm resolution, using rectangular quartz cuvettes with 1 cm path length. BDs, before the spectroscopic analysis, were suitably diluted in the corresponding extraction solvent (glycerol/ethanol 1/1 *w/w* or water/glycerol/ethanol 50/20/30 *w/w/w*) depending on the followed experimental protocol ("Protocol A" and "Protocol B", respectively). Dilution was necessary to avoid signal saturation but was subsequently considered in order to make a comparison between the different spectra achieved. BDs spectra were acquired in duplicate and then averaged. The collection was performed at room temperature (25 ± 1 °C), against a blank solution represented by the corresponding extraction solvent.

2.5. HPLC Analysis

In this study, effective HPLC–DAD methods were used for fingerprint analysis and phytochemical identification of samples. Four polyphenolic classes were considered: benzoic acids (ellagic and gallic acids), catechins ((+)catechin and (-)epicatechin), cinnamic acids (caffeic, chlorogenic, coumaric, and ferulic acids), and flavonols (hyperoside, isoquercitrin, quercetin, quercitrin, and rutin). Total bioactive compound content (TBCC) was determined as the sum of the most important bioactive compounds with positive effects on human organism ("multimarker approach") [48].

The external standard method was used for quantitative determination of bioactive compounds. Stock solutions of cinnamic acids and flavonols with a concentration of 1.0 mg·mL^{-1} were prepared in methanol: five calibration standards were prepared by dilution with methanol; stock solutions of benzoic acids and catechins with a concentration of 1.0 mg·mL^{-1} were prepared in 95% methanol and 5% water. In this case, five calibration standards were prepared by dilution with 50% methanol–water.

An Agilent 1200 High-Performance Liquid Chromatograph coupled to an Agilent UV-Vis diode array detector (Agilent Technologies, Santa Clara, CA, USA) was used for the chromatographic analysis. Four chromatographic methods were used to separate the bioactive molecules on a Kinetex C18 column (4.6 × 150 mm, 5 μm, Phenomenex, Torrance, CA, USA). Several mobile phases were used for bioactive compound identification and UV spectra were recorded at different wavelengths, based on HPLC methods, previously tested and validated [10,40], with some modifications: (i) a solution of 10 mM KH2PO4/H3PO4 (A) and acetonitrile (B) with a flow rate of 1.5 mL·min^{-1} (method A—analysis of cinnamic acids and flavonols, gradient analysis: 5% B to 21% B in 17 min + 21% B in 3 min + 2 min of conditioning time); (ii) a solution (A) of methanol/water/formic acid (5:95:0.1 *v/v/v*) and a mix (B) of methanol/formic acid (100:0.1 *v/v*) with a flow rate of 0.6 mL·min^{-1} (method B—analysis of benzoic acids and catechins, gradient analysis: 3% B to 85% B in 22 min + 85% B in 1 min + 2 min of conditioning time). UV spectra were recorded at 330 nm (A); 280 nm (B).

Biomarkers were selected for their demonstrated positive healthy properties and antioxidant capacity by literature in relation to the use of this plant-derived products. All single compounds were identified in samples by comparison and combination of their retention times and UV spectra with those of authentic standards in the same chromatographic conditions. Each sample was analyzed in triplicate and results were reported as mean value ± standard deviation to assess the repeatability of the employed methods.

2.6. Data Analysis

2.6.1. Chemometric Analysis

Multivariate data analysis has been performed by CAT (*Chemometric Agile Tool*) software, one advanced chemometric multivariate analysis tool based on R, developed by the Chemistry Group of the Italian Chemical Society [49].

PCA was applied as common multivariate statistical method of unsupervised pattern recognition. Its aim is extracting important information from the data and decreasing the high-dimensional dataset volume by maintaining the important information [50,51].

2.6.2. Data Matrices Organization

A data matrix $A_{32,601}$ consisting of 32 rows (corresponding to the BDs analyzed, 4 samples for each of the eight botanical species investigated) and 601 columns (the absorbance values in the range of 200–500 nm of the UV-Visible spectra, with 0.5 nm of resolution) was prepared and further analyzed by PCA. Standard normal variate (SNV) transform and column autoscaling were previously performed on the spectral data to remove multiplicative effects of scattering and to scale the data, respectively [52].

Available sample were divided in two different subsets: a calibration (or training) set and a test (or evaluation) set in order to build and validate the statistical model, respectively [53]. For a reliable validation strategy, it is important that data used as test set were not used to build the model in order to avoid the overestimations of the prediction ability [53]. 32 samples, previously reported in Table 2, were selected for the construction and identification of the model (Calibration set). The representative calibration data set consisted of 4 extracts (M_A, M_B, US_A, US_B) for each botanical species investigated (*Carpinus betulus* L., *Cornus mas* L., *Ficus carica* L., *Fraxinus excelsior* L., *Larix decidua* Mill., *Pinus montana* Mill., *Quercus petraea* (Matt.) Liebl., *Tilia tomentosa* Moench). Furthermore 16 BDs, obtained both by conventional maceration and ultrasound extraction respectively from the same eight vegetal species, were randomly selected and used as an independent set to test the model and assess its validity (Test set, Table 3).

Table 3. External test set. 16 BDs obtained starting from the eight vegetable species using two different methods (cold maceration M and Pulsed Ultrasound-Assisted Extraction US) and two different experimental protocol (Protocol A and B) are taken into account as independent set to test the statistical model.

	Sample Identification Code	Vegetable Species	Extraction Method	Experimental Protocol
1	Cb_TS	*Carpinus betulus*	US	Protocol A
2	Cb_TS2	*Carpinus betulus*	US	Protocol B
3	Cm_TS	*Cornus mas*	US	Protocol A
4	Cm_TS2	*Cornus mas*	US	Protocol B
5	Fc_TS	*Ficus carica*	US	Protocol A
6	Fc _TS2	*Ficus carica*	US	Protocol B
7	Fe_TS	*Fraxinus excelsior*	M	Protocol A
8	Fe_TS2	*Fraxinus excelsior*	US	Protocol A
9	Ld_TS	*Larix decidua*	US	Protocol A
10	Ld_TS2	*Larix decidua*	US	Protocol B
11	Pm_TS	*Pinus montana*	M	Protocol A
12	Pm_TS2	*Pinus montana*	US	Protocol A
13	Qp_TS	*Quercus petraea*	M	Protocol A
14	Qp_TS2	*Quercus petraea*	US	Protocol A
15	Tt_TS	*Tilia tomentosa*	US	Protocol A
16	Tt_TS2	*Tilia tomentosa*	US	Protocol B

All the pre-treated UV-Visible absorption spectra, in the range 200–500 nm, are reported in Figure 2. For each species, the four averaged spectral profiles corresponding to the Calibration set (Table 2) are highlighted in grey while in red have been reported the Test set samples (TS/TS2) belonging to the same class.

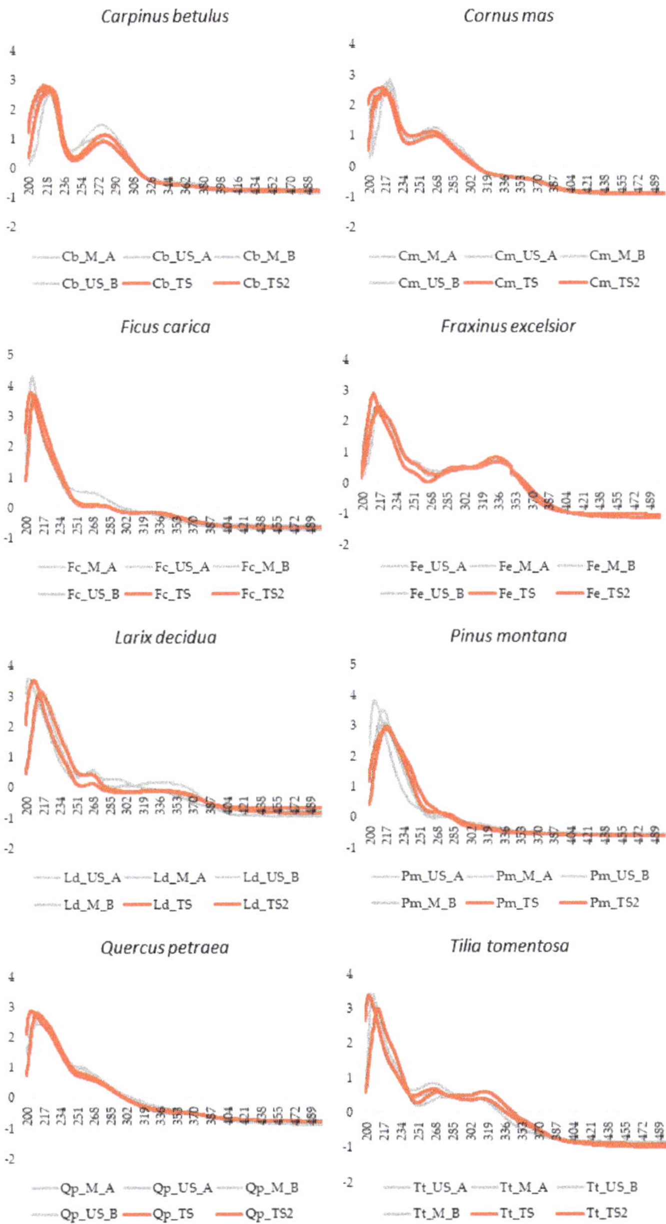

Figure 2. Averaged UV-Visible spectra of the 8 botanical species after SNV pre-treatment of data. For each species, the four averaged spectral profiles of the Calibration set (Table 2) are highlighted in grey while in red are reported the External Test set samples (Table 3).

Then, a data matrix $B_{32,620}$ consisting of 32 rows and 620 columns was prepared and analogously analyzed by PCA. $B_{32,620}$ rows correspond to the 32 BDs analyzed (Calibration set), and columns are the absorbance values of the UV-Visible spectra after SNV in the range 200–500 nm coupled to the chromatographic quantifications by HPLC (4 polyphenolic classes and 13 bioactive compounds). The data set was previously scaled by using a block scaling procedure [54], with the aim to give to the spectroscopic and chromatographic variables a comparable influence in the data analysis. In fact, this pretreatment allows to divide variables in different blocks whose values will be scaled to attain the same block-variance after pretreatment. Moreover, the variables belonging to the same block are equally weighted.

3. Results and Discussion

The quality control of vegetal material is critical both if the botanical product is to be used as a drug or as an herbal food supplement. For consumer safety and the protection of who operate in this industrial field, quality control should be applied throughout the different processing steps, from the raw material to the final product. Scientific-naturalistic illustrations of the most common buds used in BDs production (Table 1) have been realized within the Finnover project by an expert botanical graphic designer, in order to provide a useful first tool for the operators in the BDs manufacturing. In fact, this peculiar raw material is generally spontaneously collected and mistakes in the attribution of some botanical species may be possible. For this, bud illustrations could represent a preliminary control of these vegetable materials after their collection in the point of view of a controlled manufacturing chain of BDs.

Moreover, a strategy based on the untargeted UV-Visible fingerprinting coupled to chemometrics allows rapid screening of the polyphenolic BDs profile to obtain a preliminary control tool to identify the botanical species.

3.1. Bud-Derivatives Identification: UV-Visible Fingerprint

Figure 2 show the UV–Visible spectral profiles, after SNV pretreatment of the data, recorded for the eight vegetable species investigated: *Carpinus betulus* L., *Cornus mas* L., *Ficus carica* L., *Fraxinus excelsior* L., *Larix decidua* Mill., *Pinus montana* Mill., *Quercus petraea* (Matt.) Liebl., *Tilia tomentosa* Moench. The extracts were obtained by the conventional maceration and the innovative green extraction (M or US) respectively, using the two experimental protocols (A or B) as described in detail in Table 2. Ultrasounds represent one of the innovative processing techniques of officinal plants [39]. In fact, several companies already exploit innovative applications of ultrasound to obtain liquid foods, beverages, and alcoholic drinks [55,56]. Previously, the Authors described PUAE as an alternative time-saving method to the conventional maceration for the extraction of the polyphenolic fraction from buds [11]. Particularly, PUAE on a lab pilot reactor demonstrated to be an excellent approach for a rapid (20 min vs. 21 days or 3 months of maceration, depending on the Protocol applied) and efficient extraction of phenolic compounds.

Looking at Figure 2, the spectra of the different vegetable species are quite different, highlighting as the pattern of absorbances in the UV–Visible region is strictly connected with the botanical origin of the plants. On the contrary, for each botanical species the spectral differences due to the extraction method (M or US) and to the extraction solvent (Protocol A or B), are minimal. The 501–900 nm interval has been preliminarily removed because there were none interesting absorptions in this spectral region at the assayed concentrations.

PCA, an unsupervised pattern recognition technique [50,51], was applied in order to explore and to analyze the data set using a multivariate approach since the analytical information contained in each spectrum was considered as a multivariate fingerprint. Particularly, the data matrix $A_{32,601}$, whose rows are the extracts (Calibration set) and the columns are the absorbances recorded in the spectral range 200–500 nm, was considered. PCA was performed on the pretreated and autoscaled data matrix. The first two principal components (PCs) of the data set ($A_{32,601}$), which together explained the 77.9% of the total information of the data set since they visualize almost the 80% of the total variance, were firstly taken into account. Figure 3a,b shows the PCA score plots on the 1st–2nd principal components (PC1-PC2) obtained from the above-mentioned data matrix. In Figure 3a the extracts are categorized according to the vegetable species and each one is visualized with a different color (*Carpinus betulus* L.: black, *Cornus mas* L.: red, *Ficus carica* L: green., *Fraxinus excelsior* L.: blue, *Larix decidua* Mill.: brown, *Pinus montana* Mill.: light blue, *Quercus petraea* (Matt.) Liebl.: orange, *Tilia tomentosa* Moench: pink). In Figure 3b, for each vegetable class all the extracts belonging to the calibration set were indicated with their identification code (see Table 2). PC1, the direction of maximum variance which explains almost the 60% of the total information, allows good discrimination between the botanical class regardless of the extraction method (M or US) and the experimental preparation protocol (A or B). Particularly, the *Fraxinus* class (blue, lowest scores on PC1) separates from *Ficus* (green) and *Pinus* (light blue) which have higher scores on PC1. PC2, which explains the 21.1% of the remaining variance, allows to mainly separate *Larix* class (brown, highest scores on PC2) from *Quercus* (orange) and *Carpinus* (black, lowest scores on PC2).

Figure 3c,d show the PCA score plots on the PC1-PC3, which explain together the 69.3% of the total variance of the data set. A good separation among the above cited botanical classes is also highlighted except for *Larix* and *Carpinus* ones. In fact, these latter separate on PC2 (Figure 3a,b) and since PCs are orthogonal, they are uncorrelated and no duplicate information are shown in their plots [50].

In Figure 3e,f, the projections of the external test set (red samples) were reported on the PC1-PC2 and PC1-PC3 score plots respectively, showing a good correspondence with the calibration set for each botanical species.

The spectral variables having greater importance (loading values) on the first three PCs are represented by spectral areas near the following absorbances (in ascending order): 200 nm, 212 nm, 240 nm, 275 nm, 310 nm, 360 nm, 420 nm, as highlighted in the Loading plot on PC1-PC2-PC3 (Figure 4).

Several of them could be related to some secondary metabolites largely distributed in plant material (even in buds) such as tannins, whose structural variability depends on the vegetal species and even among organs of the same plant species [57]. The chemotaxonomic values of tannins have been recognized in the literature for several botanical species [58,59] and, the distribution of hydrolysable tannins has been used as chemotaxonomic markers by several authors [60].

It is well known that the different classes of tannins present characteristic absorption bands in the UV spectral region. Particularly as far as hydrolysable tannins are concerned, gallotannins show two characteristic absorption maximums, λ max around 212 nm and λ max around 275 nm, with an inflection point (λ min) around 242 nm; ellagitannins present strong absorption near 200 nm and a shoulder around 277 nm and another absorption near 360 nm. Instead condensed tannins (or proanthocyanidins), chemically defined as flavonoid polymers in which the phenolic hydroxyls are partially or totally esterified with gallic acid, present an absorption around 200 nm, a λ min between 258–259 nm and λ max between 279–281 nm [57]. Nevertheless, also other polyphenols, such as hydroxycinnamic acids and flavonoids, could contribute to the UV-Visible fingerprints, even if some of them are more ubiquitarians and lesser species-specific [61,62]. Furthermore, as far as flavonoids are concerned, it is important to underline that their absorptions in the Visible are almost negligible at the measured concentrations, which are instead useful to avoid saturation of the UV region.

The fingerprint UV-Visible, at least in a preliminary screening step, seems to discriminate the peculiar polyphenols composition of BDs and could be a simple and quick method to confirm the proper identification of the botanical source after the botanic check by a professional botanist.

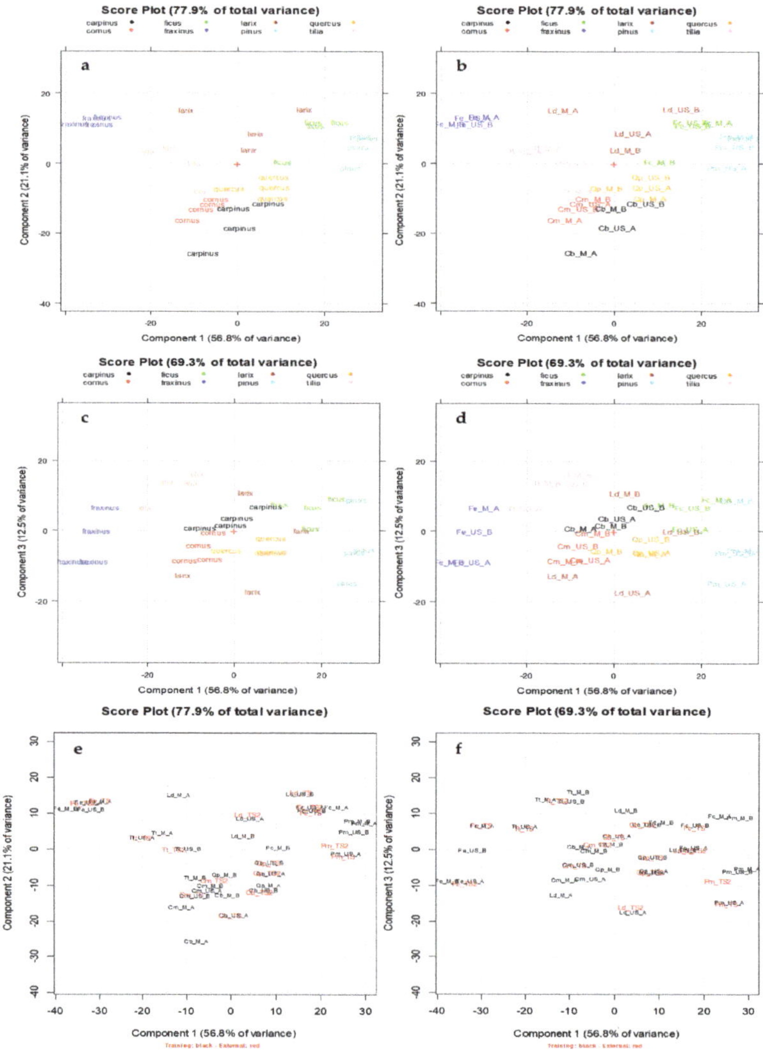

Figure 3. The scores plots of the UV–Visible absorbances data matrix $A_{32,601}$. Each vegetable species is reported with a different color (*Carpinus betulus* L.: black, *Cornus mas* L.: red, *Ficus carica* L: green., *Fraxinus excelsior* L.: blue, *Larix decidua* Mill.: brown, *Pinus montana* Mill.: light blue, *Quercus petraea* (Matt.) Liebl.: orange, *Tilia tomentosa* Moench: pink). (**a**) PC1-PC2 score plot with BDs categorized according to the vegetable species; (**b**) PC1-PC2 score plot with BDs categorized according to their identification code (Table 2); (**c**) PC1-PC3 score plot with BDs categorized according to the vegetable species; (**d**) PC1-PC3 score plot with BDs categorized according to their identification code (Table 2); (**e**) PC1-PC2 score plot obtained projecting the external test set samples (highlighted in red); (**f**) PC1-PC3 score plot obtained projecting the external test set samples (highlighted in red).

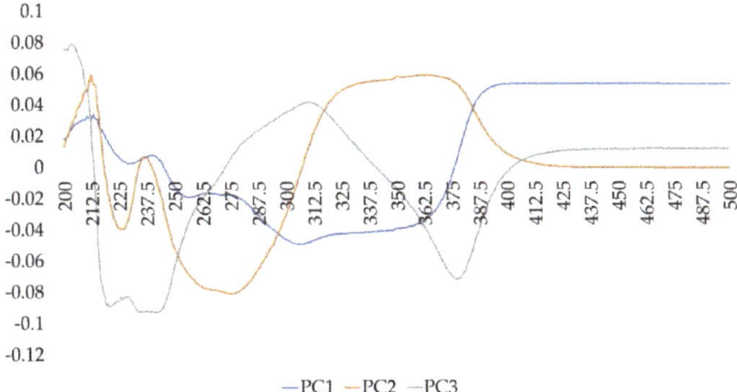

Figure 4. Loading plot on PC1, PC2, PC3.

3.2. Bud-Derivatives Identification: UV-Visible and HPLC Fingerprints

Figure 5 shows the PCA plots of the data matrix $B_{32,620}$ on PC1-PC2, which together explained the 76.2% of the total variance.

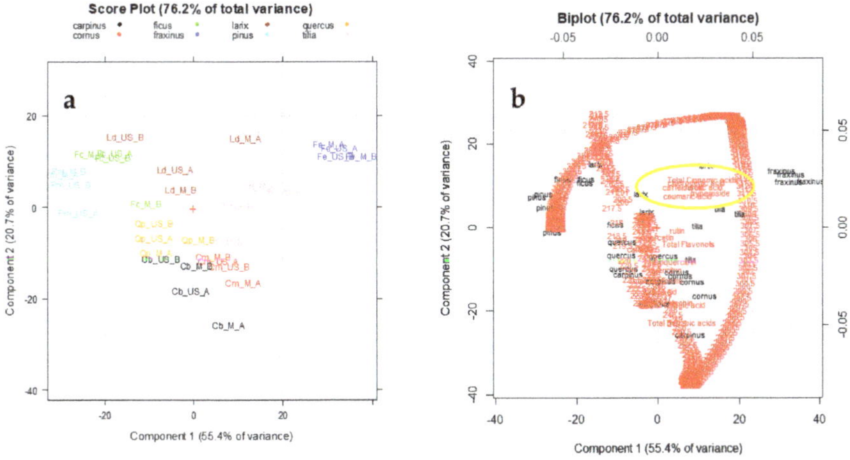

Figure 5. The PC1–PC2 plots of the UV–Visible absorbances coupled to the HPLC data (data matrix $B_{32,620}$): (**a**) Score plot; (**b**) Biplot.

PCA was performed on the pretreated and autoscaled data matrix, after the block scaling treatment in order to consider in the data analysis the same importance for the spectroscopic and chromatographic variables [63]. The PC1-PC2 score plot (Figure 5a) highlights a good separation between the vegetal species. Particularly PC1, which represents the direction of maximum variance explaining the 55.4% of the total information, allows good discrimination between *Fraxinus* class (blue, highest scores on PC1), *Ficus* (green) and *Pinus* (light blue) classes, which have lowest scores on this PC. As highlighted in the Biplot (Figure 5b) the variables having greater importance (loading value) on this separation are represented by total cinnamic acids, caffeic acid, coumaric acid and hyperoside content which are high in *Fraxinus* species and very low in *Pinus* one (as reported in Table 4). Instead PC2, which explains the 20.7% of the remaining information, allows mainly to separate *Carpinus* (black) and *Cornus* (red) classes from all the other ones. These species result particularly rich in tannins (catechins and benzoic acids).

Table 4. Bioactive classes and total phenolics in the analyzed samples.

Sample ID	Cinnamic Acids		Flavonols		Benzoic Acids		Catechins		Total Phenolics	
	Mean Value	SD	Mean Value	SD	Mean Value	SD	Mean Value	SD	Mean Value	SD
	(mg/100 gFW **)		(mg/100 gFW **)		(mg/100 gFW **)		(mg/100 gFW **)		(mg/100 gFW **)	
Tt_M_A	5.30	0.73	51.64	2.66	22.98	0.79	52.17	1.46	132.09	5.64
Tt_M_B	23.87	1.06	90.79	5.02	6.62	1.04	50.68	1.03	171.97	8.16
Tt_US_A	5.33	1.39	71.26	5.92	132.56	1.68	156.46	1.78	365.61	10.77
Tt_US_B	12.43	5.20	100.23	14.84	96.28	8.41	81.15	10.16	290.10	38.61
Pm_M_A	n.d.	/	31.13	1.45	n.d.	/	171.38	1.65	202.51	3.10
Pm_M_B	n.d.	/	n.d.	/	n.d.	/	49.36	2.29	49.36	2.29
Pm_US_A	n.d.	/	31.36	3.86	3.67	1.56	378.90	2.54	413.93	7.96
Pm_US_B	n.d.	/	38.74	4.35	n.d.	/	325.88	4.77	364.62	9.12
Ld_M_A	n.d.	/	275.15	0.91	97.07	0.31	112.09	0.67	484.31	1.88
Ld_M_B	n.d.	/	151.57	2.23	137.23	0.88	70.90	2.62	359.70	5.72
Ld_US_A	2.40	1.02	810.86	3.32	190.25	0.95	152.12	2.12	1155.63	7.42
Ld_US_B	n.d.	/	941.62	13.22	219.28	3.66	127.08	7.33	1287.98	24.21
Fe_M_A	829.03	2.26	499.08	2.52	214.49	0.69	328.25	1.68	1870.85	7.15
Fe_M_B	119.44	0.98	223.61	3.43	40.81	1.25	98.75	2.52	482.61	8.18
Fe_US_A	151.00	2.32	378.93	4.62	115.82	0.93	225.26	2.21	871.01	10.07
Fe_US_B	113.53	6.70	551.07	10.06	77.40	2.30	215.96	5.28	957.96	24.34
Cm_M_A	23.97	0.40	1055.03	1.87	577.48	0.37	104.70	0.53	1761.19	3.18
Cm_M_B	24.59	1.55	310.99	2.06	541.34	2.35	1161.65	2.48	2038.58	8.45
Cm_US_A	14.87	1.04	672.04	3.57	276.38	1.33	98.83	1.21	1062.12	7.15
Cm_US_B	n.d.	/	784.79	12.98	329.55	2.85	167.03	4.67	1281.37	20.50
Cb_M_A	47.04	0.83	442.45	2.04	286.40	1.25	523.93	1.14	1299.83	5.26
Cb_M_B	n.d.	/	203.20	1.18	418.85	2.56	248.73	2.73	870.78	6.47
Cb_US_A	n.d.	/	230.16	2.82	80.56	1.04	297.57	1.07	608.29	4.92
Cb_US_B	n.d.	/	198.98	5.89	206.42	4.05	227.60	3.00	633.00	12.95
Fc_M_A	62.21	0.84	287.89	4.35	67.29	0.89	267.35	2.16	684.74	8.25
Fc_M_B	n.d.	/	123.28	3.65	45.86	1.08	68.42	2.11	237.57	6.83
Fc_US_A	6.49	2.62	116.68	4.31	26.33	1.18	138.27	2.64	287.77	10.76
Fc_US_B	10.77	5.54	155.02	11.39	52.18	3.49	183.91	7.34	401.88	27.76
Qp_M_A	5.08	0.65	223.63	1.97	283.59	1.28	294.75	0.85	807.06	4.75
Qp_M_B	n.d.	/	59.40	2.75	84.02	2.16	109.81	2.18	253.23	7.09
Qp_US_A	1.76	1.29	55.98	4.96	223.32	2.35	253.81	2.23	534.87	10.83
Qp_US_B	n.d.	/	72.09	8.50	58.43	5.70	161.81	4.89	292.32	19.08

SD: standard deviation; ** FW: fresh weight

In the Supplementary materials an example (*Larix decidua*) of chromatographic pattern was reported. As shown in Figure 5, the addition of chromatographic variables does not greatly improve the taxonomic separation previously obtained by the only UV-Visible fingerprint (Figure 3). However, these results show that the main polyphenols evaluated could be useful markers for identifying the botanical species regardless of the extraction method and the experimental preparation protocol.

3.3. Phenolic Composition of BDs

In this study, the health-promoting compounds were grouped into four different polyphenolic classes in order to assess the contribution of each class to the phytocomplex composition of buds belonging to the eight different species: cinnamic acids (as sum of caffeic acid, chlorogenic acid, coumaric acid, ferulic acid), flavonols (as sum of hyperoside, isoquercitrin, quercetin, quercitrin and rutin), benzoic acids (ellagic and gallic acids) and catechins ((+)catechin and (-)epicatechin). The identification and quantification of each single bioactive compound, expressed in mg/100 g $_{FW}$, is reported in the Supplementary Materials (Table S2). For a better data visualization, Figure 6 shows the radar plot, made considering for each botanical species the mean values obtained from the 4 different extracts (M_A, M_B, US_A, US_B) for each marker compound quantified.

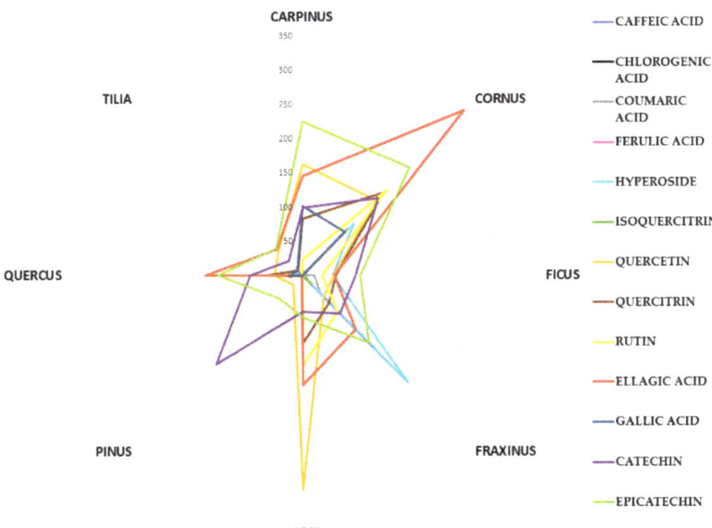

Figure 6. The mean content of each phenolic marker (caffeic acid, chlorogenic acid, coumaric acid, ferulic acid, hyperoside, isoquercitrin, quercetin, quercitrin and rutin, ellagic acid, gallic acid, (+)catechin and (-)epicatechin) for the eight botanical species investigated.

Several markers of cinnamic acids were considered but not detected in all the extracts. *Fraxinus excelsior* BDs showed the highest content in cinnamic acids (ranged from 113.53 ± 6.70 to 829.03 ± 2.26 mg/100 g $_{FW}$), and as shown in Table S2, ferulic and chlorogenic acids were the most abundant. *Cornus mas* and *Tilia tomentosa* species showed very low amounts of ferulic acid (respectively 12.14 and 11.73 mg/100 g $_{FW}$), while in the other species it was not detected. In recent years, several physiological functions of ferulic acid have been demonstrated [64]. Particularly, its free radical scavenging activity and its cholesterol-lowering activity, together with the low toxicity, suggested its chemo preventive effects on heart diseases [65]. Chlorogenic acid is also involved in beneficial effects on human health due to its anti-inflammatory, antioxidative, anti-aging and anticancer activities [66]. Chlorogenic acid was detected only in *Fraxinus excelsior* BDs (ranges from 43.88 to 489.94 mg/100 g$_{FW}$), in all the other species it was not detectable. Li et al. 2013 reported that chlorogenic acid and flavonols may be considered the main phenolic compounds responsible for in vitro anti-cancer property (i.e., against breast, colon, liver and lung cancer) [66]. As regards the total flavonol content, it was highly variable among species. The highest content was quantified in *Cornus mas* species (mean value: 705.71 mg/100 g $_{FW}$) while the lowest value in *Pinus montana* (mean value: 25.31 mg/100 g $_{FW}$). As highlighted in Figure 6, quercetin represented the phenolic marker of BDs belonged to *Larix decidua* species (in orange), while hyperoside was more abundant in *Fraxinus excelsior* ones (in light blue).

Benzoic acids are known to be very important in the human diet because of their relation to many biological and functional activities including antioxidative, anti-inflammatory, anticancer and antihepatotoxic properties [67]. Gallic acid, due to its antioxidant activity, has been shown to be effective against oxidative stress (OS), and many other properties have been reported (i.e., anti-mutagenic, anti-carcinogenic, antiviral, antibacterial, anti-inflammatory, antithrombotic and anti-atherosclerotic activities) [68]. A multi-target activity of ellagic acid, mainly ascribed to its antioxidant property and free radical trapping ability, has been reported too. In particular anti-angiogenic, anti-atherogenic, anti-carcinogenic, anti-obesity, anti-inflammatory, antioxidant, anti-thrombotic and anti-neurodegenerative properties have been demonstrated [69]. Ellagic acid was very abundant in almost all the described species (Table S2) while gallic acid was not detectable in *Ficus carica*, *Fraxinus excelsior*, *Larix decidua* and *Pinus montana* species. The highest content in ellagic acid was identified in *Cornus mas* extracts, followed by *Larix decidua* and *Quercus petraea* BDs (Figure 6, in red).

Catechins have important effects on human health thanks to its antioxidant, anti-inflammatory, antidiabetic, and antimicrobial properties [67]. The intake of foods and dietary supplements rich in catechins could have an important role in the prevention of various diseases (i.e., cardiovascular diseases), inhibition of lipid peroxidation, improvement of blood flow, elimination of several toxins and inhibition of human cancer cell line proliferation and cyclooxygenase enzymes [70]. All the vegetal species considered in this research were a good source of catechins (catechin and epicatechin) as shown in Table 3 and Table S2. Particularly, as highlighted in Figure 6, catechin represented the phenolic marker of *Pinus montana* BDs (in violet), while epicatechin was more abundant in *Carpinu betulus* and *Quercus petraea* extracts (in light green).

All BDs analyzed showed a good content of phenolics although there was a high variability both between the different vegetal species and between the extracts obtained by the different manufacturing method and experimental conditions starting from the same botanical species. Figure 7 showed the radar plots of each botanical species in order to better highlight the phenolic composition of the 4 different extracts (M_A, M_B, US_A, US_B).

As showed in Figure 7, the manufacturing methods (conventional maceration or sonication) and the experimental conditions used for the preparation of BDs (i.e., extraction solvent, extraction time, solid/solvent ratio, extraction time) strongly influenced the phenolic extraction yield despite having removed the variability of the raw material (same batch of buds for each vegetal species). Generally US_A (green line) and US_B (red line) appears more similar in terms of phenolic composition respect to M extracts (M_A and M_B), except for some species, such as *Pinus montana* and *Larix decidua*, in which there is a greater homogeneity in the polyphenolic profile of the final products. In almost all species, the M_B extract (yellow line) is the most different from the others. In example, the M_B extract of *Cornus mas* was rich in catechin which was not detected in extracts obtained by different extraction conditions (M_A, US_A, US_B). Analogously, rutin represents a phenolic marker of the M_B extract of *Fraxinus excelsior*, while it was poorly detectable in the other extracts of the same species. Surely Protocol A, according to the European Pharmacopoeia, provided a higher alcoholic concentration of the extraction solvent than protocol B and it is known that a higher solvent polarity allows a higher phenolic extraction from plant materials [71]. Moreover, Protocol A used an higher solid/solvent ratio because it is evaluated on the dry weight of the raw material while following the industrial Protocol (B), the fresh weight of buds was taken into account. Regarding the effect of ultrasounds, the implosion of cavitation bubbles on the material surface results in micro-jetting which generates several effects such as surface peeling, detexturation, erosion and cell breakdown [40]. Probably, the destruction of vegetal cells allowed to increase the extraction yield making up for the lower alcohol content of protocol B.

Due to the lack of a single regulation and an unique preparation protocol for these botanicals, very different products can be found on the market in terms of their polyphenolic fraction depending on both the raw materials (i.e., taking into account their specific agro-environmental and biological traits) and on the experimental conditions of manufacturing (method of preparation, extraction solvent, solid/solvent ratio, extraction time).

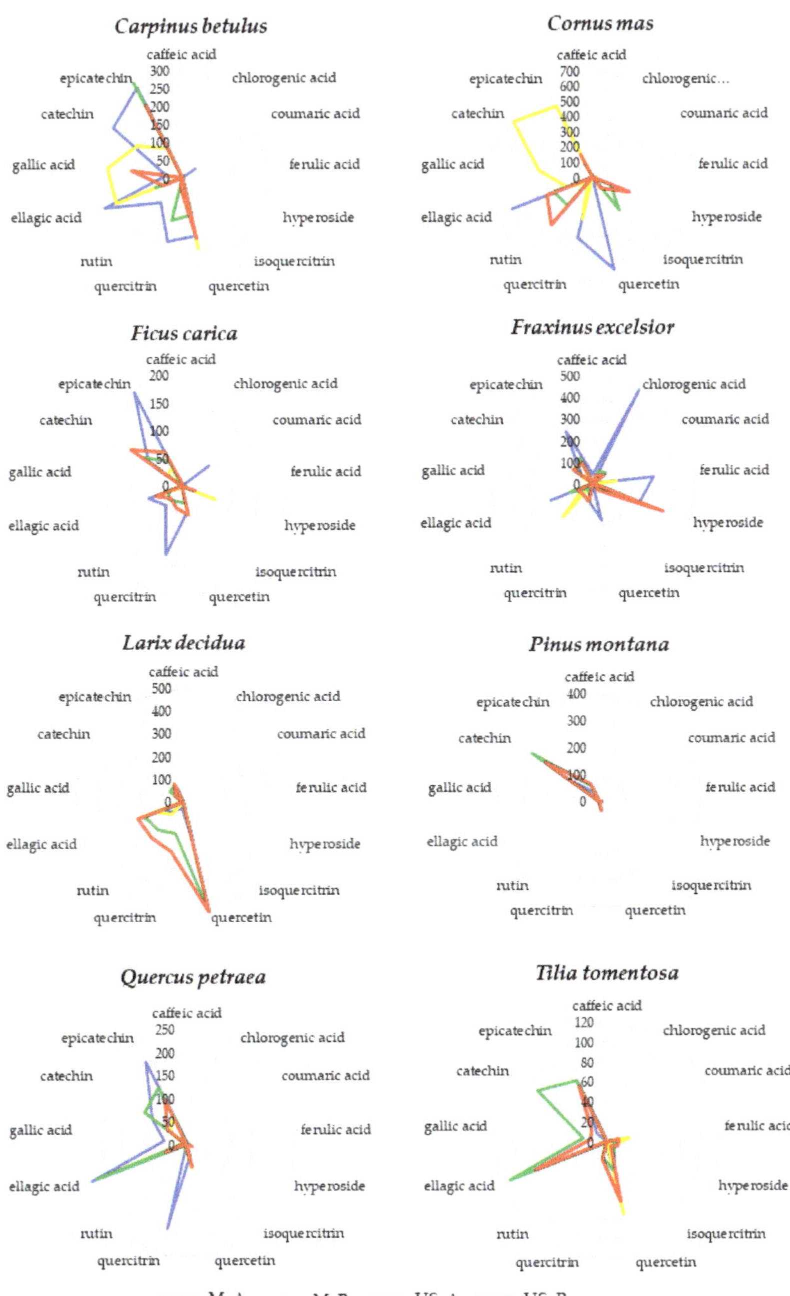

Figure 7. For each botanical species the phenolic composition (caffeic acid, chlorogenic acid, coumaric acid, ferulic acid, hyperoside, isoquercitrin, quercetin, quercitrin and rutin, ellagic acid, gallic acid, (+)catechin and (-)epicatechin) of the 4 different extracts (M_A: blue line, M_B: yellow line, US_A: green line, US_B: red line) was reported.

4. Conclusions

Although BDs have been widely used in traditional medicine because of the peculiar content of buds in phenolic compounds, nowadays they are a category of botanicals still poorly studied. The lack of detailed scientific information and a clear and unique regulation, it makes these products high risk and vulnerable for accidental mistakes in the attribution of the botanical species, but also frauds and adulterations. Moreover, the polyphenols content of BDs is strongly influenced by the manufacturing processes whose parameters are often not strictly defined (e.g., solvent ratios in the extraction mixtures, raw material/extraction mixture ratios, extraction time) and thus they affect their final compositions.

This research, within the Finnover project, aims to answer to the growing demand for efficient quality control in the BDs field to guarantee the proper attribution of the botanical source and their content. Moreover, a manufacturing process specification should be advisable to monitor the bioactive contents.

UV-Visible spectroscopy and HPLC-DAD analysis have been employed to obtain an untargeted and a targeted phytochemical fingerprint of BDs, respectively. UV-Visible coupled with an appropriate chemometric data processing is a simple, rapid and low-cost technique proved to be very useful to identify the botanical source regardless the manufacturing method and the experimental conditions used. Moreover, the targeted phytochemical fingerprint by HPLC-DAD allowed to obtain a detailed screening of the BDs polyphenolic profile which highlighted an high variability due to the different vegetal species and to the manufacturing method and protocol. The ultrasonic extraction of buds compared to conventional maceration proved less sensitive to the different extraction protocols.

The proposed strategy offers to those operating in this industrial sector an untargeted method for the identification of the bud's botanical species and a green extraction strategy (PUAE) which is more robust with respect to the different extractive protocols that can be used. The same approach, described for BDs, could be analogously applied to other botanical productions.

Supplementary Materials: The following are available online at http://www.mdpi.com/2304-8158/9/10/1343/s1, **Figure S1**: *Larix decidua* chromatographic pattern. Table S1: Purity of all the used standards for HPLC analysis of BDs. Table S2: Single phenolic compound fingerprint of BDs.

Author Contributions: Conceptualization, G.L.B., P.Z. and R.B.; Data curation, F.T. and D.D.; Formal analysis, R.B.; Funding acquisition, G.L.B., A.P. and R.B.; Investigation, F.T., D.D. and M.G.; Project administration, G.L.B. and R.B.; Software, F.T. and R.B.; Supervision, G.L.B., A.P., P.Z. and R.B.; Writing—original draft, F.T. and D.D.; Writing—review & editing, G.L.B. and R.B. All authors have read and agreed to the published version of the manuscript.

Funding: This research was funded by an European Union project called FINNOVER (n° 1198), http://www.interreg-finnover.com/.

Acknowledgments: The authors acknowledge GEAL Pharma (Bricherasio, Torino, Italy) for providing the samples of plant materials and Teresa Fior for the scientific-naturalistic illustrations of buds.

Conflicts of Interest: The authors declare no conflict of interest.

References

1. Czepielewska, E.; Makarewicz-Wujec, M.; Różewski, F.; Wojtasik, E. Kozłowska-Wojciechowska, M. Drug adulteration of food supplements: A threat to public health in the European Union? *Regul. Toxicol. Pharmacol.* **2018**, *97*, 98–102. [CrossRef]

2. Italian Ministry of Health. Available online: http://www.salute.gov.it/portale/temi/p2_5.jsp?lingua=italiano&area=Alimentiparticolarieintegratori&menu=integratori (accessed on 20 March 2020).

3. European Commission. Food Supplements. Available online: https://ec.europa.eu/food/safety/labelling_nutrition/supplements_en (accessed on 21 May 2020).

4. Colombo, F.; Restani, P.; Biella, S.; Di Lorenzo, C. Botanicals in Functional Foods and Food Supplements: Tradition, Efficacy and Regulatory Aspects. *Appl. Sci.* **2020**, *10*, 2387. [CrossRef]

5. Restani, P.; Di Lorenzo, C.; Garcia-Alvarez, A.; Frigerio, G.; Colombo, F.; Maggi, F.M.; Milà-Villarroel, R.; Serra-Majem, L. The PlantLIBRA consumer survey: Findings on the use of plant food supplements in Italy. *PLoS ONE* **2018**, *13*, e0190915. [CrossRef]

6. Deconinck, E.; Vanhamme, M.; Bothy, J.L.; Courselle, P. A strategy based on fingerprinting and chemometrics for the detection of regulated plants in plant food supplements from the Belgian market: Two case studies. *J. Pharm. Biomed. Anal.* **2019**, *166*, 189–196. [CrossRef]

7. Fibigr, J.; Šatínský, D.; Solich, P. Current trends in the analysis and quality control of food supplements based on plant extracts. *Anal. Chim. Acta* **2018**, *1036*, 1–15. [CrossRef]

8. FINNOVER Interreg Alcotra Project 2017–2020. Available online: http://www.interreg-finnover.com/ (accessed on 20 May 2020).

9. Pharmacopée Française. *Codex Medicamentarius Gallicus, Codex Français: Monographie, Préparations Homéopathiques*; Ordre National des Pharmaciens: Paris, France, 1965; Available online: http://ansm.sante.fr/ Mediatheque/Publications/Pharmacopee-francaise-Plan-Preambule-index (accessed on 21 May 2020).

10. Turrini, F.; Donno, D.; Boggia, R.; Beccaro, G.L.; Zunin, P.; Leardi, R.; Pittaluga, A.M. An innovative green extraction and re-use strategy to valorize food supplement by-products: Castanea sativa bud preparations as case study. *Food Res. Int.* **2019**, *115*, 276–282. [CrossRef]

11. Turrini, F.; Donno, D.; Beccaro, G.L.; Zunin, P.; Pittaluga, A.; Boggia, R. Pulsed Ultrasound-Assisted Extraction as an Alternative Method to Conventional Maceration for the Extraction of the Polyphenolic Fraction of Ribes nigrum Buds: A New Category of Food Supplements Proposed by The FINNOVER Project. *Foods* **2019**, *8*, 466. [CrossRef]

12. Allio, A.; Calorio, C.; Franchino, C.; Gavello, D.; Carbone, E.; Marcantoni, A. Bud extracts from Tilia tomentosa Moench inhibit hippocampal neuronal firing through GABAA and benzodiazepine receptors activation. *J. Ethnopharmacol.* **2015**, *172*, 288–296. [CrossRef]

13. Calorio, C.; Donno, D.; Franchino, C.; Carabelli, V.; Marcantoni, A. Bud extracts from Salix caprea L. inhibit voltage gated calcium channels and catecholamines secretion in mouse chromaffin cells. *Phytomedicine* **2017**, *36*, 168–175. [CrossRef]

14. Nervo, T.; Bergamini, L.; Guido, M.; Ferraro, F. Composizione e Relativo Uso Nel Trattamento Dell'endometrite Animale. U.S. Patent n. TO2015A000193, 2 April 2015.

15. Olivero, G.; Turrini, F.; Vergassola, M.; Boggia, R.; Zunin, P.; Donno, D.; Beccaro, G.L.; Grilli, M.; Pittaluga, A. The 3Rs: Reduction and refinement through a multivariate statistical analysis approach in a behavioural study tounveil anxiolytic effects of natural extracts of Tilia tomentosa. *Biomed. Sci. Eng.* **2019**, *3*, 116. [CrossRef]

16. Antonaci, I. Effetto Di Un Trattamento Fitoterapico Su Alcuni Parametri Ematologici Dell'asina Da Latte. Bachelor's Thesis, Degree-granting Produzioni e Gestione degli Animali in Allevamento e Selvatici University of Turin, Turin, Italy, 2017.

17. Guerra, C.; Nury, C. Utilizzo Di Una Soluzione Fitoterapica Per Un Trattamento Alternativo Dell'endometrite Equina. Bachelor's Thesis, Degree-Granting Medicina Veterinaria University of Turin, Turin, Italy, 2015.

18. Donno, D.; Beccaro, G.L.; Cerutti, A.K.; Mellano, M.G.; Bounous, G. Bud Extracts as New Phytochemical Source for Herbal Preparations-Quality Control and Standardization by Analytical Fingerprint. In *Phytochemicals—Isolation, Characterisation and Role in Human Health*, 1st ed.; Rao, A.V., Rao, L.G., Eds.; InTech: Rijeka, Croatia, 2015; pp. 187–218. [CrossRef]

19. Donno, D.; Mellano, M.G.; Cerutti, A.K.; Beccaro, G.L. Biomolecules and Natural Medicine Preparations: Analysis of New Sources of Bioactive Compounds from *Ribes* and *Rubus* spp. Buds. *Pharmaceuticals* **2016**, *9*, 7. [CrossRef]

20. Sanzini, E.; Badea, M.; Dos Santos, A.; Restani, P.; Sievers, H. Quality control of plant food supplements. *Food Funct.* **2011**, *2*, 740–746. [CrossRef]

21. Donno, D.; Boggia, R.; Zunin, P.; Cerutti, A.K.; Guido, M.; Mellano, M.G.; Prgomet, Z.; Beccaro, G.L. Phytochemical fingerprint and chemometrics for natural food preparation pattern recognition: An innovative technique in food supplement quality control. *J. Food Sci. Technol.* **2016**, *53*, 1071–1083. [CrossRef]

22. Directive 2002/46/EC of the European Parliament and of the Council of 10 June 2002 on the Approximation of the Laws of the Member States Relating to Food Supplements. Available online: https://eur-lex.europa.eu/ eli/dir/2002/46/2017-07-26 (accessed on 5 September 2020).

23. Decreto Legislativo 21 Maggio 2004, n.169, Attuazione Della Direttiva 2002/46/CE Relativa Agli Integratori Alimentari. Available online: https://www.gazzettaufficiale.it/eli/id/2004/07/15/004G0201/sg (accessed on 5 September 2020).

24. European Federation of Associations of Health Product Manufacturers (EHPM). Available online: https://www.ehpm.org/attachments/article/117/EHPM%20Quality%20Guide%20101214.pdf (accessed on 5 September 2020).

25. Watson, R.R. *Polyphenols in Plants: Isolation, Purification and Extract Preparation*, 2nd ed.; Academic Press: London, UK, 2018.

26. Ma, G.; Chen, Y. Polyphenol supplementation benefits human health via gut microbiota: A systematic review via meta-analysis. *J. Funct. Foods* **2020**, *66*, 103829. [CrossRef]

27. Tsao, R. Chemistry and biochemistry of dietary polyphenols. *Nutrients* **2010**, *2*, 1231–1246. [CrossRef]

28. Tresserra-Rimbau, A.; Rimm, E.B.; Medina-Remón, A.; Martínez-González, M.A.; de la Torre, R.; Corella, D.; Salas-Salvadó, J.; Gómez-Gracia, E.; Lapetra, J.; Arós, F.; et al. Inverse association between habitual polyphenol intake and incidence of cardiovascular events in the PREDIMED study. *Nutr. Metab. Cardiovasc. Dis.* **2014**, *24*, 639–647. [CrossRef]

29. Kwok, C.S.; Boekholdt, S.M.; Lentjes, M.A.H.; Loke, Y.K.; Luben, R.N.; Yeong, J.K.; Wareham, N.J.; Myint, P.K.; Khaw, K.T. Habitual chocolate consumption and risk of cardiovascular disease among healthy men and women. *Heart* **2015**, *101*, 1279–1287. [CrossRef]

30. Wang, P.Y.; Fang, J.C.; Gao, Z.H.; Zhang, C.; Xie, S.Y. Higher intake of fruits, vegetables or their fiber reduces the risk of type 2 diabetes: A meta-analysis. *J. Diabetes Investig.* **2016**, *7*, 56–69. [CrossRef]

31. Wang, S.; Moustaid-Moussa, N.; Chen, L.; Mo, H.; Shastri, A.; Su, R.; Bapat, P.; Kwun, I.; Shen, C.L. Novel insights of dietary polyphenols and obesity. *J. Nutr. Biochem.* **2014**, *25*, 1–18. [CrossRef]

32. Serra, D.; Almeida, L.M.; Dinis, T.C.P. Dietary polyphenols: A novel strategy to modulate microbiota-gut-brain axis. *Trends Food Sci. Technol.* **2018**, *78*, 224–233. [CrossRef]

33. Vauzour, D. Dietary polyphenols as modulators of brain functions: Biological actions and molecular mechanisms underpinning their beneficial effects. *Oxidative Med. Cell. Longev.* **2012**, *2012*, 914273. [CrossRef]

34. Liu, X.; Du, X.; Han, G.; Gao, W. Association between tea consumption and risk of cognitive disorders: A dose-response meta-analysis of observational studies. *Oncotarget* **2017**, *8*, 43306–43321. [CrossRef]

35. Brglez Mojzer, E.; Knez Hrnčič, M.; Škerget, M.; Knez, Ž.; Bren, U. Polyphenols: Extraction Methods, Antioxidative Action, Bioavailability and Anticarcinogenic Effects. *Molecules* **2016**, *21*, 901. [CrossRef]

36. Donno, D.; Beccaro, G.L.; Mellano, M.G.; Bonvegna, L.; Bounous, G. Castanea spp. buds as a phytochemical source for herbal preparations: Botanical fingerprint for nutraceutical identification and functional food standardization. *J. Sci. Food Agric.* **2014**, *94*, 2863–2873. [CrossRef]

37. Chemat, F.; Vian, M.A.; Cravotto, G. Green extraction of natural products: Concept and principles. *Int. J. Mol. Sci.* **2012**, *13*, 8615–8627. [CrossRef]

38. Green Chemistry's 12 Principles, United States Environmental Protection Agency. Available online: https://www.epa.gov/greenchemistry/basics-green-chemistry#twelve (accessed on 19 March 2020).

39. Ministero Delle Politiche Agricole Alimentari E Forestali. Piano Di Settore Della Filiera Delle Piante Officinali 2014–16. Available online: https://www.politicheagricole.it/flex/cm/pages/ServeBLOB.php/L/IT/IDPagina/7562 (accessed on 21 May 2020).

40. Chemat, F.; Rombaut, N.; Sicaire, A.G.; Meullemiestre, A.; Fabiano-Tixier, A.S.; Abert-Vian, M. Ultrasound assisted extraction of food and natural products. Mechanisms, techniques, combinations, protocols and applications. A review. *Ultrason. Sonochem.* **2017**, *34*, 540–560. [CrossRef]

41. Vinatoru, M.; Mason, T.J.; Calinescu, I. Ultrasonically assisted extraction (UAE) and microwave assisted extraction (MAE) of functional compounds from plant materials. *Trends Anal. Chem.* **2017**, *97*, 159–178. [CrossRef]

42. Vernès, L.; Vian, M.; Chemat, F. Chapter 12—Ultrasound and Microwave as Green Tools for Solid-Liquid Extraction. In *Handbooks in Separation Science, Liquid-Phase Extraction*, 1st ed.; Poole, C.F., Ed.; Elsevier: Amsterdam, The Netherlands, 2020; pp. 355–374. [CrossRef]

43. Boggia, R.; Turrini, F.; Anselmo, M.; Zunin, P.; Donno, D.; Beccaro, G.L. Feasibility of UV-VIS-Fluorescence Spectroscopy combined with pattern recognition techniques to authenticate a new category of plant food supplements. *J. Food Sci. Technol.* **2017**, *54*, 2422–2432. [CrossRef]

44. Donno, D.; Mellano, M.G.; Riondato, I.; De Biaggi, M.; Andriamaniraka, H.; Gamba, G.; Beccaro, G.L. Traditional and Unconventional Dried Fruit Snacks as a Source of Health-Promoting Compounds. *Antioxidants* **2019**, *8*, 396. [CrossRef]

45. Li, A.N.; Li, S.; Zhang, Y.J.; Xu, X.R.; Chen, Y.M.; Li, H.B. Resources and Biological Activities of Natural Polyphenol. *Nutrients* **2014**, *6*, 6020–6047. [CrossRef]

46. Vilkhu, K.; Mawson, R.; Simons, L.; Bates, D. Applications and opportunities for ultrasound assisted extraction in the food industry—A review. *Innov. Food Sci. Emerg. Technol.* **2008**, *9*, 161–169. [CrossRef]

47. Pan, Z.; Qu, W.; Ma, H.; Atungulu, G.G.; McHugh, T.H. Continuous and pulsed ultrasound-assisted extractions of antioxidants from pomegranate peel. *Ultrason. Sonochem.* **2011**, *18*, 1249–1257. [CrossRef]

48. Mok, D.K.W.; Chau, F.T. Chemical information of Chinese medicines: A challenge to chemist. *Chemom. Intell. Lab. Syst.* **2006**, *82*, 210–217. [CrossRef]

49. Italian Chemical Society. Division of Analytical Chemistry-Group of Chemometrics. CAT Chemometric Agile Tool. Available online: http://www.gruppochemiometria.it/index.php/software (accessed on 25 May 2020).

50. Wold, S.; Esbensen, K.; Geladi, P. Principal Component Analysis. *Chemom. Intell. Lab. Syst.* **1987**, *2*, 37–52. [CrossRef]

51. Jolliffe, I.T. *Principal Component Analysis*, 2nd ed.; Springer Series in Statistics; Springer: New York, NY, USA, 2002.

52. Barnes, R.J.; Dhanoa, M.S.; Lister, S.J. Standard normal variate transformation and de-trending of near-infrared diffuse reflectance spectra. *Appl. Spectrosc.* **1989**, *43*, 772–777. [CrossRef]

53. Oliveri, P. Class-modelling in food analytical chemistry: Development, sampling, optimisation an validation issues—A tutorial. *Anal. Chim. Acta* **2017**, *982*, 9–19. [CrossRef]

54. Wold, S.; Johansson, E.; Cocchi, M. *3D QSAR in Drug Design: Theory, Methods and Applications*; Hugo, K., Ed.; ESCOM Science Publishers: Leiden, The Netherlands, 1993; p. 523.

55. Chemat, F.; Ashokkumar, M. Preface: Ultrasound in the processing of liquid foods, beverages and alcoholic drinks. *Ultrason. Sonochem.* **2017**, *38*, 753. [CrossRef]

56. Paniwnyk, L. Applications of ultrasound in processing of liquid foods: A review. *Ultrason. Sonochem.* **2017**, *38*, 794–806. [CrossRef]

57. Falcão, L.; Araújo, M.E.M. Vegetable Tannins Used in the Manufacture of Historic Leathers. *Molecules* **2018**, *23*, 1081. [CrossRef]

58. Okuda, T.; Yoshida, T.; Hatano, T.; Iwasaki, M.; Kubo, M.; Orime, T.; Yoshizaki, M.; Naruhashi, N. Hydrolysable tannins as chemotaxonomic markers in the rosaceae. *Phytochemistry* **1992**, *31*, 3091–3096. [CrossRef]

59. Yoshida, T.; Amakura, Y.; Yoshimura, M. Structural features and biological properties of ellagitannins in some plant families of the order Myrtales. *Int. J. Mol. Sci.* **2010**, *11*, 79–106. [CrossRef]

60. Moilanen, J.; Koskinen, P.; Salminen, J.P. Distribution and content of ellagitannins in Finnish plant species. *Phytochemistry* **2015**, *116*, 188–197. [CrossRef]

61. Rocha, L.D.; Monteiro, M.C.; Teodoro, A.J. Anticancer properties of hydroxycinnamic acids—A Review. *Cancer Clin. Oncol.* **2012**, *1*, 109–121. [CrossRef]

62. Zelber-Sagi, S.; Salomone, F.; Mlynarsky, L. The Mediterranean dietary pattern as the diet of choice for non-alcoholic fatty liver disease: Evidence and plausible mechanisms. *Liver Int.* **2017**, *37*, 936–949. [CrossRef]

63. Brereton, R.G. *Applied Chemometrics for Scientists*; John Wiley & Sons, Ltd.: Chichester, UK, 2007. [CrossRef]

64. De Biaggi, M.; Donno, D.; Mellano, M.G.; Gamba, G.; Riondato, I.; Rakotoniaina, E.N.; Beccaro, G.L. Emerging species with nutraceutical properties: Bioactive compounds from Hovenia dulcis pseudofruits. *Food Chem.* **2020**, *310*, 125816. [CrossRef]

65. Kim, H.K.; Jeong, T.-S.; Lee, M.-K.; Park, Y.B.; Choi, M.-S. Lipid-lowering efficacy of hesperetin metabolites in high-cholesterol fed rats. *Clin. Chim. Acta* **2003**, *327*, 129–137. [CrossRef]

66. Li, F.; Li, S.; Li, H.-B.; Deng, G.-F.; Ling, W.-H.; Wu, S.; Xu, X.-R.; Chen, F. Antiproliferative activity of peels, pulps and seeds of 61 fruits. *J. Funct. Foods* **2013**, *5*, 1298–1309. [CrossRef]

67. Beccaro, G.L.; Donno, G.; Lione, G.G.; De Biaggi, M.; Gamba, G.; Rapalino, S.; Riondato, I.; Gonthier, P.; Mellano, M.M. Castanea spp. Agrobiodiversity Conservation: Genotype Influence on Chemical and Sensorial Traits of Cultivars Grown on the Same Clonal Rootstock. *Foods* **2020**, *9*, 1062. [CrossRef]

68. Alfei, S.; Signorello, M.G.; Schito, A.; Catena, S.; Turrini, F. Reshaped as polyester-based nanoparticles, gallic acid inhibits platelet aggregation, reactive oxygen species production and multi-resistant Gram-positive bacteria with an efficiency never obtained. *Nanoscale Adv.* **2019**, *1*, 4148–4157. [CrossRef]

69. Alfei, S.; Turrini, F.; Catena, S.; Zunin, P.; Grilli, M.; Pittaluga, A.M.; Boggia, R. Ellagic acid a multi-target bioactive compound for drug discovery in CNS? A narrative review. *Eur. J. Med. Chem.* **2019**, *183*, 111724. [CrossRef]
70. Ananingsih, V.K.; Sharma, A.; Zhou, W. Green tea catechins during food processing and storage: A review on stability and detection. *Food Res. Int.* **2013**, *50*, 469–479. [CrossRef]
71. Franco, D.; Sineiro, J.; Rubilar, M.; Sánchez, M.; Jerez, M.; Pinelo, M.; Costoya, N.; Núñez, M.J. Polyphenols from plant materials: Extraction and antioxidant power. *Electron. J. Environ. Agric. Food Chem.* **2008**, *7*, 3210–3216.

Article

Phenolic Profiles, Antioxidant, and Inhibitory Activities of *Kadsura heteroclita* (Roxb.) Craib and *Kadsura coccinea* (Lem.) A.C. Sm.

Varittha Sritalahareuthai [1], Piya Temviriyanukul [1,2], Nattira On-nom [1,2], Somsri Charoenkiatkul [1] and Uthaiwan Suttisansanee [1,2,*]

[1] Institute of Nutrition, Mahidol University, Salaya, Phuttamonthon, Nakhon Pathom 73170, Thailand; varittha.sri@hotmail.com (V.S.); piya.tem@mahidol.ac.th (P.T.); nattira.onn@mahidol.ac.th (N.O.-n.); somsri.chr@mahidol.ac.th (S.C.)

[2] Food and Nutrition Academic and Research Cluster, Institute of Nutrition, Mahidol University, Salaya, Phuttamonthon, Nakhon Pathom 73170, Thailand

* Correspondence: uthaiwan.sut@mahidol.ac.th; Tel.: +66-(0)2800-2380 (ext. 422)

Received: 4 August 2020; Accepted: 31 August 2020; Published: 2 September 2020

Abstract: *Kadsura* spp. in the Schisandraceae family are woody vine plants, which produce edible red fruits that are rich in nutrients and antioxidant activities. Despite their valuable food applications, *Kadsura* spp. are only able to grow naturally in the forest, and reproduction handled by botanists is still in progress with a very low growth rate. Subsequently, *Kadsura* spp. were listed as endangered species by the International Union for Conservation of Nature and Natural Resources (IUCN) in 2011. Two different *Kadsura* spp., including *Kadsura coccinea* (Lem.) A.C. Sm. and *Kadsura heteroclita* (Roxb.) Craib, are mostly found in northern Thailand. These rare, wild fruits are unrecognizable to outsiders, and there have only been limited investigations into its biological properties. This study, therefore, aimed to comparatively investigate the phenolic profiles, antioxidant activities, and inhibitory activities against the key enzymes involved in diabetes (α-glucosidase and α-amylase) and Alzheimer's disease (acetylcholinesterase (AChE), butyrylcholinesterase (BChE), and beta-secretase 1 (BACE-1)) in different fruit parts (exocarp, mesocarp (edible part), seed, and core) of *Kadsura coccinea* (Lem.) A.C. Sm. and *Kadsura heteroclita* (Roxb.) Craib. The results suggested that *Kadsura* spp. extracts were rich in flavonol (quercetin), flavanone (naringenin), anthocyanidins (cyanidin and delphinidin), and anthocyanins (cyanidin 3-*O*-glucoside (kuromanin), cyanidin 3-*O*-galactoside (ideain), cyanidin 3-*O*-rutinoside (keracyanin), and cyanidin 3,5-di-*O*-glucoside (cyanin)). These flavonoids were found to be responsible for the high antioxidant activities and key enzyme inhibitions detected in *Kadsura* spp. extracts. The findings of the present study can support further development of *Kadsura* spp. as a potential source of phenolics and anti-oxidative agents with health benefits against diabetes and Alzheimer's disease. Besides, exocarp and the core of *Kadsura* spp. exhibited higher phenolic contents, antioxidant activities, and key enzyme inhibitory activities compared to the mesocarp and seeds, respectively. This information can promote the use of fruit parts other than the edible mesocarp for future food applications using *Kadsura* spp. rather than these being wasted.

Keywords: *Kadsura* spp.; fruit parts; phenolics; antioxidant capacity; in vitro health properties

---

1. Introduction

For thousands of years, humans have consumed naturally indigenous plants as green medicines to treat illnesses, including vegetables, herbs, and fruits. However, the development of effective, advanced scientific technology and synthetic drugs has significantly lowered the popularity of the natural medicine approach. Besides several advantages, including ready-to-use therapeutic applications and the commercial availability of synthetic drugs, severe side effects and economic impacts are other factors of concern. These issues have brought indigenous plants back into focus. Natural products have been considered as better functional food sources for alternative approaches in the prevention of certain non-communicable diseases (NCDs), and have even been used as green medicine to treat particular diseases. Subsequently, the study of medicinal applications and functional food development from beneficial plants are currently in focus.

Kadsura spp. in the *Schisandraceae* family produce ovate-elliptical shaped leaves and solitarily unisexual flowers [1]. The globose fruit is between 14 and 20 cm in diameter and forms a hexagonal structured skin developed by each carpel [1]. The fruit shape is similar to that of sugar-apples, but with a sourer and astringent taste. While the ripened fruit is edible [2], other plant parts can also be used as traditional medicine [1]. The stem and root of *Kadsura coccinea* are used to reduce rheumatic pain in the bones, chronic enteritis, acute gastritis, and the immunologic hepatic fibrosis effect [3], while those of *Kadsura longipedunculata* are used in the treatment of rheumatic arthritis, traumatic injury, dysmenorrhea, abdominal pain, irregular menstruation, canker sores, and gastrointestinal inflammation [3]. Moreover, the *Kadsura heteroclita* stem can prevent and treat rheumatic and arthritic diseases with anti-nociceptive and anti-inflammatory effects [4,5]. Phenolics detected in *Kadsura* spp. appear to be responsible for these health properties. Flavonoids identified in *Kadsura oblongifolia*, including quercetin and kaempferol, possess potent antioxidant activities [6]. Additionally, *K. coccinea* has been previously reported to contain gallic acid as a major phenolic acid, which has strongly demonstrated DPPH-radical scavenging activity (the half maximal inhibitory concentration (IC_{50}) = 23.64 µg/mL) compared to commercial antioxidants, ascorbic acid, and 2,6-ditertbuty l-4-methylphenol (BHT) (the IC_{50} of 48.46 and 975.96 µg/mL, respectively) [7]. However, stressful conditions i.e., aphid feeding [8], light and drought intensity [9,10], or even the processing of biological material [11] can significantly affect the antioxidant properties or the phenolic profile.

The local Thai name for *Kadsura* spp. is noi-na-kreau and was first discovered in northern Thailand in 1972 [12]. *Kadsura* spp. were listed as an endangered species by the International Union for Conservation of Nature and Natural Resources (IUCN) in 2011, and since then only 11 trees have been found in the Thai provinces of Chiang Rai, Chiang Mai, and Mae Hong Son. Due to how rare they are, *Kadsura* spp. were listed by the Plant Genetic Conservation Project under the initiative of Her Royal Highness Princess Maha Chakri Sirindhorn (RSPG) in the same year of 2011, aiming to sustainably conserve and allocate plant resources to optimize beneficial utilizations [13]. Unlike more ubiquitous fruits in Thailand, *Kadsura* spp. are not well known, and due to largely being unrecognized, they are at high risk of extinction. In order to obtain the main goal of RSPG to sustainably conserve *Kadsura* spp. by expanding the knowledge on their biological properties, eventually leading to proper management of the plant, the objective of the present study is, therefore, to examine the phenolic profiles, antioxidant activities, and in vitro key enzyme inhibitory properties of *K. coccinea* (Lem.) A.C. Sm. and *K. heteroclita* (Roxb.) Craib, the two most abundant *Kadsura* spp. found in northern Thailand. These enzyme inhibitory properties are related to medicinal abilities against diabetes and Alzheimer's disease (AD) by inhibiting key enzymes, which control these diseases. Targeting the inhibition of the carbohydrate degrading enzymes, including α-glucosidase and α-amylase, is a key approach for drugs designed to control diabetes. Based on enzyme–inhibitor structural interactions, several phenolics have been proven as effective α-glucosidase and α-amylase inhibitors [5,14]. In addition, neurotransmitter degrading enzymes, acetylcholinesterase (AChE) and butyrylcholinesterase (BChE), and amyloid precursor protein (APP) degrading enzyme, beta-secretase (BACE-1), are the key enzymes, which control the occurrence of AD. It has been previously suggested that the extreme loss of cholinergic markers found in the cerebral cortex is associated with AD, while BACE-1 causes the formation of beta-amyloid peptides or senile plaques, a hallmark of AD [15]. The findings of the present study provide valuable information for the sustainable conservation of the nearly extinct *Kadsura* spp. by describing their advantageous biological properties and potential applications, which can help to establish proper management and prolong the existence of the plants.

2. Materials and Methods

2.1. Sample Collection, Preparation, and Extraction

Kadsura coccinea (Lem.) A.C. Sm. and *Kadsura heteroclita* (Roxb.) Craib were collected from forests near Hui-Nam-Guen village, Mae-Jedi-Mai sub-district, Weing-Pa-Pao district, Chiang Rai Province, Thailand (19°11′58.3″ N and 99°31′00.7″ E). The fruits of both species could be harvested only once a year at different time periods. The fruits of *Kadsura coccinea* (Lem.) A.C. Sm. were collected in October 2017, while those of *Kadsura heteroclita* (Roxb.) Craib were collected in January 2018. Both samples were identified and authenticated by Assist. Prof. Dr. Chunthana Suwanthada and Assoc. Prof. Dr. Chusri Trisonthi of the Plant Genetic Conservation Project under the royal initiative of Her Royal Highness Princess Maha Chakri Sirindhorn. The voucher specimens, including BK No. 071406 (*Kadsura coccinea* (Lem.) A.C. Sm.) and BK No. 071407 (*Kadsura heteroclita* (Roxb.) Craib), were deposited at the Bangkok Herbarium (BK), Bangkok, Thailand.

All fresh samples were cleaned with deionized water and prepared as exocarp, mesocarp (edible part), seed, and core (Supplementary Table S1). All the samples were cut (approximately 0.3 cm thick) before freeze drying by a Heto Powerdry PL9000 freeze dryer (Heto Lab Equipment, Allerod, Denmark) for 3 days. Dry samples were then ground using a grinder (Philips 600 W series from Philips Electronic Co., Ltd., Jakarta, Indonesia) into a fine powder before packing in a vacuum aluminum foil bag and storing in a freezer at −20 °C for further analysis.

A spectrophotometer (ColorFlex EZ, Hunter Associates Laboratory, Reston, VA, USA) was used to analyze the colors of the fresh and dry samples, which were expressed as CIELAB units, where L* represents dark (0) to white (100) colors, a* represents green (−) to red (+) colors, and b* represents blue (−) to yellow (+) colors. The moisture content of the powdered samples was analyzed using a Halogen moisture analyzer (HE53 series, Mettler-Toledo AG, Greifensee, Switzerland). The color data and moisture content data are presented in Supplementary Table S2.

The extraction of *Kadsura* spp. was optimized as described by a previous study [16]. Briefly, the powered samples (100 mg) were dissolved in distilled water (10 mL) and incubated in a 90 °C temperature-controlled water bath shaker (WNE45 series from Memmert GmBh, Eagle, WI, USA) for 1 h. The supernatant was then collected by centrifugation at 3800× g using a Hettich® ROTINA 38R refrigerated centrifuge (Andreas Hettich GmBH, Tuttlingen, Germany) for 15 min and filtered through a 0.45 µM PES membrane syringe filter. All the extracted samples were stored at −20 °C for further analysis.

2.2. Determination of Antioxidant Activity

The antioxidant activities, including 2,2-diphenyl-1-picrylhydrazyl (DPPH) radical scavenging, ferric ion reducing antioxidant power (FRAP), and oxygen radical absorbance capacity (ORAC) assays of the *Kadsura* spp. extracts, naringenin, and quercetin were determined using existing, well-established protocols [17–20].

The DPPH radical scavenging assay was performed using DPPH in 95% (*v/v*) aqueous ethanol as previously described [17,20]. Trolox solution (0.01–0.64 mM) was used as a standard, and the results were reported as µmol TE/g dried weight (DW).

The FRAP assay was determined using FRAP reagent as previously described [18,20]. Trolox solution (7.81–250.00 µM) was used as a standard, and the results were reported as µmol TE/g DW.

The ORAC assay was examined using fluorescein reagent as previously described [19,20]. Trolox solution (3.12–100.00 µM) was used as a standard, and the results were reported as µmol TE/g DW.

2.3. Determination of Total Phenolic Contents, Total Flavonoid Contents (TFCs), Total Anthocyanin Contents (TACs), and Phenolic Profiles

The total phenolic contents (TPCs) of the *Kadsura* spp. extracts were determined using Folin-Ciocalteu reagent as previously described [21,22]. Gallic acid (10–200 μg/mL) was used as a standard, and the TPCs were reported as the mg gallic acid equivalent (GAE)/g DW.

Total flavonoid contents (TFCs) of the *Kadsura* spp. extracts were determined using aluminum trichloride as previously described [23]. Quercetin (0–100 μg/mL) was used as a standard, and the TFCs were reported as the mg quercetin equivalent (QE)/g DW.

Total anthocyanidin contents (TACs) of the *Kadsura* spp. extracts were determined using the pH differential method as previously described [24]. Cyanidin 3-*O*-glucoside (2–63 μg/mL) was used as a standard, and the TACs were reported as the mg cyanidin 3-*O*-glucoside equivalent (C3GE)/g DW.

Phenolic profiles were analyzed using high performance liquid chromatography (HPLC) as described previously [25]. The powdered samples (0.5 g) were mixed with 62.5% (*v/v*) aqueous methanol containing 0.5 g/L tBHQ (40 mL) and 6 N HCl (10 mL) before incubating in a 80 °C temperature-controlled water bath shaker (WNE45 series from Memmert GmBh, Eagle, WI, USA) for 2 h in dark. The mixture was then cooled in ice for 5 min before sonicating in an ultrasonic cleansing bath (Branson Ultrasonics™ M series, Branson Ultrasonics Corp., Danbury, CT, USA) for a further 5 min. The mixture was filtered through a 0.22 μM PTFE membrane syringe filter. Phenolic acids in the filtrate were analyzed by HPLC utilizing a 5 μm Zorbax Eclipse XDB-C_{18} column (150 × 4.6 mm from Agilent Technologies, Santa Clara, CA, USA) on an Agilent 1100 HPLC system with a photodiode array detector from Agilent Technologies (Santa Clara, CA, USA). The gradient mobile phases comprised Milli-Q water (18.2 MΩ.cm resistivity at 25 °C) containing 0.05% (*v/v*) TFA (solvent A), methanol containing 0.05% (*v/v*) TFA (solvent B), and acetonitrile containing 0.05% (*v/v*) TFA (solvent C) with a constant flow rate of 0.6 mL/min, as shown in Table 1. The existence of the phenolic acids were visualized at 280 and 325 nm using the ChemStation software (Agilent Technologies, Santa Clara, CA, USA) by comparing retention time (t_R) and spectral fingerprint with standards including 4-hydroxybenzoic acid (>99.0% GC, T), caffeic acid (>98.0% HPLC, T), chlorogenic acid (>98.0% HPLC, T), ferulic acid (>98.0% GC, T), *p*-coumaric acid (>98.0% GC, T), sinapic acid (>99.0% GC, T), syringic acid (>97.0% T) from Tokyo Chemical Industry (Tokyo, Japan), and gallic acid (97.5–102.5% T) from Sigma-Aldrich (St. Louis, MO, USA).

Table 1. Solvent system of the phenolic acids and flavonoids using high performance liquid chromatography (HPLC) analysis.

Time (min)	Flow Rate (mL/min)	Solvent A (%)	Solvent B (%)	Solvent C (%)
0	0.6	90	6	4
5	0.6	85	9	6
30	0.6	71	17.4	11.6
60	0.6	0	85	15
61	0.6	90	6	4
66	0.6	90	6	4

Solvent A = Milli-Q water containing 0.05% (*v/v*) TFA; solvent B = methanol containing 0.05% (*v/v*) TFA; solvent C = acetonitrile containing 0.05% (*v/v*) TFA.

Flavonoid identification was performed similar to that of phenolic acids. However, the existence of flavonoids was visualized at 338 and 368 nm and evaluated by comparing t_R and spectral fingerprint with standards including apigenin (>98.0% HPLC), genistein (>98.0% HPLC), hesperidin (>90.0% HPLC, T), kaempferol (>97.0% HPLC), luteolin (>98.0% HPLC), myricetin (>97.0% HPLC), naringenin (>93.0% HPLC, T), quercetin (>98.0% HPLC, E) from Tokyo Chemical Industry (Tokyo, Japan), and isorhamnetin (≥99.0% HPLC) from Extrasynthese (Genay, France).

Anthocyanidin and anthocyanin identification was performed as previously described [26,27]. Anthocyanidins were extracted using the powdered sample (500 mg) that was dispersed in 50% (*v/v*) aqueous methanol containing 2 N HCl (5 mL). The mixture was incubated in a 100 ± 2 °C water bath (TW20 series from Julabo GmbH, Seelbach, Germany) for 1 h before filtering through a 0.22 µM PTFE membrane syringe filter. Anthocyanidins were identified by the HPLC system (an Ultimate 3000 with diode array and multiple-wavelength detectors) from Thermo Fisher Scientific (Dreieich, Germany) and a 5 µm ReproSil-Pur® ODS-3 column (250 × 4.6 mm) from Dr. Maisch GmbH (Ammerbuch, Germany). Milli-Q water (18.2 MΩ.cm resistivity) containing 0.4% (*v/v*) TFA (solvent A) and acetonitrile containing 0.4% *v/v* TFA (solvent B) were used as an isocratic mobile phase at a ratio of 82% solvent A and 18% solvent B, with a constant flow rate of 1.0 mL/min. The existence of anthocyanidins was visualized at 530 nm using a Chromeleon™ chromatography Data System (CDS) software (Thermo Fisher Scientific, Dreieich, Germany) by comparing t_R and spectral fingerprint with the standards including cyanidin (≥96.0% HPLC), delphinidin (≥97.0% HPLC), pelargonidin (≥97.0% HPLC), peonidin (≥97.0% HPLC), and petunidin (≥95.0% HPLC) from Extrasynthese (Genay, France).

Anthocyanin identification was performed similarly to the anthocyanidin identification. Anthocyanins were extracted using the powdered sample (500 mg) that was dispersed in 50% (*v/v*) aqueous methanol containing 2% (*v/v*) HCl (5 mL). The extract was then sonicated using an ultrasonic cleansing bath (Branson Ultrasonics™ M series, Branson Ultrasonics Corp., Danbury, CT, USA) for 20 min and filtered through a 0.22 µM PTFE membrane syringe filter. Anthocyanins were identified utilizing the same HPLC system as anthocyanidins. Milli-Q water containing 0.4% (*v/v*) TFA (solvent A) and acetonitrile containing 0.4% (*v/v*) TFA (solvent B) were used as gradient mobile phases with a constant flow rate 1.0 mL/min, as shown in Table 2. The existence of anthocyanins was visualized at 525 nm and compared t_R and spectral fingerprint with the standards including cyanidin 3-*O*-sophoroside (≥95.0% HPLC), cyanidin 3,5-di-*O*-glucoside (≥97.0% HPLC), cyanidin 3-*O*-galactoside (≥97.0% HPLC), cyanidin 3-*O*-rutinoside (≥96.0% HPLC), cyanidin 3-*O*-glucoside (≥96.0% HPLC), and delphinidin 3,5-di-*O*-glucoside (≥97.0% HPLC) from Extrasynthese (Genay, France).

Table 2. Solvent system of the anthocyanins using HPLC analysis.

Time (min)	Solvent A	Solvent B
0	88	12
6	88	12
8	85	15
25	85	15
25	88	12
30	88	12

Solvent A = Milli-Q water containing 0.4% (*v/v*) TFA; solvent B = acetonitrile containing 0.4% (*v/v*) TFA.

2.4. Determination of Enzyme Inhibitory Activities

The α-glucosidase inhibitory activity was determined as previously described [28]. Briefly, the assay comprising 2 mM *p*-nitrophenyl-α-D-glucopyranoside (50 µL) in a 50 mM phosphate buffer (pH 7.0), 0.1 U/mL *Saccharomyces cerevisiae* α-glucosidase (type 1, ≥10 U/mg protein, 100 µL), and sample extract (50 µL) was visualized at a wavelength of 405 nm using a microplate reader (Synergy™ HT 96-well UV-visible spectrophotometer using the Gen5 data analysis software from BioTek Instruments, Inc., Winooski, VT, USA). The inhibition percentage was then calculated as follows:

$$\% \text{ inhibition} = \left(1 - \frac{B - b}{A - a}\right) \times 100$$

where A is the initial velocity of the reaction with the enzyme, a is the initial velocity of the reaction without the enzyme, B is the initial velocity of the enzyme reaction with the extract, and b is the initial velocity of the reaction with the extract but without the enzyme. The efficiency of flavonoids against α-glucosidase was also determined using the half maximal inhibitory concentration (IC$_{50}$), and analyzed by a dose-response plot of flavonoids versus the inhibition percentage.

The α-amylase inhibitory activity was determined as previously described [29]. Briefly, the assay comprising of 30 mM *p*-nitrophenyl-α-D-maltohexaoside (50 µL) in a 50 mM phosphate buffer (pH 7.0) containing 200 mM KCl, 30 mg/mL of porcine pancreatic α-amylase (type VII, $\geq$10 unit/mg, 100 µL), and the sample extract (50 µL) was visualized at a wavelength of 405 nm using the microplate reader. The inhibition percentage was then calculated as above.

Acetylcholinesterases (AChE) inhibitory activities were determined as previously described [25]. The enzyme assay consisting of 20 ng of *Electrophorus electricus* AChE (1000 units/mg, 100 µL) in 50 mM KPB (pH 7.0), 16 mM 5,5-dithio-bis-(2-nitrobenzoic acid) (DTNB, 10 µL), 0.8 mM acetylthiocholine (40 µL) in 50 mM KPB (pH 7.0), and the extract (50 µL) was detected at a wavelength of 412 nm using the microplate reader. Butyrylcholinesterases (BChE) inhibitory activities were determined similarly to AChE. However, 100 ng equine serum BChE ($\geq$10 units/mg protein, 100 µL) in 50 mM KPB (pH 7.0) containing 1 mM $MgCl_2$ and 0.1 mM butyrylthiocholine (40 µL) in 50 mM KPB (pH 7.0) were used as the enzyme and substrate, respectively. The results were expressed as the inhibition percentage and the IC_{50} value, as mentioned above.

The beta-secretase (BACE-1) inhibitory activity was determined utilizing a BACE-1 activity detection kit (Sigma-Aldrich, St. Louis, MO, USA). The manufacturer's instructions were followed, and the results expressed as a percentage of BACE-1 inhibition, as above.

All the enzymes, chemicals, and reagents in the enzyme inhibitory assays were purchased from Sigma-Aldrich (St. Louis, MO, USA).

2.5. Statistical Analysis

All experiments were carried out in triplicate ($n = 3$) and expressed as mean $\pm$ standard deviation (SD). One-way analysis of variance (ANOVA) followed by Duncan's multiple comparison test and unpaired *t*-test were performed to determine the significant differences between values with $p < 0.05$.

3. Results

3.1. Total Phenolic Contents (TPCs), Total Flavonoid Contents (TFCs), Total Anthocyanin Contents (TACs), and Phenolic Profiles

The total phenolic contents (TPCs) of *K. coccinea* (Lem.) A.C. Sm. extracts (KCE) ranged between 0.83 and 43.61 mg gallic acid equivalent (GAE)/g dry weight (DW), while *K. heteroclita* (Roxb.) Craib extract (KHE) exhibited the TPCs ranging from 1.13 to 103.72 mg GAE/g DW (Table 3). Among the different fruit parts, the exocarp of KCE exhibited the significantly highest TPCs, followed by the core, mesocarp, and seed, respectively. However, for KHE, its core exhibited significantly higher TPCs than exocarp, mesocarp, and seed, respectively. Furthermore, significantly higher TPCs were observed in all the fruit parts of KHE compared to KCE.

Similarly, the total flavonoid contents (TFCs) in KCE ranged between 1.45 and 60.52 mg quercetin equivalent (QE)/g DW, with the exocarp providing the highest TFCs, followed by the core, while mesocarp and seed exhibited the lowest TFCs (Table 3). The core of KHE exhibited the highest TFCs, followed by the exocarp, mesocarp, and seed, respectively. In line with the TPC results, higher TFCs were detected in all the fruit parts of KHE (ranging from 1.40 to 206.92 mg QE/g DW) compared to KCE, in exception of seed.

Interestingly, the total anthocyanin contents (TACs) of KCE were only detected in the exocarp and mesocarp (0.02–0.27 mg cyanidin 3-*O*-glucoside equivalent (C3GE)/g DW), while TACs in KHE were found in the exocarp, mesocarp, and core (0.14–0.28 mg C3GE/g DW) (Table 3). The mesocarp of KCE exhibited higher TACs than its exocarp. Meanwhile, the exocarp and mesocarp of KHE exhibited insignificantly different TACs, but were significantly higher than that of the core.

Among all the phenolic acids and flavonoids investigated utilizing high performance liquid chromatography (HPLC), quercetin and naringenin were the only flavonoids identified in *Kadsura* spp. extracts (Table 4, Supplementary Figures S1 and S2). The most abundant flavonoid, naringenin, was detected in all fruit parts of KCE (ranging from 1395.22 to 1972.65 mg/100 g DW) and KHE (ranging from 1739.91 to 1835.46 mg/100 g DW). Meanwhile, quercetin was only found in the exocarp of KCE (17.94 mg/100 g DW), and in the exocarp and mesocarp of KHE (51.19 and 12.59 mg/100 g DW, respectively). Among all the fruit parts, the exocarp and seed of KCE exhibited significantly higher naringenin content than the core and mesocarp, respectively. However, insignificant naringenin content differences were detected in all the fruit parts of KHE.

Table 3. Quantification of the total phenolic contents (TPCs), total flavonoid contents (TFCs), and total anthocyanin contents (TACs) in different fruit parts of *Kadsura coccinea* (Lem.) A.C. Sm. extract (KCE) and *Kadsura heteroclita* (Roxb.) Craib extract (KHE).

Kadsura spp.	Total Phenolic Contents (mg GAE/g DW)	Total Flavonoid Contents (mg QE/g DW)	Total Anthocyanin Contents (mg C3GE/g DW)
KCE			
Exocarp	43.61 ± 0.65 [a,*]	60.52 ± 5.51 [a,*]	0.02 ± 0.00 [b,*]
Mesocarp	6.20 ± 0.15 [c,*]	1.45 ± 0.52 [c,*]	0.27 ± 0.02 [a]
Seed	0.83 ± 0.08 [d,*]	3.48 ± 1.06 [c,*]	ND
Core	17.71 ± 0.07 [b,*]	43.23 ± 1.18 [b,*]	ND
KHE			
Exocarp	54.00 ± 2.11 [B]	113.36 ± 4.87 [B]	0.28 ± 0.02 [A]
Mesocarp	21.20 ± 0.54 [C]	33.02 ± 1.55 [C]	0.26 ± 0.01 [A]
Seed	1.13 ± 0.14 [D]	1.40 ± 0.10 [D]	ND
Core	103.72 ± 1.14 [A]	206.92 ± 4.08 [A]	0.14 ± 0.01 [B]

All data are expressed as mean ± standard deviation (SD) of triplicate experiments ($n = 3$). GAE: gallic acid equivalent; QE: quercetin equivalent; C3GE: cyanidin 3-*O*-glucoside equivalent; DW: dry weight; ND: not detected. Lower case and upper case letters indicate significant differences ($p < 0.05$) in different fruit parts of KCE and KHE, respectively, using one-way analysis of variance (ANOVA) and Duncan's multiple comparison test; * showed significant difference ($p < 0.05$) in the same fruit part when comparing between KCE and KHE using the unpaired *t*-test.

Anthocyanidins detected in *Kadsura* spp. extracts were identified as cyanidin and delphinidin (Table 4 and Supplementary Figure S3). Only the exocarp, mesocarp, and core of both *Kadsura* spp. extracts were found to contain these anthocyanidins, while none was detected in the seed. KCE exhibited cyanidin contents in the range of 0.34–1.03 µg/100 g DW, with its exocarp and mesocarp containing higher cyanidin content compared to its core. Similar results were observed in KHE (cyanidin content ranging from 0.03 to 1.00 µg/100 g DW), in which its exocarp and mesocarp provided higher cyanidin contents than its core. Delphinidin in KCE ranged between 0.42 and 4.02 µg/100 g DW, while KHE exhibited delphinidin contents in the range of 0.31–0.50 µg/100 g DW. The exocarp in both *Kadsura* spp. extracts provided higher delphinidin content than the mesocarp and core, respectively.

Anthocyanins (or glycosylated anthocyanidins) detected in *Kadsura* spp. extracts were identified as cyanidin 3,5-di-*O*-glucoside (cyanin), cyanidin 3-*O*-galactoside (ideain), cyanidin 3-*O*-glucoside (kuromanin), and cyanidin 3-*O*-rutinoside (keracyanin) (Table 4 and Supplementary Figure S4). Only the mesocarp of KCE contained cyanin (9.94 µg/100 g DW) and ideain (10.16 µg/100 g DW), while none was detected in the other fruit parts. Meanwhile, KHE exhibited ideain (2.11–53.36 µg/100 g DW), kuromanin (0.27–26.06 µg/100 g DW), and keracyanin (4.16–112.54 µg/100 g DW) in the exocarp, mesocarp, and core, while none was detected in the seed. The exocarp contained higher content for all three anthocyanins compared to the mesocarp and core, respectively. Besides, kuromanin and keracyanin were only detected in KHE, while none was found in KCE. On the other hand, cyanin was only found in KCE, while none was detected in KHE.

Table 4. Phenolic profiles in different fruit parts of *Kadsura coccinea* (Lem.) A.C. Sm. extract (KCE) and *Kadsura heteroclita* (Roxb.) Craib extract (KHE).

Kadsura spp.	Flavonoids (mg/100 g DW)		Anthocyanidins (µg/100 g DW)			Anthocyanins (µg/100 g DW)		
	Quercetin	Naringenin	Cyanidin	Delphinidin	Cyanin	Ideain	Kuromanin	Keracyanin
KCE								
Exocarp	17.94 ± 0.96 *	1972.65 ± 135.07 [a]	1.03 ± 0.13 [a,*]	4.02 ± 0.91 [a]	ND	ND	ND	ND
Mesocarp	ND	1395.22 ± 50.23 [c,*]	0.99 ± 0.38 [b]	0.42 ± 0.19 [a]	9.94 ± 0.20	10.16 ± 0.19 *	ND	ND
Seed	ND	1861.15 ± 31.35 [ab]	ND	ND	ND	ND	ND	ND
Core	ND	1752.54 ± 0.70 [b]	0.34 ± 0.05 [b]	0.60 ± 0.29 [b,*]	ND	ND	ND	ND
KHE								
Exocarp	51.19 ± 1.92 [A]	1811.02 ± 15.28 [A]	1.00 ± 0.07 [A]	0.50 ± 0.02 [A]	ND	53.36 ± 2.40 [A]	26.06 ± 4.08 [A]	112.54 ± 9.65 [A]
Mesocarp	12.59 ± 1.17 [B]	1812.07 ± 88.25 [A]	0.41 ± 0.07 [B]	0.43 ± 0.11 [AB]	ND	6.83 ± 0.49 [B]	2.68 ± 0.98 [B]	12.56 ± 0.65 [B]
Seed	ND	1835.46 ± 54.82 [A]	ND	ND	ND	ND	ND	ND
Core	ND	1739.91 ± 55.90 [A]	0.03 ± 0.01 [C]	0.31 ± 0.09 [B]	ND	2.11 ± 0.08 [C]	0.27 ± 0.07 [B]	4.16 ± 0.56 [B]

All data are expressed as mean ± standard deviation (SD) of triplicate experiments ($n = 3$). Cyanidin 3,5-di-*O*-glucoside: cyanin; cyanidin 3-*O*-galactoside: ideain; cyanidin 3-*O*-glucoside: kuromanin; cyanidin 3-*O*-rutinoside: keracyanin; DW: dry weight; ND: not detected. Lower case and upper case letters indicate significant differences ($p < 0.05$) in different fruit parts of KCE and KHE, respectively, using one-way analysis of variance (ANOVA) and Duncan's multiple comparison test; * showed significant difference ($p < 0.05$) in the same fruit part when comparing between KCE and KHE using unpaired *t*-test.

3.2. Antioxidant Activities

Antioxidant activities of *Kadsura* spp. extracts were determined using 2,2-diphenyl-1-picrylhydrazyl (DPPH) radical scavenging activity, ferric ion reducing antioxidant power (FRAP), and oxygen radical absorbance capacity (ORAC) assays (Table 5). KCE exhibited DPPH radical scavenging activities in the range of 0.04–2.73 µmol trolox equivalent (TE)/g DW, FRAP activities in the range of 3.14–300.44 µmol TE/g DW, and ORAC activities in the range of 41.93–957.80 µmol TE/g DW. Among all fruit parts, the exocarp provided significantly higher antioxidant activities than the core, mesocarp, and seed, respectively. This data corresponds with the TPCs, in which the highest TPCs were detected in the exocarp of KCE and the lowest in the seed. Similarly, KHE exhibited DPPH radical scavenging activities in the range of 0.04–6.48 µmol TE/g DW, FRAP activities in the range of 2.91–900.60 µmol TE/g DW, and ORAC activities in the range of 15.84–1330.23 µmol TE/g DW. Among all fruit parts, the core exhibited significantly higher antioxidant activities, followed by the exocarp, mesocarp, and seed, respectively. This data corresponds to the TPCs and TFCs, in which the highest TPCs and TFCs were detected in the core of KHE and the lowest in the seed. Interestingly, considering the abundant flavonoids detected in *Kadsura* spp., naringenin and quercetin, the results suggested that quercetin is a stronger antioxidant with higher DPPH radical scavenging activity of 8.07 µmol TE/g DW, FRAP activity of 13,519.54 µmol TE/g DW, and ORAC activity of 30,112.47 µmol TE/g DW than naringenin with lower DPPH radical scavenging activity of 0.02 µmol TE/g DW, FRAP activity of 30.98 µmol TE/g DW, and ORAC activity of 25,592.45 µmol TE/g DW.

Table 5. Antioxidant activities in the different fruit parts of *Kadsura coccinea* (Lem.) A.C. Sm. extract (KCE) and *Kadsura heteroclita* (Roxb.) Craib extract (KHE) and their detected flavonoids, naringenin and quercetin.

Kadsura spp.	Antioxidant Activities (µmol TE/g DW)		
	DPPH Radical Scavenging Assay	**FRAP Assay**	**ORAC Assay**
KCE			
Exocarp	2.73 ± 0.08 [a]	300.44 ± 12.09 [a,*]	957.80 ± 77.76 [a,*]
Mesocarp	0.23 ± 0.00 [c,*]	25.80 ± 1.04 [c,*]	120.20 ± 10.88 [c,*]
Seed	0.04 ± 0.00 [d]	3.14 ± 0.18 [d,*]	41.93 ± 4.17 [d,*]
Core	1.00 ± 0.03 [b,*]	100.19 ± 1.69 [b,*]	440.35 ± 33.60 [b,*]
KHE			
Exocarp	2.75 ± 0.10 [B]	351.48 ± 18.79 [B]	812.66 ± 77.67 [B]
Mesocarp	1.02 ± 0.03 [C]	143.23 ± 12.29 [C]	260.91 ± 23.48 [C]
Seed	0.04 ± 0.00 [D]	2.91 ± 0.24 [D]	15.84 ± 1.41 [D]
Core	6.48 ± 0.22 [A]	900.60 ± 6.22 [A]	1330.23 ± 49.67 [A]
Flavonoids			
Naringenin	0.02 ± 0.00	30.98 ± 0.69	25,592.45 ± 495.92
Quercetin	8.07 ± 0.25	13,519.54 ± 242.20	30,112.47 ± 1631.24

All data are expressed as mean ± standard deviation (SD) of triplicate experiments ($n = 3$). TE: trolox equivalent; DW: dry weight. Lower case and upper case letters indicate significant differences ($p < 0.05$) in different fruit parts of KCE and KHE, respectively, using one-way analysis of variance (ANOVA) and Duncan's multiple comparison test; * showed significant difference ($p < 0.05$) in the same fruit part when comparing between KCE and KHE using unpaired *t*-test.

3.3. In Vitro Enzyme Inhibitory Activities

Kadsura spp. extracts inhibited the key enzymes relevant to diabetes, including α-glucosidase and α-amylase, with different degrees of inhibition (Table 6). The α-glucosidase inhibitory activities in both *Kadsura* spp. extracts were detected in all fruit parts except for the seed. KCE exhibited α-glucosidase inhibitory activities in the range of 60.04–98.23% inhibitions using an extract concentration of 0.5 mg/mL, while those in KHE ranged between 89.72% and 91.84% inhibitions with the same extract concentration. It is also found that the exocarp and core of both *Kadsura* spp. extracts exhibited significantly higher α-glucosidase inhibitory activities than mesocarp, as indicated by the half maximal inhibitory concentration (IC_{50}) (Table 6). Lower IC_{50} values indicated greater effectiveness of enzyme inhibitions. The exocarp and core of KCE exhibited significantly lower IC_{50} values (0.13 and 0.45 mg/mL, respectively) than mesocarp (0.56 mg/mL). Similar results were observed with KHE, in which its exocarp and core exhibited significantly lower IC_{50} values (0.06 mg/mL) than its mesocarp (the IC_{50} value of 0.64 mg/mL). Interestingly, the exocarp and core of KHE exhibited significantly lower IC_{50} values compared to those of KCE, while insignificantly different IC_{50} values were observed in the mesocarp of both *Kadsura* spp. extracts.

Likewise, KCE exhibited α-amylase inhibitory activities in the range of 12.18–51.85% inhibitions using an extract concentration of 2.5 mg/mL, while those in KHE ranged between 33.93% and 44.75% inhibitions when using the same extract concentration. When comparing the different fruit parts of KCE, its exocarp exhibited significantly higher inhibitory activities than its core, mesocarp, and seed, respectively. Similar results were observed in KHE, in which its exocarp and core possessed significantly higher inhibitory activities than its mesocarp, but no inhibitory activity was observed in its seed. Due to low inhibitory activities, the IC_{50} value against α–amylase of *Kadsura* spp. extracts is unavailable.

Kadsura spp. extracts were able to inhibit acetylcholinesterase (AChE), butyrylcholinesterase (BChE), and beta-secretase (BACE-1), the key enzymes relevant to AD with different degrees of inhibition. The AChE inhibitory activities in KCE ranged between 20.40% and 87.44% inhibitions, while those of KHE ranged between 29.98% and 91.74% inhibitions using the extract concentration of 2.0 mg/mL. The exocarp and core of both *Kadsura* spp. extracts possessed significantly higher AChE inhibitory activities than mesocarp, while no inhibitory activities were detected in the seed. Due to the low inhibitory activities in the mesocarp, the IC_{50} values against AChE were only detected in exocarp and core. KCE exhibited the IC_{50} values of 0.88 mg/mL in the exocarp and 1.54 mg/mL in the core, while KHE exhibited the IC_{50} values of 0.52 mg/mL in the exocarp and 0.41 mg/mL in the core (Table 6). It is also suggested that the KHE exocarp and core exhibited significantly lower IC_{50} values than those of KCE.

Similar results were observed with BChE inhibition, in which only the exocarp, mesocarp, and core of both *Kadsura* spp. extracts possessed inhibitory activities, while none was detected in the seed. KCE exhibited BChE inhibitory activities in the range of 41.27–96.81% inhibitions, while those of KHE ranged between 63.67% and 99.15% inhibitions using the same extract concentration of 2.0 mg/mL. The exocarp and core of KHE exhibited significantly lower IC_{50} values (0.20 and 0.16 mg/mL, respectively) than its mesocarp (1.32 mg/mL) (Table 6). However, KCE exhibited significantly lower IC_{50} values in its exocarp (0.49 mg/mL) than its core (0.67 mg/mL). Interestingly, the exocarp and core of KHE exhibited significantly lower IC_{50} values than those of KCE.

Additionally, KCE exhibited BACE-1 inhibitory activity in the range of 4.72–56.34% inhibitions, while those of KHE ranged between 12.70% and 35.26% inhibitions using the same extract concentration of 2.0 mg/mL. In KCE, the mesocarp exhibited a significantly higher BACE-1 inhibitory activity than the exocarp, core, and seed, respectively. Similar results were observed in KHE, in which the mesocarp exhibited the significantly higher BACE-1 inhibitory activity, followed by the exocarp and seed, respectively. No inhibitory activity was observed in the core of KHE.

Table 6. In vitro enzyme inhibitory activities in different fruit parts of *Kadsura coccinea* (Lem.) A.C. Sm. extract (KCE) and *Kadsura heteroclita* (Roxb.) Craib extract (KHE).

Kadsura spp.	α-Glucosidase		α-Amylase	AChE		BChE		BACE1
	%Inhibition [1]	IC$_{50}$ (mg/mL)	%Inhibition [2]	%Inhibition [3]	IC$_{50}$ (mg/mL)	%Inhibition [3]	IC$_{50}$ (mg/mL)	%Inhibition [3]
KCE								
Exocarp	92.32 ± 7.04 [a]	0.13 ± 0.01 [a,*]	51.58 ± 3.52 [a,*]	87.44 ± 8.67 [a]	0.88 ± 0.01 [a,*]	96.81 ± 1.26 [a,*]	0.49 ± 0.02 [a,*]	41.39 ± 2.56 [b,*]
Mesocarp	60.04 ± 5.17 [b,*]	0.56 ± 0.03 [c]	40.17 ± 3.31 [c,*]	20.40 ± 1.60 [c,*]	N/A	41.27 ± 1.97 [c,*]	N/A	56.34 ± 2.12 [a,*]
Seed	ND	N/A	12.18 ± 0.71 [d]	ND	N/A	ND	N/A	4.72 ± 2.85 [d,*]
Core	98.23 ± 1.49 [a,*]	0.45 ± 0.00 [b,*]	44.78 ± 2.72 [b]	62.86 ± 3.21 [b,*]	1.54 ± 0.10 [b,*]	88.86 ± 0.65 [b,*]	0.67 ± 0.04 [b,*]	26.36 ± 1.59 [c]
KHE								
Exocarp	91.84 ± 3.76 [A]	0.06 ± 0.01 [A]	44.75 ± 3.52 [A]	88.64 ± 2.34 [A]	0.52 ± 0.05 [A]	99.15 ± 1.53 [A]	0.20 ± 0.04 [A]	16.29 ± 1.16 [B]
Mesocarp	90.11 ± 2.60 [A]	0.64 ± 0.03 [B]	33.93 ± 2.00 [B]	29.98 ± 3.01 [B]	N/A	63.67 ± 5.56 [B]	1.32 ± 0.08 [B]	35.26 ± 2.61 [A]
Seed	ND	N/A	ND	ND	N/A	ND	N/A	12.70 ± 0.93 [C]
Core	89.72 ± 3.68 [A]	0.06 ± 0.00 [A]	44.12 ± 3.20 [A]	91.74 ± 0.71 [A]	0.41 ± 0.00 [A]	97.99 ± 0.22 [A]	0.16 ± 0.00 [A]	ND

All data are expressed as mean ± standard deviation (SD) of triplicate experiments (n = 3). [1] Extract concentration = 0.5 mg/mL; [2] Extract concentration = 2.5 mg/mL; [3] Extract concentration = 2.0 mg/mL; N/A: not available; ND: not detected. The extract concentrations in each enzyme assay was chosen in attempt to differentiate the inhibitory activities among different fruit parts. Lower case and upper case letters indicate significant differences ($p < 0.05$) in different fruit parts of KCE and KHE, respectively, using one-way analysis of variance (ANOVA) and Duncan's multiple comparison test; * showed significant difference ($p < 0.05$) in the same fruit part when comparing between KCE and KHE using unpaired t-test.

4. Discussion

Two species of *Kadsura* spp., *K. coccinea* (Lem.) A.C. Sm. and *K. heteroclita* (Roxb.) Craib, are found abundantly in Thailand. *Kadsura* spp. has been reportedly used as a folk medicine for several decades. Despite their valuable medicinal applications, *Kadsura* spp. can only be naturally populated in the forest, which makes them unrecognizable for outsiders, while knowledge of their biological properties is also limited. The present research is the first comparative and comprehensive study of two *Kadsura* spp. in terms of their phytochemicals (total phenolic contents (TPCs), total flavonoid contents (TFCs), total anthocyanin contents (TACs), and phenolic profiles), and their health properties (in vitro antioxidant, anti-diabetic, and anti-Alzheimer's properties). The researchers found that (i) *Kadsura* spp. extracts were rich in flavonol (quercetin), flavanone (naringenin), anthocyanidins (cyanidin and delphinidin), and anthocyanins (cyanidin 3-*O*-glucoside (kuromanin), cyanidin 3-*O*-galactoside (ideain), cyanidin 3-*O*-rutinoside (keracyanin), and cyanidin 3,5-di-*O*-glucoside (cyanin)); (ii) the antioxidant activities of *Kadsura* spp. extracts were related to their phenolic contents; (iii) *Kadsura* spp. extracts exhibited strong inhibitory activities against key diabetic enzymes, including α-glucosidase and α-amylase; (iv) *Kadsura* spp. extracts provided effective in vitro anti-Alzheimer properties through acetylcholinesterase (AChE), butyrylcholinesterase (BChE), and beta-secretase (BACE-1) inhibitions. This information supports the further development of *Kadsura* spp. as a potential source of phenolics with health benefits against the occurrence of diabetes and AD.

Flavonoids are diphenylpropanes-structured phytochemicals, which are mostly found in plant-based diets. The HPLC outcomes reveal that the identified flavonol and flavanone in *Kadsura* spp. extracts were quercetin (12.59–51.19 mg/100 g DW) and naringenin (1395.22–1972.65 mg/100 g DW), respectively. Naringenin is the most abundant flavonoid in *Kadsura* spp. extracts and is also widely distributed in citrus fruits such as kumquats, grapefruits, yuzu, and pomelo [30]. Quercetin is the most common flavonol and is found ubiquitously in mulberry, apricot, apple, and onion [31]. The amounts of quercetin and naringenin in the *Kadsura* spp. extracts corresponded with the TPCs and TFCs, in which the exocarp appeared to exhibit higher TPCs and TFCs than mesocarp. Similar results were observed in *K. coccinea* collected in China, which suggested that the TPCs of exocarp was double those detected in mesocarp [7]. Besides, anthocyanidins, the pigmented water-soluble compounds widely found in red to purplish-blue colored plants, were also identified in *Kadsura* spp. extracts as cyanidin and delphinidin. This data corresponds to the anthocyanins found in most vegetables and fruits, in which cyanidin (50%) is the most abundant anthocyanidin, followed by delphinidin (12%), peonidin (12%), pelargonidin (12%), malvidin (7%), and petunidin (7%) [32]. Furthermore, glycosylated anthocyanidins, namely anthocyanins, were identified in *Kadsura* spp. extracts as kuromanin, ideain, keracyanin, and cyanin. These findings correspond to a previous study, suggesting that since cyanidin is the most abundant anthocyanidin found in plants, its glycosylated form, kuromanin, is also the most common form of glycoside derivatives in colored plants [32]. Besides *Kadsura* spp., these anthocyanins are also abundantly detected in most berries. Cyanin is mostly found in mulberries, pomegranate, and wild blackberry [26,27,33,34], while ideain is found in blueberries and cranberries [35,36]. Keracyanin is detected in mulberries, raspberries, blackcurrants, and redcurrants [26,27,35]. In addition, kuromanin is abundantly found in strawberries, mulberries, blackcurrants, and raspberries [26,27,35,37].

These flavonoids appear to be responsible for the high antioxidant activities detected in *Kadsura* spp. extracts. The results on antioxidant activities of individual flavonoid suggested that quercetin is a stronger antioxidant than naringenin. However, due to high content of detected naringenin, this flavonoid might be responsible for high antioxidant activities in *Kadsura* spp. extracts. When comparing fruit parts, the exocarp and core of *Kadsura* spp. extracts exhibited higher antioxidant activities than the mesocarp and seed, respectively. These results correspond with the previous study, finding that antioxidants detected in exocarp of *K. coccinea* collected in China were two times higher than its mesocarp [38]. Comparably, the ORAC values of our *Kadsura* spp. extracts (675–2943 μmol TE/100 g FW) are similar to those of high antioxidant containing fruits, including fuji apple (2589 μmol TE/100 g FW), gala apple (2828 μmol TE/100 g FW), apricot (1110 μmol TE/100 g FW), hass avocado

(1922 µmol TE/100 g FW), grapefruits (1640 µmol TE/100 g FW), grapes (red, black, white, and green as 1018–1837 µmol TE/100 g FW), white-flesh guava (2550 µmol TE/100 g FW), lemon (1346 µmol TE/100 g FW), mangosteen (2510 µmol TE/100 g FW), oranges (2103 µmol TE/100 g FW), peach (1922 µmol TE/100 g FW), and green-cultivar pear (2201 µmol TE/100 g FW) [39]. Additionally, the antioxidant activities are related to the TPCs and TFCs. The phenolics are proven to be the strong anti-oxidative agents [40,41]. Naringenin, the most abundant flavonoid detected in *Kadsura* spp. extracts, can strongly inhibit oxidative stress [42] and induce endogenous antioxidants [43]. Besides, it can suppress lipid peroxidation in rat liver tissue induced by hydroxyl and peroxyl radicals [44]. Many previous studies report that the effectiveness of anti-oxidative agents is related to their structures. The antioxidant properties activated by flavonoids are due to the existence of functional hydroxyl moieties, for instance, an active 4′ hydroxyl moiety of naringenin is responsible for its antioxidant activity [45]. Since these phenolics are responsible for the antioxidant activities in *Kadsura* spp. extracts, the exocarp and core with higher TPCs and TFCs than the mesocarp and seed also exhibited higher antioxidant activities. This data suggests that *Kadsura* spp. extracts have a strong antioxidant potential, leading to health promoting biological effects.

Besides being rich sources of anti-oxidative agents, *Kadsura* spp. extracts also exhibit strong inhibitory activities against α-glucosidase and α-amylase, the key enzymes in controlling diabetes. These enzymes play a significant role in hydrolyzing carbohydrates (polysaccharides) into glucose subunits (monosaccharide) before being absorbed into the small intestine. Various plant extracts have been examined previously and reported to possess α-amylase and α-glucosidase inhibitions [46]. Moreover, epidemiological studies support that phenolics, mainly flavonoids and phenolic acids, contribute to the prevention of DM through the α-amylase and α-glucosidase inhibitory effects. The most abundant phenolic detected in *Kadsura* spp., naringenin, can act as a mixed, close to non-competitive inhibitor with the IC_{50} value of 75 µM against α-glucosidase [28,47]. Additionally, naringenin inhibited α-glucosidase activity in diabetic rats, resulting in delayed carbohydrate absorption and decreased postprandial blood glucose level [48]. However, its action against α-amylase is much lower than that of α-glucosidase, in which its IC_{50} value on α-amylase inhibition was found to be greater than 500 µM [47]. Since KHE contained higher quantities of phenolics than KCE, it also had higher α-amylase and α-glucosidase inhibitory activities than the latter. Besides, these phenolics were more effective against α-glucosidase than α-amylase, while *Kadsura* spp. extracts exhibited potentially higher α-glucosidase inhibitions than α-amylase inhibitions.

Kadsura spp. extracts also provide effective in vitro anti-AD properties through AChE, BChE, and BACE-1 inhibitions. Under normal conditions, acetylcholine (ACh), a neurotransmitter working as a molecular messenger between neurons, is hydrolyzed mainly by AChE and co-regulated by BChE. In the AD hypothesis, these neurotransmitters decrease as a result of overexpression of AChE and BChE, leading to cognitive impairment [16]. Recent AD treatments are based on increasing cerebral acetylcholine levels by cholinesterase inhibitors [49]. Synthetic drugs used widely to treat AD symptoms, such as tacrine and donepezil, and can introduce both effective outcomes and cause side effects such as hepatotoxicity and gastrointestinal complaints [20]. Thus, natural therapeutics such as plant extracts are of interest, since some plant phenolics have been proven to provide a strong contribution in the prevention of neurodegenerative diseases [50]. The major flavonoid contained in *Kadsura* spp. extract, naringenin, has been previously reported to exhibit AChE inhibitory activity with the IC_{50} value of 42.66 µM and BChE inhibitory activity with the IC_{50} value of over 100 µM [51]. It has also been suggested that naringenin could introduce neuroprotective effects in an in vivo study, in which naringenin was found to suppress the activity of AChE, resulting in elevated synaptic cholinergic neurotransmitter level, improved cognitive functions, and prevented memory extinction [52]. Another pathogenesis of AD is amyloids cascade formation. Amyloid beta-peptides (Aβ) are overproduced and gradually accumulated in the brain. This aggregation leads to amyloid plaques, which cause oxidative stress and neurotoxicity. Aβ is derived from proteolytic cleavage by endogenous BACE-1, an enzyme that is increasingly found in AD patients [53]. BACE-1 is, therefore, another key enzyme for AD progression. Naringenin was previously reported to inhibit BACE-1 activity with the IC_{50} value of 30.31 µM [51]. Additionally, an oral administration of naringenin could also improve memory deficits in an Aβ-induced mouse model of AD [54]. Thus, AChE, BChE, and BACE-1 inhibitory activities detected in *Kadsura* spp. extracts may be the result of the biological function of this flavonoid, which acts as effective enzyme inhibitor.

In conclusion, *Kadsura* spp., including *Kadsura coccinea* (Lem.) A.C. Sm. and *Kadsura heteroclita* (Roxb.) Craib, contained high phenolic and flavonoid contents with naringenin being the most abundant compound. This flavonoid was responsible for the antioxidant activities and inhibitory activities against the key enzymes controlling diabetes (α-glucosidase and α-amylase) and AD (AChE, BChE, and BACE-1). This supports their future application as potential sources of phenolics with health benefits against the occurrence of diabetes and AD. Interestingly, by comparing different fruit parts within the same *Kadsura* spp., some fruit parts (exocarp of KCE, and exocarp and core of KHE) appeared to exhibit higher TPCs, TFCs, and antioxidant activities than its edible mesocarp. Besides, the exocarp and core of both *Kadsura* spp. also exhibited greater enzyme inhibitory activities than mesocarp, suggesting the potential food application of other fruit parts rather than these being wasted. Most importantly, this information can also lead to the sustainable conservation and establish proper agricultural management of the nearly extinct *Kadsura* spp. by providing knowledge of their advantageous biological properties and potential applications.

Supplementary Materials: The following are available online at http://www.mdpi.com/2304-8158/9/9/1222/s1, Table S1: Images of whole fruit, sectioned fruit, exocarp, mesocarp (edible part), seed, and core of *Kadsura coccinea* (Lem.) A.C. Sm. and *Kadsura heteroclita* (Roxb.) Craib; Table S2: Color (where L* describes darkness (−) to lightness (+), a* describes green (−) to red (+) colors, and b* describes indigo (−) to yellow (+)) and the percentage (%) of moisture content of fresh and freeze-dried *Kadsura* spp. Samples; Figure S1: High-performance liquid chromatograms of (A.) naringenin and *Kadsura* spp. samples including (B.) exocarp, (C.) mesocarp, (D.) seed, and (E.) core of *Kadsura coccinea* (Lem.) A.C. Sm. and (F.) exocarp, (G.) mesocarp, (H.) seed, and (I.) core of *Kadsura heteroclita* (Roxb.) Craib. Retention times (R_t) of phenolics in *Kadsura* spp. extracts are indicated at a wavelength of 280 nm; Figure S2: High-performance liquid chromatograms of (A.) quercetin and *Kadsura* spp. samples including (B.) exocarp, (C.) mesocarp, (D.) seed, and (E.) core of *Kadsura coccinea* (Lem.) A.C. Sm. and (F.) exocarp, (G.) mesocarp, (H.) seed, and (I.) core of *Kadsura heteroclita* (Roxb.) Craib. Retention times (R_t) of phenolics in *Kadsura* spp. extracts are indicated at a wavelength of 368 nm; Figure S3: High-performance liquid chromatograms of (A.) cyanidin, (B.) delphinidin, and *Kadsura* spp. samples including (C.) exocarp, (D.) mesocarp, (E.) seed, and (F.) core of *Kadsura coccinea* (Lem.) A.C. Sm. and (G.) exocarp, (H.) mesocarp, (I.) seed, and (J.) core of *Kadsura heteroclita* (Roxb.) Craib. Retention times (R_t) of phenolics in *Kadsura* spp. extracts are indicated at a wavelength of 530 nm; Figure S4: High-performance liquid chromatograms of (A.) cyanidin 3,5-di-O-glucoside (cyanin), (B.) cyanidin 3-O-glucoside (kuromanin), (C.) cyanidin 3-O-galactoside (ideain), (D.) cyanidin 3-O-rutinoside (keracyanin), and *Kadsura* spp. samples including (E.) exocarp, (F.) mesocarp, (G.) seed, and (H.) core of *Kadsura coccinea* (Lem.) A.C. Sm. and (I.) exocarp, (J.) mesocarp, (K.) seed, and (L.) core of *Kadsura heteroclita* (Roxb.) Craib. Retention times (R_t) of phenolics in *Kadsura* spp. are indicated at a wavelength of 525 nm.

Author Contributions: V.S. performed experiments, interpreted the results, generated the figures and tables, and wrote the manuscript. P.T. and N.O.-n. designed the research, suggested, and reviewed the manuscript. S.C. suggested and reviewed the manuscript. U.S. designed the research, supervised, performed experiments, interpreted the results, generated the figures and tables, wrote the manuscript, and reviewed the manuscript. All authors have read and agreed to the published version of the manuscript.

Funding: This study was supported by the National Research Council of Thailand (NRCT) (grant No. 57/2561).

Acknowledgments: The information on *Kadsura* spp. was kindly provided by Chunthana Suwanthada and Chusri Trisonthi, the board of the Plant Genetic Conservation Project under the Royal Initiative of Her Royal Highness Princess Maha Chakri Sirindhorn (RSPG). Besides, we would like to express our gratitude to Viparat Tepkaew, the RSPG coordinator, and Boonlert Ittiponjun, the RGSP staff in Lam-pang province, who provided the information and samples of *Kadsura* spp. utilized in this study.

Conflicts of Interest: The authors declare no conflict of interest.

References

1. Trisonthi, C.; Trisonthi, P. A new description of *Kadsura ananosma* Kerr (Schisandraceae). In *Thai Forest Bulletin (Botany)*; Forest Herbarium: Bangkok, Thailand, 1999; Volume 27, pp. 31–35.

2. Tepkaew, V.; Sangkaew, W.; Mahawan, P.; Sanguanpong, V.; Thainurak, H.; Ittiponjun, B.; Klongklaew, B.; Worapratheep, K.; Suwanthada, C.; Trisonthi, P.; et al. Surveying of RSPG Conserved Plant Genus Kadsura (Schisandraceae) in Northern Thailand. In Proceedings of the 7th Conference of RGSP Committee, Khon Haen University, Khon Kaen, Thailand, 24–26 March 2016; pp. 315–318.

3. Mulyaningsih, S.; Youns, M.; El-Readi, M.Z.; Ashour, M.L.; Nibret, E.; Sporer, F.; Herrmann, F.; Reichling, J.; Wink, M. Biological activity of the essential oil of *Kadsura longipedunculata* (Schisandraceae) and its major components. *J. Pharm. Pharmacol.* **2010**, *62*, 1037–1044. [CrossRef]

4. Yao, Y.; Yang, X.; Tian, J.; Liu, C.; Cheng, X.; Ren, G. Antioxidant and antidiabetic activities of black mung bean (*Vigna radiata* L.). *J. Agric. Food Chem.* **2013**, *61*, 8104–8109. [CrossRef] [PubMed]

5. Xiang, Y.; Zhang, T.; Yin, C.; Zhou, J.; Huang, R.; Gao, S.; Zheng, L.; Wang, X.; Manyande, A.; Tian, X.; et al. Effects of the stem extracts of *Schisandra glaucescens* Diels on collagen-induced arthritis in Balb/c mice. *J. Ethnopharmacol.* **2016**, *194*, 1078–1086. [CrossRef] [PubMed]

6. Liu, H.T.; Xu, L.J.; Peng, Y.; Li, R.T.; Xiao, P.G. Chemical study on ethyl acetate soluble portion of *Kadsura oblongifolia*. *China J. Chin. Mater. Med.* **2009**, *34*, 864–866.

7. Sun, J.Y.; Jiang, I.; Amin, Z.; Li, K.N.; Prasad, X.; Duan, B.Y.; Xu, L. An exotic fruit with high nutritional value: *Kadsura coccinea* fruit. *Int. Food Res.* **2011**, *18*, 651–657.

8. Banaszczak, E.W.; Radzikowska, D.; Ratajczak, K. Chemical profile and antioxidant activity of *Trollius europaeus* under the influence of feeding aphids. *Open Life Sci.* **2018**, *13*, 312–318. [CrossRef]

9. Kowalczewski, P.L.; Radzikowska, D.; Ivanišová, E.; Szwengiel, A.; Kačániová, M.; Sawinska, Z. Influence of Abiotic Stress Factors on the Antioxidant Properties and Polyphenols Profile Composition of Green Barley (*Hordeum vulgare* L.). *Int. J. Mol. Sci.* **2020**, *21*, 397. [CrossRef]

10. Vanacker, H.; Guichard, M.; Bohrer, A.S.; Issakidis-Bourguet, E. Redox regulation of monodehydroascorbate reductase by thioredoxin y in plastids revealed in the context of water stress. *Antioxidants* **2018**, *7*, 183. [CrossRef]

11. Kowalczewski, P.L.; Olejnik, A.; Białas, W.; Kubiak, P.; Siger, A.; Nowicki, M.; Lewandowicz, G. Effect of thermal processing on antioxidant activity and cytotoxicity of waste potato juice. *Open Life Sci.* **2019**, *14*, 150–157. [CrossRef]

12. Keng, H. Schisandraceae. In *Flora of Thailand*; Smitinand, T., Larsen, K., Eds.; ASRCT Press: Bangkok, Thailand, 1972; Volume 2, pp. 112–114.

13. Plant Genetic Conservation Project Under The Royal Initiative of Her Royal Highness Princess Maha Chakri Sirindhorn (RSPG) Webmaster. Available online: www.rspg.or.th (accessed on 30 December 2017).

14. Qi, X.Z.; Liu, J.B.; Chen, J.B.; Li, S. Lignans and triterpenoids from roots of *Kadsura longipedunculata*. *China Trad. Herb. Drug* **2017**, *48*, 2164–2171.

15. Chen, D.F.; Zhang, S.X.; Chen, K.; Zhou, B.N.; Wang, P.; Cosentino, L.M.; Lee, K.H. Two new lignans, interiotherins A and B, as anti-HIV principles from *Kadsura interior*. *China J. Nat. Prod.* **1996**, *59*, 1066–1068. [CrossRef] [PubMed]

16. Ferreira-Vieira, T.H.; Guimaraes, I.M.; Silva, F.R.; Ribeiro, F. Alzheimer's disease: Targeting the cholinergic system. *Curr. Neuropharmacol.* **2016**, *14*, 101–115. [CrossRef] [PubMed]

17. Fukumoto, L.; Mazza, G. Assessing antioxidant and prooxidant activities of phenolic compounds. *J. Agric. Food Chem.* **2000**, *48*, 3597–3604. [CrossRef] [PubMed]

18. Benzie, I.F.F.; Strain, J.J. The ferric reduction ability of plasma (FRAP) as a measure of antioxidant power: The FRAP assay. *Anal. Biochem.* **1996**, *239*, 70–76. [CrossRef] [PubMed]

19. Ou, B.; Hampsch-Woodill, M.; Prior, R.L. Development and validation of an improved oxygen radical absorbance capacity assay using fluorescein as the fluorescent probe. *J. Agric. Food Chem.* **2001**, *49*, 4619–4926. [CrossRef] [PubMed]

20. Atukeren, P.; Cengiz, M.; Yavuzer, H.; Gelisgen, R.; Altunoglu, E.; Oner, S.; Erdenen, F.; Yuceakın, D.; Derici, H.; Cakatay, U.; et al. The efficacy of donepezil administration on acetylcholinesterase activity and altered redox homeostasis in Alzheimer's disease. *Biomed. Pharmacother.* **2017**, *90*, 786–795. [CrossRef] [PubMed]

21. Ainsworth, E.A.; Gillespie, K.M. Estimation of total phenolic content and other antioxidant substrates in plant tissues using Folin-Ciocalteu reagent. *Nat. Protoc.* **2007**, *2*, 875–877. [CrossRef]

22. Sripum, C.; Kukreja, R.K.; Charoenkiatkul, S.; Kriengsinyos, W.; Suttisansanee, U. The effect of extraction conditions on antioxidant activities and total phenolic contents of different processed Thai Jasmine rice. *Int. Food Res. J.* **2017**, *24*, 1644–1650.

23. Zhishen, J.; Mengcheng, T.; Jianming, W. The determination of flavonoid contents in mulberry and their scavenging effects on superoxide radicals. *Food Chem.* **1999**, *64*, 555–559. [CrossRef]

24. Lee, J. Determination of total monomeric anthocyanin pigment content of fruit juices, beverages, natural colorants, and wines by the pH differential method: Collaborative Study. *J. AOAC Int.* **2005**, *88*, 1269–1278. [CrossRef]

25. Thuphairo, K.; Sornchan, P.; Suttisansanee, U. Bioactive compounds, antioxidant activity and inhibition of key enzymes relevant to Alzheimer's disease from sweet pepper (*Capsicum annuum*) extracts. *Prev. Nutr. Food Sci.* **2019**, *24*, 327–337. [CrossRef] [PubMed]

26. Temviriyakul, P.; Sritalahareuthai, V.; Na Jom, K.; Jongruaysup, B.; Tabtimsri, S.; Pruesapan, K.; Thangsiri, S.; Inthachat, W.; Siriwan, D.; Charoenkiatku, S.; et al. Comparison of phytochemicals, antioxidant, and *in vitro* anti-Alzheimer properties of twenty-seven *Morus* spp. cultivated in Thailand. *Molecules* **2020**, *25*, 2600. [CrossRef] [PubMed]

27. Suttisansanee, U.; Charoenkiatkul, S.; Jongruaysup, B.; Tabtimsri, S.; Siriwan, D.; Temviriyanukul, P. Mulberry fruit cultivar 'Chiang Mai' prevents beta-amyloid toxicity in PC12 neuronal cells and in a *Drosophila* model of Alzheimer's disease. *Molecules* **2020**, *25*, 1837. [CrossRef] [PubMed]

28. Promyos, N.; Temviriyanukul, P.; Suttisansanee, U. Evaluation of α-glucosidase inhibitory assay using different sub-classes of flavonoids. *Curr. Appl. Sci. Technol. J.* **2017**, *17*, 172–180.

29. Pongkunakorn, T.; Watcharachaisoponsiri, T.; Chupeerach, C.; On-nom, N.; Suttisansanee, U. Inhibitions of key enzymes relevant to obesity and diabetes of Thai local mushroom extracts. *Curr. Appl. Sci. Technol. J.* **2017**, *17*, 181–190.

30. Kawaii, S.; Tomono, Y.; Katase, E.; Ogawa, K.; Yano, M. Quantitation of flavonoid constituents in citrus fruits. *J. Agric. Food Chem.* **1999**, *47*, 3565–3571. [CrossRef]

31. Hakkinen, S.H.; Karenlampi, S.O.; Heinonen, M.; Mykkanen, H.M.; Torronen, A.R. Content of the flavonols quercetin, myricetin and kaempferol in 25 edible berries. *J. Agric. Food Chem.* **1999**, *47*, 2274–2279. [CrossRef]

32. Kong, J.M.; Chia, L.S.; Goh, N.K.; Chia, T.F.; Brouillard, R. Analysis and biological activities of anthocyanins. *Phytochemistry* **2003**, *64*, 923–933. [CrossRef]

33. Tulipani, S.; Mezzetti, B.; Capocasa, F.; Bompadre, S.; Beekwilder, J.; de Vos, C.; Capanoglu, E.; Bovy, A.; Battino, M. Antioxidants, phenolic compounds, and nutritional quality of different strawberry genotypes. *J. Agric. Food Chem.* **2008**, *56*, 696–704. [CrossRef]

34. Mertz, C.; Cheynier, V.; Gunata, Z.; Brat, P. Analysis of phenolic compounds in two blackberry species (*Rubus glaucus* and *Rubus adenotrichus*) by high-performance liquid chromatographywith diode array detection and electrospray ion trap mass spectrometry. *J. Agric. Food Chem.* **2007**, *55*, 8616–8624. [CrossRef]

35. Borges, G.; Degeneve, A.; Mullen, W.; Crozier, A. Identification of flavonoid and phenolic antioxidants in black currants, blueberries, raspberries, red currants, and cranberries. *J. Agric. Food Chem.* **2010**, *58*, 3901–3909. [CrossRef] [PubMed]

36. Fredericks, C.H.; Fanning, K.J.; Gidley, M.J.; Netzel, G.; Zabaras, D.; Herrington, M.; Netzel, M. High-anthocyanin strawberries through cultivar selection. *J. Sci. Food Agric.* **2013**, *93*, 846–852. [CrossRef] [PubMed]

37. Wang, E.; Yina, Y.; Xuc, C.; Liu, J. Isolation of high-purity anthocyanin mixtures and monomersfrom blueberries using combined chromatographic techniques. *J. Chromatogr. A* **2014**, *1327*, 39–48. [CrossRef] [PubMed]

38. Sun, J.; Yao, J.Y.; Huang, S.X.; Long, X.; Wang, J.B.; Garcia-Garcia, E. Antioxidant activity of polyphenol and anthocyanin extracts from fruits of *Kadsura coccinea* (Lem.) A.C. Smith. *Food Chem.* **2009**, *117*, 276–281. [CrossRef]

39. USDA Database for the Oxygen Radical Absorbance Capacity (ORAC) of Selected Foods. Available online: www.orac-info-portal.de/download/ORAC_R2.pdf (accessed on 23 March 2018).

40. Lodovici, M.; Guglielmi, F.; Meoni, M.; Dolara, P. Effect of natural phenolic acids on DNA oxidation *in vitro*. *Food Chem. Toxicol.* **2001**, *39*, 1205–1210. [CrossRef]

41. Heim, K.E.; Tagliaferro, A.R.; Bobilya, D.J. Flavonoid antioxidants: Chemistry, metabolism and structure-activity relationships. *J. Nutr. Biochem.* **2002**, *13*, 572–584. [CrossRef]

42. Manchope, M.F.; Campos, C.C.; Silva, L.C.; Zarpelon, A.C.; Ribeiro, F.A.P.; Georgetti, S.R.; Baracat, M.M.; Casagrande, R.; Verri, W.A., Jr. Naringenin inhibits superoxide anion-induced inflammatory pain: Role of oxidative stress, cytokines, Nrf-2 and the NO–cGMP–PKG–KATP channel signaling pathway. *PLoS ONE* **2016**, *11*, e0153015. [CrossRef]

43. Raza, S.S.; Khan, M.M.; Ahmad, A.; Ashafaq, M.; Islam, F.; Wagner, A.P.; Safhi, M.M.; Islam, F. Neuroprotective effect of naringenin is mediated through suppression of NF-kB signaling pathway in experimental stroke. *Neuroscience* **2013**, *230*, 157–171. [CrossRef]

44. Cavia-Saiz, M.; Busto, M.D.; Pilar-Izquierdo, M.C.; Ortega, N.; Perez-Mateos, M.; Muniz, P. Antioxidant properties, radical scavenging activity and biomolecule protection capacity of flavonoid naringenin and its glycoside naringin: A comparative study. *J. Sci. Food Agric.* **2010**, *90*, 1238–1244. [CrossRef]

45. Zheng, Y.Z.; Deng, G.; Guo, R.; Chen, D.F.; Fu, Z.M. DFT studies on the antioxidant activity of naringenin and its derivatives: Effects of the substituents at C3. *Int. J. Mol. Sci.* **2019**, *20*, 1450. [CrossRef]

46. Adisakwattana, S.; Jiphimai, P.; Prutanopajai, P.; Chanathong, B.; Sapwarobol, S.; Ariyapitipan, T. Evaluation of α-glucosidase, α-amylase and protein glycation inhibitory activities of edible plants. *Int. J. Food. Sci. Nutr.* **2010**, *61*, 295–305. [CrossRef] [PubMed]

47. Tadera, K.; Minami, Y.; Takamatsu, K.; Matsuoka, T. Inhibition of alpha-glucosidase and alpha-amylase by flavonoids. *J. Nutr. Sci. Vitaminol. (Tokyo)* **2006**, *52*, 149–153. [CrossRef] [PubMed]

48. Priscilla, D.H.; Roy, D.; Suresh, A.; Kumar, V.; Thirumurugan, K. Naringenin inhibits α-glucosidase activity: A promising strategy for the regulation of postprandial hyperglycemia in high fat diet fed streptozotocin induced diabetic rats. *Chem. Biol. Interact.* **2014**, *210*, 77–85. [CrossRef] [PubMed]

49. Ingkaninan, K.; de Best, C.M.; van der Heijden, R.; Hofte, A.J.P.; Karabatak, B.; Irth, H.; Tjaden, U.R.; van der Greef, J.; Verpoorte, R. High-performance liquid chromatography with on-line coupled UV, mass spectrometric and biochemical detection for identification of acetylcholinesterase inhibitors from natural products. *J. Chromatogr. A* **2000**, *872*, 61–73. [CrossRef]

50. Singhai, A.K.; Naithani, V.; Bangar, O.P. Medicinal plants with a potential to treat Alzheimer and associated symptoms. *Int. J. Nutr. Pharmacol. Dis.* **2012**, *2*, 84–91. [CrossRef]

51. Lee, S.; Youn, K.; Lim, G.; Lee, J.; Jun, M. *In silico* docking and *in vitro* approaches towards BACE1 and cholinesterases inhibitory effect of citrus flavanones. *Molecules* **2018**, *23*, 1509. [CrossRef] [PubMed]

52. Liaquat, L.; Batool, Z.; Sadir, S.; Rafiq, S.; Shahzad, S.; Perveen, T.; Haider, S. Naringenin-induced enhanced antioxidant defence system meliorates cholinergic neurotransmission and consolidates memory in male rats. *Life Sci.* **2018**, *194*, 213–223. [CrossRef]

53. Querfurth, H.W.; Laferla, F.M. Mechanisms of disease: Alzheimers disease. *N. Engl. J. Med.* **2010**, *362*, 329–344. [CrossRef]

54. Ghofrani, S.; Joghataei, M.-T.; Mohseni, S.; Baluchnejadmojarad, T.; Bagheri, M.; Khamse, S.; Roghani, M. Naringenin improves learning and memory in an Alzheimer's disease rat model: Insights into the underlying mechanisms. *Eur. J. Pharmacol.* **2015**, *764*, 195–201. [CrossRef]

Article

In Vivo Anti-Inflammatory Potential of Viscozyme®-Treated Jujube Fruit

Yoonsu Kim [1,†], Jisun Oh [2,†], Chan Ho Jang [1], Ji Sun Lim [2], Jeong Soon Lee [3] and Jong-Sang Kim [1,2,*]

[1] School of Food Science and Biotechnology (BK21PLUS), Kyungpook National University, Daegu 41566, Korea; yunsu531@gmail.com (Y.K.); cksghwkd7@gmail.com (C.H.J.)

[2] Institute of Agriculture Science and Technology, Kyungpook National University, Daegu 41566, Korea; j.oh@knu.ac.kr (J.O.); lzsunny@daum.net (J.S.L.)

[3] Forest Resources Development Institute of Gyeongsangbuk-do, Andong 36605, Korea; ljs7942@korea.kr

*   Correspondence: vision@knu.ac.kr; Tel.: +82-53-950-5752; Fax: +82-53-950-6750

†   These authors contributed equally to this work.

Received: 8 July 2020; Accepted: 30 July 2020; Published: 1 August 2020

Abstract: The fruit of *Ziziphus jujuba*, commonly called jujube, has long been consumed for its health benefits. The aim of this study was to examine the protective effect of dietary supplementation of enzymatically hydrolyzed jujube against lung inflammation in mice. The macerated flesh of jujube was extracted with aqueous ethanol before and after Viscozyme treatment. The extract of enzyme-treated jujube, called herein hydrolyzed jujube extract (HJE), contained higher levels of quercetin, total phenolics, and flavonoids, and exhibited more effective radical-scavenging abilities in comparison to non-hydrolyzed jujube extract (NHJE). HJE treatment decreased production of inflammation-associated molecules, including nitric oxide and pro-inflammatory cytokines from activated Raw 264.7 or differentiated THP-1 cells. HJE treatment also reduced expression of nuclear factor-κB and its downstream proteins in A549 human lung epithelial cells. Moreover, oral supplementation of 1.5 g of HJE per kg of body weight (BW) attenuated histological lung damage, decreased plasma cytokines, and inhibited expression of inflammatory proteins and oxidative stress mediators in the lungs of mice exposed to benzo(*a*)pyrene at 50 mg/kg BW. Expression levels of antioxidant and cytoprotective factors, such as nuclear factor erythroid-derived 2-related factor 2 and heme oxygenase-1, were increased in lung and liver tissues from mice treated with HJE, compared to mice fed NHJE. These findings indicate that dietary HJE can reduce benzo(*a*)pyrene-induced lung inflammation by inhibiting cytokine release from macrophages and promoting antioxidant defenses in vivo.

Keywords: Jujube; hydrolysis; anti-inflammation; lung; NF-κB; Nrf2; HO-1

---

1. Introduction

Lung inflammation can be triggered by pathogens, toxins, and pollutants [1–3]. In response to exogenous inflammatory factors, the innate immune defense system is acutely activated to rehabilitate affected tissue and minimize systemic inflammation [4]. If the inflammatory response is not sufficient to restore tissue homeostasis, lung damage progresses and causes pulmonary dysfunction and respiratory failure [5].

Inflammation in the lungs is regulated by a variety of molecular mediators, including pro- and anti-inflammatory chemokines and cytokines, which are produced by damaged pulmonary epithelial cells, as well as various types of activated leukocytes [6,7]. Mediators related to lung inflammation include tumor necrosis factor α (TNFα), interleukin-6 (IL-6), and interleukin-1β (IL-1β), which regulate maturation of dendritic cells and induce recruitment of neutrophils at inflammatory sites [7,8]. Cyclooxygenase (COX) is an enzyme that metabolizes arachidonic acid into endoperoxides and is closely associated with inflammation [9]. Of the two isoforms, COX-2 is expressed in most cells in an inducible manner at the local inflammation site, whereas COX-1 is constitutively expressed. COX-2 expression is controlled by nuclear factor κB (NF-κB) which is a ubiquitous transcription factor that regulates expression of multiple proteins involved in inflammatory and immune responses [10].

Nuclear factor erythroid-derived 2-related factor 2 (Nrf2) is a key regulator of a set of phase-2 defense and antioxidant enzymes, including heme oxygenase-1 (HO-1) [11]. Nrf2 and its downstream enzymes play a protective role against pro-inflammatory and detrimental oxidizing effects [12]. Several studies have demonstrated that antioxidative phytochemicals in fruits and vegetables can upregulate the Nrf2 signaling pathway, downregulate the NF-κB pathway, and consequently modulate inflammation-associated gene expression in macrophages and lung epithelial cells [13,14].

The fruit of *Ziziphus jujuba* var. inermis Rehder (commonly called jujube) is used widely as a dietary ingredient and traditional remedy for digestive and respiratory dysfunctions [15–18]. Jujube contains not only antioxidative phytochemicals, including phenolics and flavonoids, but also polysaccharides with immunomodulatory potential [19–21]. These compounds were found to exert protective effects against gastrointestinal, hepatic, and pulmonary damage [19–21]. A couple of studies have demonstrated that crude mixtures containing jujube extract can alleviate symptoms of nasal or gastrointestinal inflammation in mice [22,23]. Some polysaccharides obtained from jujube by ultrasonic-assisted extraction reportedly suppressed pro-inflammatory cytokine production in activated Raw 264.7 cells [24,25]. In addition, a research group has shown that jujuboside B, a saponin found in jujube ethanolic extract, lowered the number of immune cells in bronchoalveolar lavage fluid, and decreased levels of allergic phenotype-regulating cytokines in lung homogenate from an ovalbumin-induced allergic asthma mouse model [26]. These findings suggest that jujube ethanolic extract contains a variety of bioactive substances and may relieve lung disease caused by exogenous inflammatory stimuli.

In an attempt to liberate the bioactive components existing in inactive glycoside forms, jujube fruit was preheated and enzymatically treated using Viscozyme, which possesses primarily endo-1,3(4)-β-glucanase and collaterally cellulase and hemicellulase activities, to break any constraining linkage and thereby enhance the bioavailability of the compounds. The fruit was then subjected to extraction with aqueous ethanol. The hydrolyzed jujube fruit extract was evaluated for its biologically beneficial effects in an acute lung inflammation mouse model using benzo(*a*)pyrene (B(a)P).

2. Materials and Methods

2.1. Preparation of Hydrolyzed Jujube Extracts

The commercially available fresh jujube fruits that were harvested from Gyeongsan region, S. Korea in October 2019 [16,27] were obtained from a local market. The seeds were removed and the flesh was cut into 100 g samples. The samples were homogenized with 10 volumes of distilled water, pretreated at 95 °C for 10 min, followed by enzymatic hydrolysis using five units of Viscozyme (Novozyme, Bagsvaerd, Denmark) in 1 M acetate buffer at a pH of 5.0 and a temperature of 55 °C, shaken at 250 rpm for 1 h. Non-hydrolyzed jujube flesh homogenate was subjected to the same procedure as HJE but without enzyme treatment. Both samples were then mixed with the same volume of absolute ethanol to make the final concentration of 50% (*v/v*) ethanol in solution, and extracted in a shaking incubator at 55 °C and 250 rpm for 1 h. Both extracts prepared in the absence and presence of Viscozyme were referred to as non-hydrolyzed jujube extract (NHJE) and hydrolyzed jujube extract (HJE), respectively. They were filtered using 8-μm filter papers (Whatman, Little Chalfont, UK), vacuum-evaporated (EYELA N-1000, Tokyo, Japan), and freeze-dried. The NHJE and HJE powders were dissolved in dimethyl sulfoxide (DMSO) at the concentration of 500 mg/mL for further experiments.

2.2. Quantification of Total Phenolics and Flavonoids

Total phenolic and flavonoid contents in the jujube extracts were determined as reported previously [28–30]. For quantification of the total phenolic content, briefly, each sample was mixed with Folin reagent and 10% sodium carbonate. After incubation at room temperature for 1 h, the absorbance was measured at 725 nm using a microplate reader. For measurement of the total flavonoid content, samples were mixed with aluminum nitrate, potassium acetate, and ethanol. After incubation at room temperature for 40 min, the absorbance was acquired at 415 nm. Total phenolic compound content and total flavonoid content were estimated from a standard curve prepared using gallic acid and quercetin (both from Sigma-Aldrich, St. Louis, MO, USA), respectively.

2.3. High-Performance Liquid Chromatography Analysis

Quantitative determination of rutin and quercetin in NHJE and HJE was performed using high-performance liquid chromatography (Shimadzu Corporation, Tokyo, Japan) equipped with a photodiode array (PDA) detector and an M1 column (SunFire, C18, 4.6 × 250 mm, 5-μm particle size) with the mobile phase consisting of a mixture of 40% (*v/v*) methanol and 15% (*v/v*) acetonitrile in water containing 1% (*v/v*) acetic acid. The flow rate was 1 mL/min. The column temperature was maintained at 40 °C for 30 min during the analysis and the injection volume was 10 μL. Detection was made with a PDA detector at 280 nm. The quantities of those compounds were determined by extrapolating the corresponding peak areas from the calibration curves of standards, rutin, and quercetin (both from Sigma-Aldrich).

2.4. Cell Culture and Treatment

Raw 264.7 murine macrophages, THP-1 human monocytes, and A549 human lung epithelial cells were obtained from the Korean Cell Line Bank (Seoul, Korea). All cell lines were grown in Dulbecco's modified Eagle medium (DMEM, Invitrogen) containing 10% heat-inactivated fetal bovine serum (FBS) and 1% penicillin-streptomycin (Welgene, Gyeongsan, Gyeongbuk) and placed in a humidified incubator that provided 5% CO_2 at 37 °C.

For cell viability assays, cells were dispensed into a 96-well plate at a density of 5×10^3 cells/well, treated with jujube extract at the concentrations of 0, 62.5, 125, 250, and 500 μg/mL for 24 h, and assayed using the Cell Counting Kit (CCK-8; Dojindo Laboratories, Kumamoto, Japan) as previously described [31].

For measurement of nitric oxide (NO) levels in culture medium, Raw 264.7 cells were seeded at a density of 2×10^3 cells per well of a 96-well plate and activated by treatment with lipopolysaccharide (LPS) at 1 μg/mL for 24 h. The culture medium was collected and reacted with the Griess reagent (Promega, Madison, WI, USA).

THP-1 cells were passaged at a density of 4×10^6 cells/mL in a 100-mm plate and maintained in DMEM containing 10% FBS. The cells were activated and differentiated by applying 200 nM of 12-*O*-tetradecanoylphorbol-13-acetate (TPA). For indirect co-culture of THP-1 and A549 cells, the cultured medium (herein called 'conditioned medium') from THP-1 cells was collected and transferred to A549 cells.

2.5. Antioxidant Response Element–Luciferase Reporter Assay

To determine the transcriptional activity of the antioxidant response element (ARE), a luciferase reporter gene assay was conducted on a HepG2 cell line carrying the ARE-luciferase construct using a luciferase assay system (Promega, Madison, WI, USA) [32]. Sulforaphane (Sigma-Aldrich, St. Louis, MO, USA) was used as a positive control.

2.6. Animal Treatment

The animal study was approved by, and conducted according to the guidelines of, the Institutional Animal Care and Use Committee of the Kyungpook National University (approval number: KNU 2019-0157). A total of 48 adult C57BL/6J mice (six weeks old males, 20–25 g BW) were purchased from Hyochang Science (Daegu, Korea). After one week of adaptation in a controlled laboratory environment (temperature, 22 ± 2 °C; humidity, 50 ± 5%; 12-h light-dark cycle) with free access to water and AIN-76A chow (Hyochang Science, Daegu, Korea), the mice were randomly assigned to six groups (eight mice per group). The experimental groups were as follows: (1) control, a group treated only with vehicle, (2) a group treated with B(a)P alone, (3) a group treated with NHJE at 0.75 g/kg BW and B(a)P, (4) a group treated with NHJE at 1.5 g/kg BW and B(a)P, (5) a group treated with HJE at 0.75 g/kg BW and B(a)P, and (6) a group treated with HJE at 1.5 g/kg BW and B(a)P. The NHJE and HJE samples were dissolved in a vehicle composed of 5% (*v/v*) Tween-80, 5% ethanol, and 90% sterilized drinking water. The freshly dissolved samples were administered by oral gavage in a volume of 0.2 mL/mouse on a daily basis for two weeks. Lung inflammation was induced by intraperitoneal injection with 50 mg/kg BW of B(a)P dissolved in corn oil (Sigma-Aldrich, St. Louis, MO, USA) 24 h prior to sacrifice. The blood, lung, and liver tissues were dissected and used for further biochemical analyses.

2.7. Histopathological Analysis

The entire lungs were dissected from the sacrificed. All lobes were fixed in a 10% (*v/v*) formalin solution and embedded in paraffin [33]. Tissue blocks were then coronally sectioned into 5-μm-thick slices using a microtome (RM-2125 Rt; Leica, Nussloch, Germany), and stained with hematoxylin and eosin (H&E, Sigma-Aldrich, St. Louis, MO, USA). Histological damage in the tissue sections was observed under a microscope (Eclipse 80i; Nikon, Tokyo, Japan).

2.8. Enzyme-Linked Immunosorbent Assay

An enzyme-linked immunosorbent assay (ELISA) was performed using commercially available kits for the quantitation of inflammatory cytokines (TNFα, IL-1β, and IL-6) in plasma (mouse ELISA sets for each cytokine, BD Biosciences, San Jose, CA, USA). Lung and liver tissues were collected from the mice and homogenized in phosphate buffered saline including protease inhibitor cocktail (Roche, Mannheim, Germany). After centrifugation at 15,000× g for 10 min, the supernatants were used for the following analyses. Lung tissue homogenate was subjected to measurement of IL-10 (ELISA MAX standard set, Biolegend, San Diego, CA, USA) and prostaglandin E_2 (PGE_2) (Cayman Chemical, Ann Arbor, MI, USA) as per the manufacturer's instructions. Liver tissue homogenate was used to determine the levels of 8-hydroxy-2'-deoxyguanosine (8-OHdG) (Cat # ADI-EKS-350; Enzo Life Sciences, Farmingdale, NY, USA) and malondialdehyde (MDA) (Cat # ALX-850-287; Enzo Life Sciences, Inc., Farmingdale, NY, USA) as previously described [34]. The obtained values were normalized to the total amount of proteins.

2.9. Western Blot Analysis

Cytoplasmic and nuclear fractions of cultured cells and mouse tissues were prepared using the NE-PER Nuclear and Cytoplasmic Extraction Reagents (Thermo Fisher Scientific, Rockford, IL, USA). The proteins in each fraction were blotted and their relative expression levels were determined as previously described [34,35]. Briefly, the primary antibodies used in this study were immunoglobulins against COX-2 (Cell Signaling Technology, Danvers, MA, USA), NF-κB (Bioworld Technology, St. Luis. MN, USA), iNOS (Enzo Life Sciences, Farmingdale, NY, USA), HO-1 (Abcam, Cambridge, UK), Nrf2 (Abcam, Cambridge, UK), β-actin, and Lamin B1 (Santa Cruz Biotechnology, Dallas, TX, USA). They were used at a dilution ratio of 1:1000 in 1% bovine serum albumin (Bio Basic Inc., Markham, ON, Canada) in tris-buffered saline (TBS). After allowing the appropriate secondary antibody (horse radish peroxidase–conjugated) at 1:2000 in TBS to interact with the primary antibody, protein bands were visualized using SuperSignal West Pico Chemiluminescent Substrate (Pierce, Cheshire, United Kingdom) and LAS4000 Mini (GE Healhcare Life Sciences, Little Chalfont, UK). The digitalized blot images were then densitometrically analyzed using Image-Studio Lite version 5.2 (LI-COR Biotechnology, Lincoln, NE, USA).

2.10. Statistical Analysis

All statistical analyses were performed using SPSS version 23.0 (SPSS Inc., Chicago, IL, USA). Statistical differences among means were tested by paired *t*-tests or one-way analysis of variance followed by Duncan's multiple range test. The *p* values less than 0.05 or 0.1 were considered significant. Different alphabetical letters indicate statistically significant difference between values.

3. Results

3.1. Enzyme Hydrolysis of Jujube Fruit Increased Total Phenolic Content in Ethanolic Extract

For the preparation of HJE, the concentration of ethanol in the extraction solvent was determined to be 50% (*v/v*) in water, based on the contents of total phenolics and flavonoids that were potent bioactive compounds (Supplementary Table S1). For hydrolysis of jujube fruit, Viscozyme was chosen among multiple food industrial enzymes as the highest total phenolic content was acquired in a 50% ethanol extract after enzymatic reaction (Supplementary Table S2). HJE exhibited higher antioxidant capacities assessed by a 2,2-diphenyl-1-picrylhydrazyl radical-scavenging assay and ferric reducing antioxidant power values (Supplementary Figure S1), and also contained approximately 1.9-fold higher total phenolic content and 10.9-fold higher quercetin content than its non-hydrolyzed counterpart, NHJE (Table 1).

Table 1. Total phenolics, rutin, and quercetin contents in jujube extracts [1].

	NHJE	**HJE**
Total phenolics (mg GAE [2]/g dry weight)	6.73 ± 0.83 [a]	12.50 ± 1.69 [b]
Rutin (mg/g dry weight)	12.68 ± 6.06 [a]	11.94 ± 4.99 [a]
Quercetin (mg/g dry weight)	0.34 ± 0.59 [a]	3.72 ± 2.54 [b]

[1] Values are expressed as means ± standard error of the mean (SEM) from three independent experimental sessions (*N* = 3). Different alphabetical letters by the values (a–b) indicate statistically significant difference from each other (*p* < 0.05). [2] GAE, gallic acid equivalent.

3.2. HJE Decreased NO Production and Increased ARE Transcription Activity in Cultured Cells

The anti-inflammatory and antioxidant effects of HJE and NHJE were compared in both Raw 264.7 cells and HepG2-ARE cells at concentrations of ≤500 μg/mL, which were non-toxic to both cell lines (Supplementary Figure S2). HJE decreased production of NO in a dose-dependent manner in LPS-stimulated Raw 264.7 cells (Figure 1A) and significantly increased ARE transcription activity in HepG2-ARE cells (Figure 1B), whereas treatment with NHJE had no significant effect on NO levels and ARE-transcriptional activity of these cells.

3.3. HJE Inhibited Secretions of Pro-Inflammatory Cytokines from TPA-Challenged THP-1 Cells and Expressions of Inflammation-Related Proteins in A549 Cells

To investigate the role of HJE in interactions between macrophages and lung epithelial cells, human monocyte THP-1 cells were differentiated by TPA treatment and incubated in the absence or presence of either HJE or NHJE, and the resulting cultured media were added to lung epithelial A549 cell culture, followed by measurements of the expression of pro-inflammatory proteins, such as iNOS, COX-2, and NF-κB.

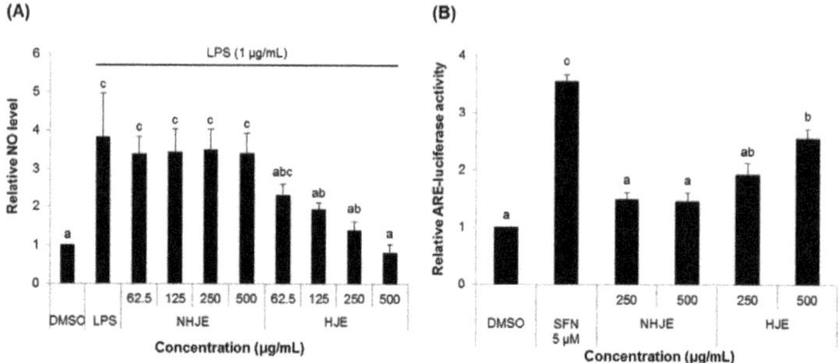

Figure 1. HJE decreased nitric oxide (NO) production and increased antioxidant response element (ARE) transcription activity in cultured cells. (**A**) Raw 264.7 cells were stimulated by LPS and treated with NHJE and HJE for 24 h. The culture media were collected and measured for the levels of NO produced from the cells. (**B**) HepG2-ARE cells carrying the luciferase reporter gene linked to the ARE sequence were treated with NHJE and HJE for 24 h. ARE transcription activity was assessed by the activity of luciferase. SFN, sulforaphane. $N = 3$; error bars, mean ± SEM. Different alphabetical letters on the bars (a–c) indicate statistically significant difference from each other ($p < 0.05$).

Quantification of pro-inflammatory cytokines in the THP-1-cultured medium (the conditioned medium) revealed that HJE considerably reduced production of TNFα and IL-6 from THP-1 cells, while NHJE had no effect (Figure 2A–C). A549 cells maintained in the conditioned media from HJE-treated THP-1 cells expressed relatively lower levels of NF-κB, cytoplasmic iNOS and COX-2 proteins compared with the cells kept in NHJE-treated conditioned media (Figure 2D–G).

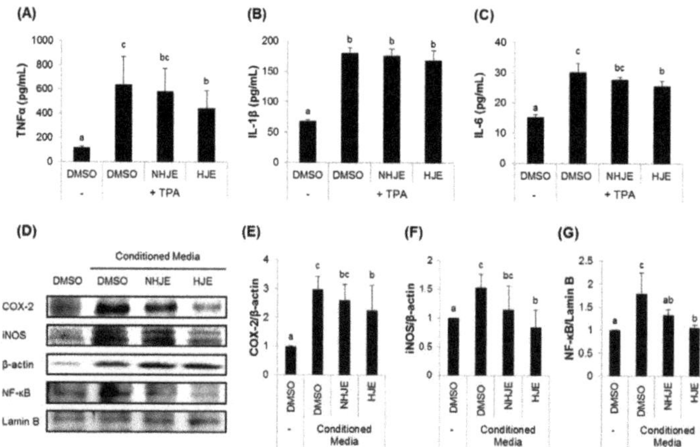

Figure 2. HJE decreased secretions of pro-inflammatory cytokines from differentiated THP-1 cells and expression of inflammation-related proteins in A549 cells. (**A–C**) THP-1 human monocytes were differentiated with tetradecanoylphorbol-13-acetate (TPA) (200 nM) and treated with either NHJE or HJE at a concentration of 500 μg/mL for 24 h. The pro-inflammatory cytokines, including TNFα (**A**), IL-1β (**B**), and IL-6 (**C**), in the culture media were quantified by enzyme-linked immunosorbent assay (ELISA). (**D–G**) A549 human lung epithelial cells were cultured in conditioned media collected from TPA-differentiated THP-1 cells treated with either NHJE or HJE. After 24 h, A549 cells were collected and subjected to western blot analysis (**D**). The relative expression levels of inflammation-related proteins, including COX-2 (**E**), iNOS (**F**), and NF-κB (**G**), were densitometrically determined. $N = 3$; error bars, mean ± SEM. Different alphabetical letters (a–c) presented on the bars indicate statistically significant difference from each other ($p < 0.05$).

3.4. Oral Supplementation of HJE Alleviated B(a)P-Induced Lung Injury in Mice

To further evaluate the anti-inflammatory effect of HJE in vivo, acute lung inflammation in mice was induced by a single intraperitoneal injection of B(a)P 24 h prior to sacrifice after 14 days of oral administration of either NHJE or HJE (Figure 3A). The changes in the average BW of mice were insignificant among the groups during the experimental period (Figure 3B). Histological observation of lungs indicated that supplementation with HJE or NHJE at a high dose (1.5 g/kg BW) attenuated B(a)P-induced lung injury (Figure 3C). In addition, H&E-stained lung tissue sections displayed that B(a)P treatment caused disruption of lung architecture [36–38], including poor arrangement of epithelial cells lining bronchiole and abnormal morphology of alveolar structure (inflated alveolar sacs and thickened or ruptured interalveolar septa). However, the severity of B(a)P-induced histological damage in the lungs was reduced in mice supplemented with HJE or NHJE-H (Figure 3D).

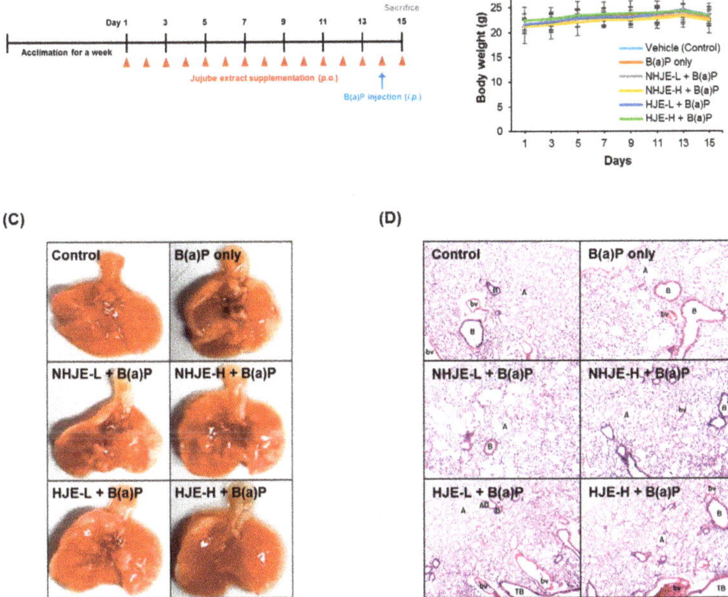

Figure 3. Oral administration of HJE alleviated B(a)P-induced lung injury in mice. C57BL/6J mice were fed either NHJE or HJE at two different doses, 0.75 g/kg BW or 1.5 g/kg BW, every day for 14 days. One day prior to sacrifice, B(a)P was given by intraperitoneal injection. (**A**) Experimental scheme. (**B**) Changes in average body weight during the experimental period. Values are means ± standard deviation (SD) (*n* = 8 mice per group). (**C**) Representative photographs of the dissected lungs. (**D**) Representative tissue sections stained with H&E (magnification, 40×). A, alveolar sac; AD, alveolar duct; B, respiratory bronchiole; TB, terminal bronchiole; bv, blood vessel. NHJE, non-hydrolyzed jujube extract; NHJE-L, NHJE at a low dose (0.75 g/kg BW); NHJE-H, NHJE at a high dose (1.5 g/kg BW); HJE, hydrolyzed jujube extract; HJE-L, HJE at a low dose (0.75 g/kg BW); HJE-H, HJE at a high dose (1.5 g/kg BW).

3.5. Oral Supplementation of HJE Lowered Plasma Pro-Inflammatory Cytokine Levels in B(a)P-Injected Mice

After supplementation with the extract for 14 days and then intraperitoneal injection with B(a)P, blood samples were collected from mice and subjected to ELISA for quantification of TNFα, IL-1β, and IL-6 (Figure 4). The group fed HJE at a high dose (1.5 g/kg BW) showed significantly lower levels of TNFα and IL-1β in plasma, compared with the group injected with B(a)P without jujube extract supplementation.

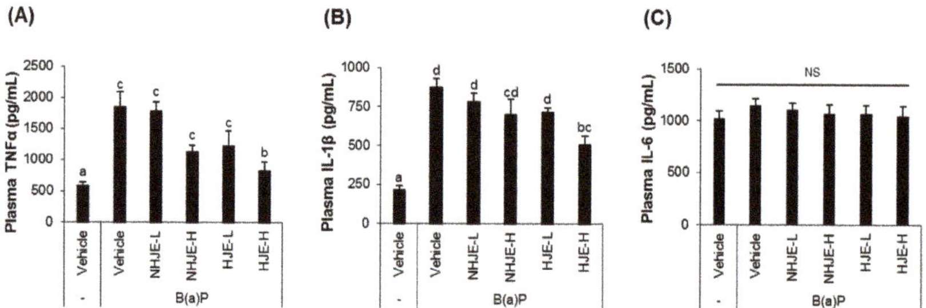

Figure 4. Oral administration of HJE lowered plasma levels of pro-inflammatory cytokines in B(a)P-injected mice. (**A–C**) Levels of pro-inflammatory cytokines, TNFα (**A**), IL-1β (**B**), and IL-6 (**C**), in the mouse blood samples were measured by ELISA. Values are means ± SD ($n = 8$). Different alphabetical letters presented on the bars (a–d) indicate statistically significant difference from each other ($p < 0.05$). NS, not significant.

3.6. Oral Supplementation of HJE Suppressed the Expression of Inflammation-Related Proteins and Increased the Expression of Antioxidant Proteins in Lung Tissue

Lung tissue homogenates were analyzed for the expression levels of inflammatory proteins, such as NF-κB, cytoplasmic iNOS, and COX-2 (Figure 5A–D). Treatment with HJE (0.75 and 1.5 g/kg BW) or NHJE (1.5 g/kg BW) significantly suppressed protein expression of NF-κB, cytoplasmic iNOS, and COX-2, which were increased by B(a)P injection. Moreover, the tissue level of PGE$_2$, which limits inflammation and promotes lung tissue repair [39], was reduced in lungs of mice treated with B(a)P alone but restored by co-treatment with 0.75 g HJE/kg BW or 1.5 g NHJE/kg BW, indicating that HJE is more effective than NHJE (Figure 5E).

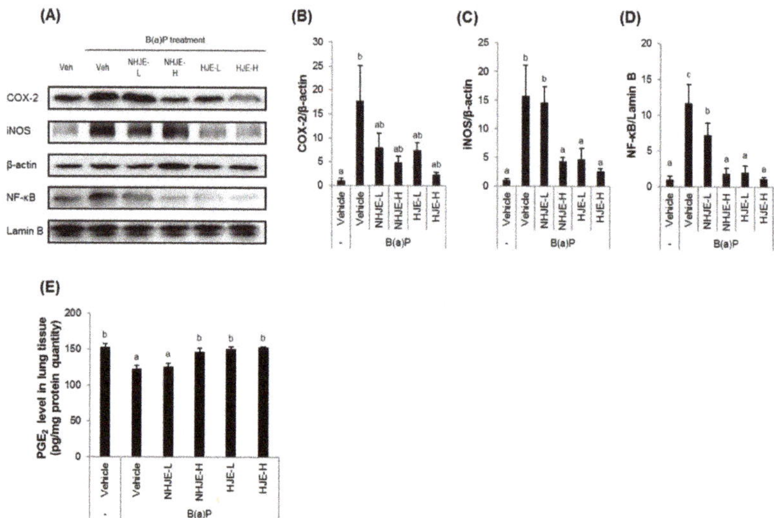

Figure 5. Oral administration of HJE suppressed expression of inflammation-related proteins in lung tissue. Expression levels of inflammation marker proteins in the lung tissues were examined. (**A**) Representative western blot images. (**B–D**) Quantitative data for cytosolic COX-2 (**B**), cytosolic iNOS (**C**), and nuclear NF-κB (**D**). (**E**) PGE$_2$ protein levels in dissected lung tissues were quantified using ELISA. Values are mean ± SD ($n = 5$). Different alphabetical letters on the bars (a–c) indicate statistically significant difference from each other ($p < 0.05$).

A key transcription factor of antioxidant and defense response, Nrf2, and one of its downstream proteins, HO-1, were highly expressed in B(a)P-treated lung tissues (Figure 6). Their expressions were increased by HJE more effectively than by NHJE. In addition, oral administration of HJE but not NHJE in mice increased the ratio of reduced to oxidized glutathione (GSH/GSSG) in lung homogenates (Supplementary Figure S3).

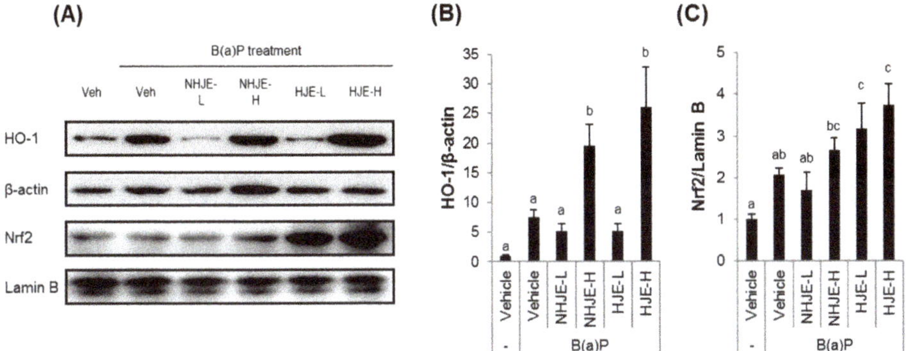

Figure 6. Oral administration of HJE increased Nrf2 and HO-1 expressions in lungs. Expression levels of a key antioxidant/defense transcription factor Nrf2 and its downstream protein HO-1 in the lung tissues were examined. (**A**) Representative western blot images. (**B–C**) Quantitative data for cytosolic HO-1 (**B**) and nuclear Nrf2 (**C**). Values are mean ± SD ($n = 5$). Different alphabetical letters on the bars (a–c) indicate statistically significant difference from each other ($p < 0.05$).

3.7. Oral Administration of HJE to Mice Increased Nrf2 and HO-1 Expression and Reduced Oxidative Stress in Liver Tissue

The relative quantities of nuclear Nrf2 and cytoplasmic HO-1 proteins were significantly higher in liver tissue from HJE-H-fed mice than from NHJE-fed or control mice (Figure 7A–C). In addition, oral administration of HJE reduced levels of oxidative stress markers, such as malondialdehyde (MDA) and 8-OHdG, in liver homogenates (Figure 7D–E).

4. Discussion

Enzymatically hydrolyzed jujube extract was evaluated for its anti-inflammatory effect in a B(a)P-induced lung inflammation mouse model. Our findings demonstrated that (1) Viscozyme-mediated hydrolysis increased total phenolic content in an ethanolic extract of jujube fruit, (2) HJE inhibited TNFα production in TPA-differentiated THP-1 cells and inflammatory protein expression in A549 cells exposed to THP-1-conditioned media more effectively than did NHJE, and (3) oral supplementation with HJE for two weeks protected the lungs from B(a)P-induced morphological and histological damage, with decreased levels of inflammation-associated proteins in the lungs and increased levels of antioxidant/defense proteins in the lungs and liver. A graphical summary of the findings is presented in Figure 8.

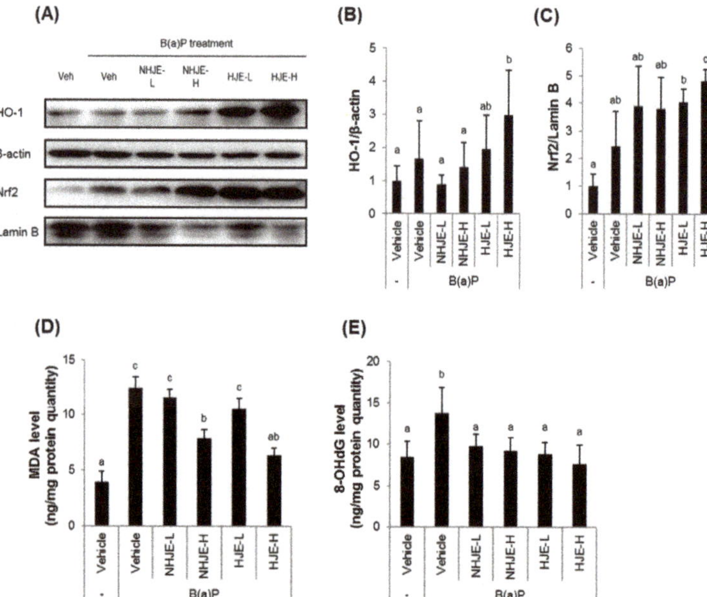

Figure 7. Oral administration of HJE increased Nrf2 and HO-1 expression and reduced oxidative stress in hepatic tissue. Liver tissue homogenates were used for analyses of expression levels of antioxidant proteins and oxidative stress markers. (**A**) Representative western blot images. (**B,C**) Quantitative data for cytosolic HO-1 (**B**) and nuclear Nrf2 (**C**). (**D,E**) Oxidative stress markers including MDA (**D**) and 8-OHdG (**E**) were quantified by ELISA. MDA, malondialdehyde. 8-OHdG, 8-hydoxydeoxyguanosine. Values are mean ± SD ($n = 8$). Different alphabetical letters presented on the bars (a–c) indicate statistically significant difference from each other ($p < 0.05$).

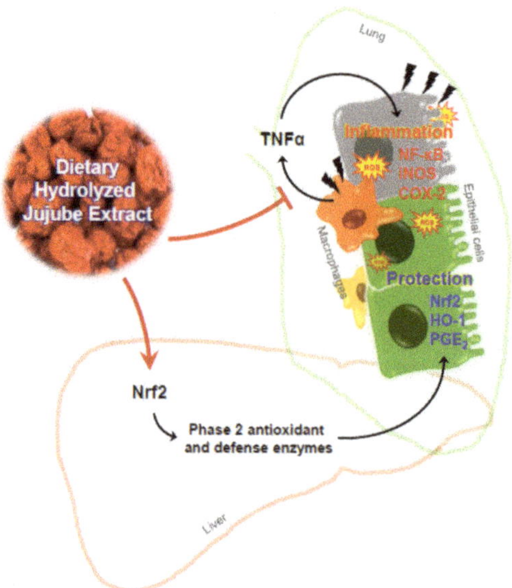

Figure 8. Schematic illustration of a potential mechanism for the anti-inflammatory activity of HJE.

Regardless of hydrolysis, jujube extracts used in this study were found to contain considerable amounts of total phenolics and flavonoids and demonstrate radical-scavenging abilities, which is consistent with previous reports [17,19,21]. HJE significantly promoted the transcription activity of the ARE-luciferase reporter gene and decreased production of NO, a signaling molecule regulating inflammatory processes, as shown by cell-based assays. In addition, HJE, but not NHJE, effectively inhibited TNFα secretion from TPA-differentiated THP-1 cells and suppressed inflammatory protein expression in A549 cells kept in THP-1-conditioned media. This suggests that the pre-treatment of jujube fruit with Viscozyme improved biological activity by increasing the bioavailability of bioactive compounds in jujube, an effect that may also be applicable to in vivo conditions.

To examine the in vivo anti-inflammatory effect of HJE compared with NHJE, a B(a)P-induced acute lung inflammation mouse model was employed [2,36,40]. B(a)P is the most common and toxic polycyclic aromatic hydrocarbon, an environmental pollutant, and reportedly the cause of health problems, such as redox imbalance, respiratory diseases, and cancer [41]. Benzo[a]pyrenediol-epoxide, which is metabolized from B(a)P by cytochrome P450 enzymes, provokes pulmonary inflammation by stimulating the NF-κB-mediated pathway in human lung fibroblasts [42]. Thus, B(a)P is widely used to generate a lung inflammation in mice [43–45]. Consistently, observations from this study demonstrated that a single administration of B(a)P to mice resulted in lung injury at morphological and histological levels. The elevation of pro-inflammatory cytokine levels in the blood samples was consistent with previous reports that showed B(a)P-induced deleterious consequences in multiple organs [41,45]. Moreover, oral supplementation of jujube extracts, both HJE and NHJE, reduced levels of NF-κB and its downstream cytoplasmic proteins, iNOS and COX-2, in the lungs while increasing expressions of Nrf2 and cytoplasmic HO-1 in lung and liver tissues. This suggests that dietary jujube extract exerts dual protective effects on both hepatic and pulmonary tissues from B(a)P-induced damage.

Intriguingly, our unpublished data demonstrated that Nrf2 and HO-1 protein expressions in A549 cells kept in THP-1-conditioned media were marginally altered by treatment with either HJE or NHJE. It implies that the stimulation of Nrf2-mediated antioxidant response in the lungs of the mice supplemented with jujube extract may be indirectly induced possibly through intestinal or hepatic metabolites. Specification and identification of biologically active metabolites produced from jujube extract in vivo will require additional study.

One of the critical findings of this study was that the biological effect of jujube extract was further enhanced when HJE was given to the mice compared with NHJE. It indicates that Viscozyme-mediated hydrolysis improved the beneficial potential of jujube fruits, likely by converting bioactive compounds in glycoside forms to aglycones and thereby increasing their bioavailability. A line of evidence has demonstrated that jujube contains a variety of bioactive substances [19,20]. Some sugars, terpenes, and phenolic acids have been reported as biologically functional components [21,24,26]. Rutin (quercetin-3-*O*-rutinoside), one of the major phenolics in jujube, has been demonstrated to possess antioxidative and anti-inflammatory activities [46–48].

Viscozyme includes a broad range of carbohydrase activities including β-glucanase, arabanase, and polygalacturonase, as described in previous reports [49,50]. Its cellulase and hemicellulase activities can degrade the polysaccharide components of plant cell walls [51,52]. HJE undergone Viscozyme-mediated hydrolysis was found to contain quercetin at levels 10.9-fold higher and total phenolic content at levels 1.9-fold higher than those of NHJE. These results are not surprising because rutin with β (1→6) glycosidic linkage can be broken by the enzyme and converted to quercetin. Furthermore, it is presumed that the β-glucanase and polygalaturonase activities have degraded cell walls and facilitated the extraction of phenolic compounds bound to cell-wall components. Considering the activity of gut microbiota and the protective effects of quercetin and rutin in the liver [53,54] and lungs against exogenous inflammatory agents [55–60], it is reasonable to assume that the hepatic and pulmonary protective effects of HJE from B(a)P insult are at least partially attributable to aglycones, including quercetin, released from glycosides during enzymatic treatment. However, further study is needed to identify the component(s) responsible for the anti-inflammatory activity of HJE. Whether the anti-inflammatory effect of HJE or its active components is mediated by the Nrf2/ARE signaling pathway or the NF-kB signaling pathway, or through crosstalk between both signaling pathways, is a fascinating question to address in the future.

5. Conclusions

This study demonstrated that dietary HJE can significantly ameliorate B(a)P-mediated lung inflammation and injury, possibly by inhibiting macrophage activity, suppressing NF-κB-mediated inflammatory protein expression in lung tissue, and stimulating Nrf2-mediated antioxidant/defense responses in the lung and liver. The findings would have implications for development of jujube-based functional foods beneficial to lung health, especially in situations where the respiratory system can be affected.

Supplementary Materials: The following are available online at http://www.mdpi.com/2304-8158/9/8/1033/s1. Table S1: Total phenolic and flavonoid contents in jujube ethanolic extracts using various concentrations of ethanol in water. Table S2: Total phenolic content in 50% ethanol extract of jujube hydrolyzed with various enzymes. Figure S1: DPPH radical scavenging activity (A) and FRAP (B) of jujube extracts. Figure S2: Cytotoxicity of jujube extracts in THP-1 human monocytes (A) and A549 human lung epithelial cells (B). Figure S3. Dietary HJE increased the ratio of reduced to oxidized glutathione (GSH/GSSG) in lung homogenates.

Author Contributions: Y.K. performed the experiments, processed and analyzed the data, and wrote the manuscript. J.O. designed the experiments, interpreted the data, and wrote the manuscript. C.H.J. analyzed and interpreted the data. J.S.L. (Ji Sun Lim) analyzed and interpreted the data. J.S.L. (Jeong Soon Lee) financially supported and interpreted the data. J.-S.K. conceptualized the project, financially supported, and finally approved the manuscript. All authors have read and agreed to the published version of the manuscript.

Funding: This study was funded by the Forest Resources Development Institute of Gyeongsangbuk-do, Andong, Korea (2019).

Acknowledgments: The authors would like to thank Enago for the English language review and Hak Young Lee, MD, PhD (INVIVO Co., Ltd., Nonsan, Chungcheongnam-do) for pathological examination of lung tissue.

Conflicts of Interest: The authors have declared no conflict of interest.

References

1. Inoue, K.-I.; Takano, H.; Yanagisawa, R.; Hirano, S.; Sakurai, M.; Shimada, A.; Yoshikawa, T. Effects of airway exposure to nanoparticles on lung inflammation induced by bacterial endotoxin in mice. *Environ. Health Perspect.* **2006**, *114*, 1325–1330. [CrossRef] [PubMed]

2. Qamar, W.; Khan, A.Q.; Khan, R.; Lateef, A.; Tahir, M.; Rehman, M.U.; Ali, F.; Sultana, S. Benzo(a)pyrene-induced pulmonary inflammation, edema, surfactant dysfunction, and injuries in rats: Alleviation by farnesol. *Exp. Lung Res.* **2012**, *38*, 19–27. [CrossRef] [PubMed]

3. Vlahos, R.; Bozinovski, S.; Jones, J.; Powell, J.; Gras, J.; Lilja, A.; Hansen, M.J.; Gualano, R.C.; Irving, L.; Anderson, G. Differential protease, innate immunity, and NF-κB induction profiles during lung inflammation induced by subchronic cigarette smoke exposure in mice. *Am. J. Physiol. Lung Cell Mol. Physiol.* **2006**, *290*, L931–L945. [CrossRef] [PubMed]

4. Liu, J.; Cao, X.T. Cellular and molecular regulation of innate inflammatory responses. *Cell. Mol. Immunol.* **2016**, *13*, 711–721. [CrossRef] [PubMed]

5. Ware, L.B.; Matthay, M.A. The acute respiratory distress syndrome. *N. Engl. J. Med.* **2000**, *342*, 1334–1349. [CrossRef]

6. Aghasafari, P.; George, U.; Pidaparti, R. A review of inflammatory mechanism in airway diseases. *Inflamm. Res.* **2019**, *68*, 59–74. [CrossRef] [PubMed]

7. Moldoveanu, B.; Otmishi, P.; Jani, P.; Walker, J.; Sarmiento, X.; Guardiola, J.; Saad, M.; Yu, J. Inflammatory mechanisms in the lung. *J. Inflamm. Res.* **2009**, *2*, 1–11.

8. Benihoud, K.; Esselin, S.; Descamps, D.; Jullienne, B.; Salone, B.; Bobe, P.; Bonardelle, D.; Connault, E.; Opolon, P.; Saggio, I.; et al. Respective roles of TNF-alpha and IL-6 in the immune response-elicited by adenovirus-mediated gene transfer in mice. *Gene Ther.* **2007**, *14*, 533–544. [CrossRef]

9. Brock, T.G.; McNish, R.W.; Peters-Golden, M. Arachidonic acid is preferentially metabolized by cyclooxygenase-2 to prostacyclin and prostaglandin E-2. *J. Biol. Chem.* **1999**, *274*, 11660–11666. [CrossRef]

10. Kirkby, N.S.; Chan, M.V.; Zaiss, A.K.; Garcia-Vaz, E.; Jiao, J.; Berglund, L.M.; Verdu, E.F.; Ahmetaj-Shala, B.; Wallace, J.L.; Herschman, H.R.; et al. Systematic study of constitutive cyclooxygenase-2 expression: Role of NF-kappa B and NFAT transcriptional pathways. *Proc. Natl. Acad. Sci. USA* **2016**, *113*, 434–439. [CrossRef]

11. Kensler, T.W.; Wakabayash, N.; Biswal, S. Cell survival responses to environmental stresses via the Keap1-Nrf2-ARE pathway. *Annu. Rev. Pharmacol. Toxicol.* **2007**, *47*, 89–116. [CrossRef] [PubMed]

12. Kobayashi, E.H.; Suzuki, T.; Funayama, R.; Nagashima, T.; Hayashi, M.; Sekine, H.; Tanaka, N.; Moriguchi, T.; Motohashi, H.; Nakayama, K.; et al. Nrf2 suppresses macrophage inflammatory response by blocking proinflammatory cytokine transcription. *Nat. Commun.* **2016**, *7*, 11624. [CrossRef] [PubMed]

13. Ahmed, S.M.; Luo, L.; Namani, A.; Wang, X.J.; Tang, X. Nrf2 signaling pathway: Pivotal roles in inflammation. *Biochim. Biophys. Acta Mol. Basis Dis.* **2017**, *1863*, 585–597. [CrossRef]

14. Li, W.; Khor, T.O.; Xu, C.; Shen, G.; Jeong, W.-S.; Yu, S.; Kong, A.-N. Activation of Nrf2-antioxidant signaling attenuates NFκB-inflammatory response and elicits apoptosis. *Biochem. Pharmacol.* **2008**, *76*, 1485–1489. [CrossRef] [PubMed]

15. Hudina, M.; Liu, M.; Veberic, R.; Stampar, F.; Colaric, M. Phenolic compounds in the fruit of different varieties of Chinese jujube (Ziziphus jujuba Mill.). *J. Hortic. Sci. Biotechnol.* **2008**, *83*, 305–308. [CrossRef]

16. Choi, S.-H.; Ahn, J.-B.; Kozukue, N.; Levin, C.E.; Friedman, M. Distribution of free amino acids, flavonoids, total phenolics, and antioxidative activities of jujube (Ziziphus jujuba) fruits and seeds harvested from plants grown in Korea. *J. Agric. Food Chem.* **2011**, *59*, 6594–6604. [CrossRef]

17. Zhang, H.; Jiang, L.; Ye, S.; Ye, Y.; Ren, F. Systematic evaluation of antioxidant capacities of the ethanolic extract of different tissues of jujube (Ziziphus jujuba Mill.) from China. *Food Chem. Toxicol.* **2010**, *48*, 1461–1465. [CrossRef]

18. Shahrajabian, M.H.; Khoshkharam, M.; Zandi, P.; Sun, W.; Cheng, Q. Jujube, a super-fruit in traditional Chinese medicine, heading for modern pharmacological science. *J. Med. Plants. Stud.* **2019**, *7*, 173–178.

19. Gao, Q.H.; Wu, C.S.; Wang, M. The jujube (Ziziphus jujuba Mill.) fruit: A review of current knowledge of fruit composition and health benefits. *J. Agric. Food Chem.* **2013**, *61*, 3351–3363. [CrossRef]

20. Sobhani, Z.; Nikoofal-Sahlabadi, S.; Amiri, M.S.; Ramezani, M.; Emami, S.A.; Sahebkar, A. Therapeutic effects of Ziziphus jujuba Mill. fruit in traditional and modern medicine: A review. *Med. Chem.* **2019**. [CrossRef]

21. Liu, N.; Yang, M.; Huang, W.Z.; Wang, Y.J.; Yang, M.; Wang, Y.; Zhao, Z.X. Composition, antioxidant activities and hepatoprotective effects of the water extract of Ziziphus jujuba cv. Jinsixiaozao. *RSC Adv.* **2017**, *7*, 6511–6522. [CrossRef]

22. Kimura, M.; Kimura, I.; Guo, X.; Luo, B.; Kobayashi, S. Combined Effects of Japanese-Sino Medicine Kakkon-to-Ka-Senkyu-Shin-I and Its Related Combinations and Component Drugs on Adjuvant-Induced Inflammation in Mice. *Phytother. Res.* **1992**, *6*, 209–216. [CrossRef]

23. Chen, P.D.; Zhou, X.; Zhang, L.; Shan, M.Q.; Bao, B.H.; Cao, Y.D.; Kang, A.; Ding, A.W. Anti-inflammatory effects of Huangqin tang extract in mice on ulcerative colitis. *J. Ethnopharmacol.* **2015**, *162*, 207–214. [CrossRef] [PubMed]

24. Ji, X.L.; Peng, Q.; Li, H.Y.; Liu, F.; Wang, M. Chemical Characterization and Anti-inflammatory Activity of Polysaccharides from Zizyphus jujube cv. Muzao. *Int. J. Food Eng.* **2017**, *13*, 20160382. [CrossRef]

25. Ji, X.; Peng, Q.; Yuan, Y.; Shen, J.; Xie, X.; Wang, M. Isolation, structures and bioactivities of the polysaccharides from jujube fruit (Ziziphus jujuba Mill.): A review. *Food Chem.* **2017**, *227*, 349–357. [CrossRef]

26. Ninave, P.B.; Patil, S.D. Antiasthmatic potential of Zizyphus jujuba Mill and Jujuboside B.—Possible role in the treatment of asthma. *Respir. Physiol. Neurobiol.* **2019**, *260*, 28–36. [CrossRef]

27. Kwon, J.H.; Won, S.J.; Moon, J.H.; Kim, C.W.; Ahn, Y.S. Control of Fungal Diseases and Increase in Yields of a Cultivated Jujube Fruit (Zizyphus jujuba Miller var. inermis Rehder) Orchard by Employing Lysobacter antibioticus HS124. *Forests* **2019**, *10*, 1146. [CrossRef]

28. Singleton, V.L.; Orthofer, R.; Lamuela-Raventos, R.M. Analysis of total phenols and other oxidation substrates and antioxidants by means of Folin-Ciocalteu reagent. *Method. Enzymol.* **1999**, *299*, 152–178.

29. Wojdylo, A.; Oszmianski, J.; Czemerys, R. Antioxidant activity and phenolic compounds in 32 selected herbs. *Food Chem.* **2007**, *105*, 940–949. [CrossRef]

30. Averilla, J.N.; Oh, J.; Wu, Z.; Liu, K.H.; Jang, C.H.; Kim, H.J.; Kim, J.S.; Kim, J.S. Improved extraction of resveratrol and antioxidants from grape peel using heat and enzymatic treatments. *J. Sci. Food Agric.* **2019**, *99*, 4043–4053. [CrossRef]

31. Woo, Y.; Lee, H.; Jeong, Y.S.; Shin, G.Y.; Oh, J.G.; Kim, J.S.; Oh, J. Antioxidant Potential of Selected Korean Edible Plant Extracts. *BioMed Res. Int.* **2017**, *2017*, 7695605. [CrossRef] [PubMed]

32. Woo, Y.; Oh, J.; Kim, J.S. Suppression of Nrf2 Activity by Chestnut Leaf Extract Increases Chemosensitivity of Breast Cancer Stem Cells to Paclitaxel. *Nutrients* **2017**, *9*, 760. [CrossRef] [PubMed]

33. Seo, H.; Oh, J.; Hahn, D.; Kwon, C.S.; Lee, J.S.; Kim, J.S. Protective Effect of Glyceollins in a Mouse Model of Dextran Sulfate Sodium-Induced Colitis. *J. Med. Food* **2017**, *20*, 1055–1062. [CrossRef] [PubMed]

34. Byeon, S.; Oh, J.; Lim, J.S.; Lee, J.S.; Kim, J.S. Protective Effects of Dioscorea batatas Flesh and Peel Extracts against Ethanol-Induced Gastric Ulcer in Mice. *Nutrients* **2018**, *10*, 1680. [CrossRef]

35. Lim, J.S.; Oh, J.; Byeon, S.; Lee, J.S.; Kim, J.S. Protective Effect of Dioscorea batatas Peel Extract Against Intestinal Inflammation. *J. Med. Food* **2018**, *21*, 1204–1217. [CrossRef]

36. Barnwal, P.; Vafa, A.; Afzal, S.M.; Shahid, A.; Hasan, S.K.; Alpashree; Sultana, S. Benzo(a)pyrene induces lung toxicity and inflammation in mice: Prevention by carvacrol. *Hum. Exp. Toxicol.* **2018**, *37*, 752–761. [CrossRef]

37. Almatroodi, S.A.; Alrumaihi, F.; Alsahli, M.A.; Alhommrani, M.F.; Khan, A.; Rahmani, A.H. Curcumin, an Active Constituent of Turmeric Spice: Implication in the Prevention of Lung Injury Induced by Benzo(a) Pyrene (BaP) in Rats. *Molecules* **2020**, *25*, 724. [CrossRef]

38. Guzel, A.; Kanter, M.; Aksu, B.; Basaran, U.N.; Yalcin, O.; Guzel, A.; Uzun, H.; Konukoglu, D.; Karasalihoglu, S. Preventive effects of curcumin on different aspiration material-induced lung injury in rats. *Pediatr. Surg. Int.* **2009**, *25*, 83–92. [CrossRef]

39. Vancheri, C.; Mastruzzo, C.; Sortino, M.A.; Crimi, N. The lung as a privileged site for the beneficial actions of PGE(2). *Trends Immunol.* **2004**, *25*, 40–46. [CrossRef]

40. Shahid, A.; Ali, R.; Ali, N.; Hasan, S.K.; Barnwal, P.; Afzal, S.M.; Vafa, A.; Sultana, S. Methanolic bark extract of Acacia catechu ameliorates benzo(a)pyrene induced lung toxicity by abrogation of oxidative stress, inflammation, and apoptosis in mice. *Environ. Toxicol.* **2017**, *32*, 1566–1577. [CrossRef]

41. Singh, L.; Varshney, J.G.; Agarwal, T. Polycyclic aromatic hydrocarbons' formation and occurrence in processed food. *Food Chem.* **2016**, *199*, 768–781. [CrossRef] [PubMed]

42. Kasala, E.R.; Bodduluru, L.N.; Barua, C.C.; Sriram, C.S.; Gogoi, R. Benzo (a) pyrene induced lung cancer: Role of dietary phytochemicals in chemoprevention. *Pharmacol. Rep.* **2015**, *67*, 996–1009. [CrossRef] [PubMed]

43. Podechard, N.; Lecureur, V.; Le Ferrec, E.; Guenon, I.; Sparfel, L.; Gilot, D.; Gordon, J.R.; Lagente, V.; Fardel, O. Interleukin-8 induction by the environmental contaminant benzo(a)pyrene is aryl hydrocarbon receptor-dependent and leads to lung inflammation. *Toxicol. Lett.* **2008**, *177*, 130–137. [CrossRef] [PubMed]

44. Hecht, S.S.; Isaacs, S.; Trushin, N. Lung-Tumor Induction in a/J Mice by the Tobacco-Smoke Carcinogens 4-(Methylnitrosamino)-1-(3-Pyridyl)-1-Butanone and Benzo[a]Pyrene—A Potentially Useful Model for Evaluation of Chemopreventive Agents. *Carcinogenesis* **1994**, *15*, 2721–2725. [CrossRef] [PubMed]

45. Deng, C.; Dang, F.; Gao, J.; Zhao, H.; Qi, S.; Gao, M. Acute benzo[a]pyrene treatment causes different antioxidant response and DNA damage in liver, lung, brain, stomach and kidney. *Heliyon* **2018**, *4*, e00898. [CrossRef] [PubMed]

46. Li, Y.; Yao, J.; Han, C.; Yang, J.; Chaudhry, M.T.; Wang, S.; Liu, H.; Yin, Y. Quercetin, inflammation and immunity. *Nutrients* **2016**, *8*, 167. [CrossRef]

47. Lee, S.-J.; Lee, S.Y.; Ha, H.J.; Cha, S.H.; Lee, S.K.; Hur, S.J. Rutin attenuates lipopolysaccharide-induced nitric oxide production in macrophage cells. *J. Food Nutr. Res.* **2015**, *3*, 202–205. [CrossRef]

48. Tian, R.; Yang, W.; Xue, Q.; Gao, L.; Huo, J.; Ren, D.; Chen, X. Rutin ameliorates diabetic neuropathy by lowering plasma glucose and decreasing oxidative stress via Nrf2 signaling pathway in rats. *Eur. J. Pharmacol.* **2016**, *771*, 84–92. [CrossRef]

49. Sun, T.; Tang, J.; Powers, J.R. Effect of pectolytic enzyme preparations on the phenolic composition and antioxidant activity of asparagus juice. *J. Agric. Food Chem.* **2005**, *53*, 42–48. [CrossRef]

50. Sun, T.; Powers, J.R.; Tang, J. Effect of enzymatic macerate treatment on rutin content, antioxidant activity, yield, and physical properties of asparagus juice. *J. Food Sci.* **2007**, *72*, S267–S271. [CrossRef]

51. Xu, C.M.; Yagiz, Y.; Borejsza-Wysocki, W.; Lu, J.; Gu, L.W.; Ramirez-Rodrigues, M.M.; Marshall, M.R. Enzyme release of phenolics from muscadine grape (Vitis rotundifolia Michx.) skins and seeds. *Food Chem.* **2014**, *157*, 20–29. [CrossRef] [PubMed]

52. Hosni, K.; Hassen, I.; Chaabane, H.; Jemli, M.; Dallali, S.; Sebei, H.; Casabianca, H. Enzyme-assisted extraction of essential oils from thyme (Thymus capitatus L.) and rosemary (Rosmarinus officinalis L.): Impact on yield, chemical composition and antimicrobial activity. *Ind. Crop. Prod.* **2013**, *47*, 291–299. [CrossRef]

53. Pingili, R.B.; Challa, S.R.; Pawar, A.K.; Toleti, V.; Kodali, T.; Koppula, S. A systematic review on hepatoprotective activity of quercetin against various drugs and toxic agents: Evidence from preclinical studies. *Phytother. Res.* **2020**, *34*, 5–32. [CrossRef] [PubMed]

54. Hosseinzadeh, H.; Nassiri-Asl, M. Review of the protective effects of rutin on the metabolic function as an important dietary flavonoid. *J. Endocrinol. Investig.* **2014**, *37*, 783–788. [CrossRef]

55. Araujo, N.P.D.; de Matos, N.A.; Mota, S.L.A.; de Souza, A.B.F.; Cangussu, S.D.; de Menezes, R.C.A.; Bezerra, F.S. Quercetin Attenuates Acute Lung Injury Caused by Cigarette Smoke Both In Vitro and In Vivo. *Copd* **2020**. [CrossRef]

56. Zhang, X.C.; Cai, Y.L.; Zhang, W.; Chen, X.H. Quercetin ameliorates pulmonary fibrosis by inhibiting SphK1/S1P signaling. *Biochem. Cell Biol.* **2018**, *96*, 742–751. [CrossRef]

57. Farazuddin, M.; Mishra, R.; Jing, Y.X.; Srivastava, V.; Comstock, A.T.; Sajjan, U.S. Quercetin prevents rhinovirus-induced progression of lung disease in mice with COPD phenotype. *PLoS ONE* **2018**, *13*, e0199612. [CrossRef]

58. Rogerio, A.P.; Kanashiro, A.; Fontanari, C.; da Silva, E.V.G.; Lucisano-Valim, Y.M.; Soares, E.G.; Faccioli, L.H. Anti-inflammatory activity of quercetin and isoquercitrin in experimental murine allergic asthma. *Inflamm. Res.* **2007**, *56*, 402–408. [CrossRef]

59. Wang, W.; Sun, C.; Mao, L.; Ma, P.; Liu, F.; Yang, J.; Gao, Y. The biological activities, chemical stability, metabolism and delivery systems of quercetin: A review. *Trends Food Sci. Technol.* **2016**, *56*, 21–38. [CrossRef]

60. Jan, A.T.; Kamli, M.R.; Murtaza, I.; Singh, J.B.; Ali, A.; Haq, Q. Dietary flavonoid quercetin and associated health benefits—An overview. *Food Rev. Int.* **2010**, *26*, 302–317. [CrossRef]

 *foods*

Article

Spray-Drying Microencapsulation of High Concentration of Bioactive Compounds Fragments from *Euphorbia hirta* L. Extract and Their Effect on Diabetes Mellitus

Ngan Tran [†]**, Minh Tran** [†]**, Han Truong**

School of Biotechnology, International University—Vietnam National University, Ho Chi Minh 700000, Vietnam; kimnganchemistry@gmail.com (N.T.); tramthiminh@gmail.com (M.T.); hantruong029@gmail.com (H.T.)
* Correspondence: ly.le@hcmiu.edu.vn
† Ngan Tran and Minh Tran made an equal contribution to the work.

Received: 5 June 2020; Accepted: 30 June 2020; Published: 4 July 2020

Abstract: The present study was performed to spray-dry the high concentration of bioactive compounds from *Euphorbia hirta* L. extracts that have antidiabetic activity. The total phenolic content (TPC) and total flavonoid content (TFC) of four different extracts (crude extract, petroleum ether extract, chloroform extract and ethyl acetate extract) from the dried powder of *Euphorbia hirta* L. were determined using a spectrophotometer. After that, the fragment containing a high number of bioactive compounds underwent spray-dried microencapsulation to produce powder which had antidiabetic potential. The total phenolic content values of the crude extract, petroleum ether extract, chloroform extract and ethyl acetate extract were 194.55 ± 0.82, 51.85 ± 3.12, 81.56 ± 1.72 and 214.21 ± 2.53 mg/g extract, expressed as gallic acid equivalents. Crude extract, petroleum ether extract, chloroform extract and ethyl acetate extracts showed total flavonoids 40.56 ± 7.27, 29.49 ± 1.66, 64.99 ± 2.60 and 91.69 ± 1.67 mg/g extract, as rutin equivalents. Ethyl acetate extract was mixed with 20% maltodextrin in a ratio of 1:10 to spray-dry microencapsulation. The results revealed that the moisture content, bulk density, color characteristic, solubility and hygroscopicity of the samples were 4.9567 ± 0.00577%, 0.3715 ± 0.01286 g/mL, 3.7367 ± 0.1424 Hue, 95.83 ± 1.44% and 9.9890 ± 1.4538 g H_2O/100 g, respectively. The spray powder was inhibited 51.19% α-amylase at 10 mg/mL and reduced 51% in fast blood glucose (FBG) after 4 h treatment. Furthermore, the administration of spray powder for 15 days significantly lowered the fast blood glucose level in streptozotocin-diabetic mice by 23.32%, whereas, acarbose—a standard antidiabetic drug—and distilled water reduced the fast blood glucose level by 30.87% and 16.89%. Our results show that obtained *Euphorbia hirta* L. powder has potential antidiabetic activity.

Keywords: *Euphorbia hirta* L.; bioactive compounds; in vitro α-amylase inhibition; streptozotocin-induced diabetic mice; diabetes mellitus

---

1. Introduction

Diabetes mellitus is a chronic disease and the result of metabolic disorders in pancreas β-cells that have hyperglycemia [1,2]. Hyperglycemia is caused by a deficiency of insulin production by pancreatic (Type 1 diabetes mellitus) or the insufficiency of insulin production in the face of insulin resistance (Type 2 diabetes mellitus) [2,3]. Hyperglycemia causes damage to eyes, kidneys, nerves, heart and blood vessels [4]. According to the ninth edition, in 2019, of the IDF Diabetes Atlas released by the International Diabetes Federation (IDF), as of 2019, the total adult population living with diabetes in the age group of 20–79 years stands at 463 million, which is set to increase to 578 million by 2030 [5]. The present treatment of diabetes mellitus is focused on controlling and lowering the blood glucose levels in the vessel to a normal level [6]. Currently, there are six main classes of modern medicines used across the world for controlling blood glucose levels and two classes of injections [7]. The tablets are known as biguanides (metformin), sulfonylureas, thiazolidinediones (glitazones), meglitinides (glinides), *alpha*-glucosidase inhibitors and DPP-4 inhibitors. The classes of medications given by injection are incretin mimetics and insulin. However, most modern drugs have many side effects causing some serious medical problems during the period of treating. For instance, the main side effects of metformin are gastrointestinal on the initial state, including dyspepsia, nausea and diarrhea. The most common adverse effects with thiazolidinediones are weight gain and fluid retention, leading to peripheral edema and a twofold increased risk for congestive heart failure. Besides modern therapies, traditional medicines have been used for a long time and play an important role as alternative medicines [8]. According to the WHO, a plant-based traditional system of medicine is still the chief support of about 75–80% of the world's population, mainly in developing countries having a diversity of plants [9]. Several patients in type 2 diabetes mellitus have used functional foods to reduce their blood glucose levels, such as olive leaf extract, turmeric and fenugreek.

Euphorbia hirta L. (*E. hirta*) is a common herb that belongs to the *Euphorbia* genus of the Euphorbiaceae family. *Euphorbia hirta* L. is found in pan-tropic, partly sub-tropic areas and worldwide including Australia, Western Australia, Northern Australia, Queensland, New South Wales, Central America, Africa, Indonesia, Malaysia, Philippines, China and India [10]. In Vietnam, *Euphorbia hirta* L. is commonly distributed in many provinces of the southern area. *Euphorbia hirta* L. was used as a traditional medicine in the treatment of diabetes a long time ago [10,11]. Furthermore, this plant contains a large number of phytochemicals including flavonoids, terpenoids, phenols, essential oil and other compounds containing antidiabetic potential [12]. Therefore, producing an ingredient containing *E. hirta* extract is necessary. However, phytochemicals such as flavonoids, terpenoids, and phenols are very sensitive to environmental conditions such as temperature or oxygen, thence, microencapsulation was supposed to protect these components [10,13].

The inhibition of the enzyme involved in the hydrolyzing carbohydrates such as α-amylase is important to approach for reducing hyperglycemia [14,15]. *Alpha*-amylase (E.C.3.2.1.1) is a potential protein tending to be a possibly applied inhibitor for anti-diabetic treatment. In human beings, *alpha*-amylase is a prominent enzyme which hydrolyses the *alpha* bonds of large, *alpha*-linked polysaccharides, such as starch and glycogen, yielding glucose and maltose [16–18]. The inhibition of α-amylase has minimized the absorption of glucose into the blood by delaying the digestion of carbohydrates. In recent years, several investigations have demonstrated the efficiency of α-amylase inhibitors in the treatment of diabetes mellitus such as Acarbose, Miglitol and Voglibose [17,18].

The spray-drying process has been one of the popular techniques for decades to encapsulate food ingredients because of its low operational cost and available equipment [19,20]. It has been used to produce powder from a liquid form by using carrier agent materials. Nowadays, maltodextrin is a popular carrier agent material to produce powder because it has many functionalities, such as wall materials, flavor carrier, bulking agents, reducing stickiness and improving product stability [21].

Therefore, this study was carried out to determine fragments containing bioactive compounds from various extracts (crude extract, petroleum ether extract, chloroform extracts and ethyl acetate extract) for spray-drying microencapsulation and evaluate the antidiabetic ability of spray-dried powder on in vitro α-amylase inhibitory activity test and the streptozotocin-induced diabetic mice model.

2. Materials and Methods

2.1. Chemicals and Reagents

The chemicals used including absolute ethanol, petroleum ether, chloroform, ethyl acetate, methanol, purchased from Chemsol Company, Vietnam. Aluminum chloride ($AlCl_3$), and sodium carbonate (Na_2CO_3), Folin–Ciocalteu reagent, dimethyl sulfoxide, starch, and 5-dinitro salicylic acid (DNS), were purchased from Merck, Germany. Gallic acid, streptozotocin (STZ), nicotinamide (NAD), enzyme α-amylase, acarbose, p-nitrophenyl glucopyranoside (pNPG) were products of Sigma-Aldrich (St. Louis, MO, USA). The mouse was purchased from The Institute of Drug Quality Control–Ho Chi Minh City (IDQC–HCMC).

2.2. Plant Material

E. hirta was identified by the Institute of Tropical Biology under the Vietnam Academy of Science and Technology. The fresh *E. hirta* plant was harvested in Thu Duc District, Ho Chi Minh City, Vietnam, in October 2015. The whole plant without root was washed with tap water to remove all contamination and soaked in ethanol 70° to prevent the presence microorganisms. Then, the samples were dried in an oven at 60 °C for 8 h to remove the water content. After that, the sample was ground into powder and stored in a plastic bag for further uses.

2.3. Preparation of E. hirta Extracts

The plant powder (1.5 kg) was macerated with a methanol solvent by the ratio of 1:2 (*w/w*) in 2 weeks [22]. Then, the aqueous phase was filtered through Whatman No.1 filter paper, and the residue was added by fresh methanol. This process was repeated seven times until a clear colorless solution was obtained. Then, the extract solution was concentrated by a rotary evaporator at 60 °C to give 184.56 g of crude extract. Subsequently, the crude extract was mixed with distilled water by ratio 1:1 (*w/w*) and soaked with petroleum ether, chloroform, and ethyl acetate, respectively, to produce the fractional extract. The rotary evaporator was used to evaporate all the solvents to take the concentrated extracts.

The percentage of the yield of the extract was calculated as

$$\% \text{ Yield } (\%Y) = \frac{\text{weight of dried extract (g)}}{\text{weight of sample (g)}} \times 100 \tag{1}$$

2.4. Determination of Bioactive Compounds

2.4.1. Determination of Total Phenolic Content (TPC)

The determination of the total phenolic content was based on the Folin–Ciocalteu assay method with some modification [23]. The reaction mixture consists of 0.2 mL of extract samples dissolving in methanol at 1 mg/mL. One milliliter of 10% Folin–Ciocalteu reagent was treated into the mixture and shaken well. After 5 min, 1.5 mL of 5% Na_2CO_3 solution was added into the mixture. Gallic acid was used as the standard solutions (20, 40, 40, 60, 80 and 100 µg/mL). The absorbance was measured at 750 nm by a spectrophotometer. The total phenol content was expressed as mg of gallic acid equivalent (GAE)/g of the extract [24].

2.4.2. Determination of Total Flavonoid Content (TFC)

The total flavonoid content was determined by the spectrophotometric method [25]. The reaction mixture consists of 5 mL of extract samples dissolving in methanol at 0.4 mg/mL. Then, 5 mL of 2% $AlCl_3$ solution was treated into the mixture. Rutin was used as a standard solution at different concentrations (20, 40, 40, 60, 80 and 100 µg/mL). The absorbance was measured at 415 nm by a spectrophotometer. The content of flavonoids in extracts was expressed in terms of rutin equivalent (mg of RU/g of extract) [24].

2.5. Spray-Drying Microencapsulation

Spray drying has been used to protect the ingredients that are sensitive to light, heat or oxygen. In this technique, a wall material protects the bioactive compounds [26,27]. The extracted sample was mixed with 20% maltodextrin at a ratio of 1:10, and the resulting mixtures were homogenized before spray dryer. The powder was obtained using LabPlant SD-06A spray dryer, serial number 485. The inlet temperatures were 180 °C, the outlet temperatures varied according to the inlet temperatures [28], the feed flow rate of the extracts was 10 rpm. Then, the *E. hirta* powder was used to determine the physical properties as the moisture content, color characteristic, buck density, solubility and hygroscopicity.

2.5.1. Determination of Moisture Content

The moisture content was determined by moisture balance, type MOC-120H, No. D207302059. Triplicate samples of *E. hirta* powder (5 mg) were weighed and dried in the oven at 105 °C until its weight was constant [21].

2.5.2. Color Characteristic

Color characteristic was measured by the Hue method. Distilled water-diluted triplicate samples of *E. hirta* powder to ratio 10:1 (*w/w*). The absorbance of this solution was measured at 510 nm and 610 nm and calculated the Hue:

$$\text{Hue} = 10 \times \log\left(\frac{Abs_{510nm}}{Abs_{610nm}}\right) \tag{2}$$

The Hue Index usually ranges from 3 (a greenish-yellow or olive hue) to 7.5 (amber red-brown) for caramel colors.

2.5.3. Solubility

The solubility of the samples was determined according to the reported procedures with some modification [29]. One gram of samples was added to 100 mL of distilled water, then the mixture was stirred at 600 rpm for 5 min. After that, 20 mL of the supernatant was transferred into petri-dishes and dried in an oven at 70 °C until the weight was constant. The solubility was calculated by weight difference and expressed in dry basis, considering the moisture content of each sample.

2.5.4. Hygroscopicity

The hygroscopicity of the *E. hirta* powder was determined according to [28] with some modifications. The *E. hirta* powder was stored at room temperature in desiccators containing saturated sodium chloride solutions. The samples were weighed after one week, and the hygroscopicity was expressed in grams of the absorbed moisture per 100 g of dry solids.

2.6. In Vitro α-Amylase Inhibitory Activity

The α-amylase inhibitory activity was measured by the dinitrosalicylic acid method [16]. The 0.5 mL of the extract samples dissolving in dimethyl sulfoxide at different concentrations were pre-incubated with α-amylase 2 U/mL for 15 min. Then, 0.5 mL of the 1% *w/v* starch solution was added to the mixture, which was further incubated at 37 °C for 10 min. Then, the reaction was stopped by adding 1 mL DNS reagent and heated in a boiling water bath for 5 min. Acarbose was used as a positive control and the absorbance was measured at 540 nm. Percentage inhibition is calculated as

$$\%\text{Inhibition} = \frac{\left(Abs_{control} - Abs_{extract}\right)}{Abs_{control}} \times 100 \tag{3}$$

The α-amylase inhibitory activity was expressed as the IC50 according to the percentage inhibition [30,31].

2.7. Acute Toxicity Testing

Spray powder safety evaluation was conducted by pyramiding single-dose (acute) toxicity testing [32]. A group of two mice (25–30 g) was fed with *E. hirta* powder with an increasing dosage on alternate days as 10, 30, 100, 300, 1000, 3000 and 5000 mg kg^{-1}, with the dosing continuing until death.

The toxicity test was found to be safe up to the dose 5000 mg/kg body weight; hence 1/10 of the dose was taken as an effective dose (500 mg/kg body weight).

2.8. Preliminary Anti-Hyperglycemic Test on STZ/NAD-Induced Mice

2.8.1. Induction of STZ/NAD-Diabetes Mice

The model of the STZ/NAD-induced mice followed the guideline of Masiello et al. [33] with modification. Male mice weighing 22c25 received an intraperitoneally administered dose of 200 mg/kg bw of NAD, and after 15 min received another intraperitoneally administered dose of 100 mg/kg bw of STZ. After the intraperitoneal administration of STZ/NAD, the mice were observed for 30 min to examine for any abnormal signals. STZ/NAD-induced mice were fed with a daily special diet with fat milk and lard to make them obese. The blood glucose levels of mice were measured and recorded to check for the change in blood glucose. After 21 days, a model of diabetic mice was completed.

2.8.2. Anti-Hyperglycemic Test

Diabetic mice (glucose level > 200 mg/dL) were divided into three groups of 6 mice each and orally administered as below:

Group 1: negative control—orally administered with distilled water;
Group 2: positive control—orally administered with acarbose (100 mg/kg bw);
Group 3: sample group—orally administered with *E. hirta* powder (500 mg/kg bw).

The administration was continued for 15 days, once daily. Then, the blood glucose level was measured and recorded after 0, 1, 2 and 4 h on the first day and on days 1, 2, 6, 10 and 15 of administration.

2.9. Statistical Analysis

Statistical analysis was performed using SPSS 22.0. One-way variance analysis (one-way ANOVA) was applied to determine the significant difference between the samples at p-value < 0.05.

3. Results

3.1. Euphorbia hirta L. Extraction Yield

Nine and a half kilograms of fresh *E. hirta* produced around 1.5 kg of dried powder, 184.56 g of crude extract, 61.25 g of petroleum ether extract, 2.1 g of chloroform extract and 32.95 g of ethyl acetate extract. The results are shown in Figure 1. Different extracts were obtained from the *E. hirta* from the different solvents (petroleum ether, chloroform, and ethyl acetate) employed.

3.2. Phytochemical Analysis

According to the previous investigations, *Euphorbia hirta* L. contain a large amount of phenolic and flavonoid components [10]. In the present study, the total phenolic and flavonoid contents were shown in Table 1. The total phenolic content in the examined extracts was expressed in terms of gallic acid equivalent (mg of GAE/g extract) by the standard curve equation: $y = 8.659x - 0.0153$, $R^2 = 0.998$. The value of the phenolic content was 194.55 ± 0.82, 51.85 ± 3.12, 81.56 ± 1.72 and 214.21 ± 2.53 mg of GAE/g extract for the crude extract, petroleum ether extract, chloroform extract and the ethyl acetate extract, respectively. The concentration of flavonoids was expressed as rutin (mg of RU/g extract) by the standard curve equation: $y = 12.905x - 0.0515$, $R2 = 0.9888$. The concentration of flavonoids in various plant extracts were 40.56 ± 7.27, 29.49 ± 1.66, 64.99 ± 2.60 and 91.69 ± 1.67 mg of RU/g extract. The highest phenolic and flavonoid concentrations were measured in ethyl acetate extract, and the lowest phenolic and flavonoid concentrations were measured in petroleum ether extract. The phenolic and flavonoid concentrations in the plant extracts depend on the solvent polarity which used in the extract preparation [24]. These data suggested that the total phenolic and flavonoid compounds were best extracted via ethyl acetate solvent from *E. hirta*. Therefore, the ethyl acetate extract was suggested to further investigate and determine the bioactive compounds and identify their antidiabetic activity.

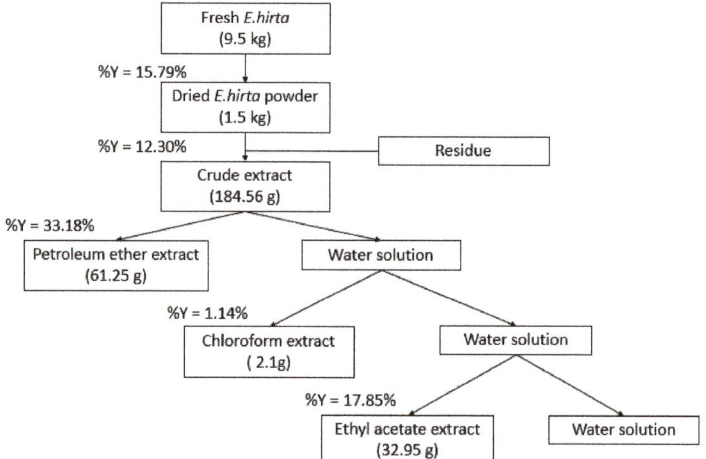

Figure 1. Percentage of extracts from *Euphorbia hirta* L.

Table 1. Total phenolic and flavonoid contents in the *E. hirta* extracts.

Extract	Total Phenolic Content (mg of GAE/g Extract)	Total Flavonoid Content (mg of RU/g Extract)
Crude extract	194.55 ± 0.82 [1]	40.56 ± 7.27
Petroleum ether extract	51.85 ± 3.12	29.49 ± 1.66
Chloroform extract	81.56 ± 1.72	64.99 ± 2.60
Ethyl acetate extract	214.21 ± 2.53	91.69 ± 1.67

[1] Each value is the average of three analyses ± standard deviation ($n = 3$).

3.3. Spray-Drying Microencapsulation

The physical properties of spray powder were shown to clarify the quality of the *E. hirta* powder (Table 2). The moisture content effects on the shelf-life of dried materials, and the shelf-life of powder can extend in moisture under than 10% [34]. The moisture content in the *E. hirta* powder is around 4.9567 ± 0.00577%, which is less than 5% so that it is suitable for storage for a long time. Bulk density is a property of food powder, the bulk density of *E. hirta* powder ranges from 0.36 to 0.48 g/mL, and the high bulk density increases its sticky or less free-flowing nature. In this research, the *E. hirta* powder had a bulk density of 0.3715 ± 0.01286 g/mL, as it adapted the requirement. Solubility shows the ability of the powder to be dissolved in water, and the solubility of the *E. hirta* powder is around 95.83 ± 1.44%. Color is a major quality parameter in a dried food product; color was presented in the Hue Index value. Table 2 shows that the Hue index value around 3.7367 ± 0.1424 means that the color of spray sample is yellow. Hygroscopicity is the ability of food powder to absorb moisture from a high relative humidity environment. The result from Table 2 showed that *E. hirta* powder could absorb 9.9890 ± 1.4538 g H_2O in a 100 g sample.

Table 2. Physical properties of the spray power.

Yield (%)	Moisture (%)	Bulk Density (g/mL)	Solubility (%)	Hygroscopicity (g H_2O/100 g)	Hue Index
21.034	4.9567 ± 0.00577 [1]	0.3715 ± 0.01286	95.83 ± 1.44	9.9890 ± 1.4538	3.7367 ± 0.1424

[1] Each value is the average of three analyses ± standard deviation ($n = 3$).

3.4. In Vitro α-Amylase Inhibitory Activity Essay

The inhibition of α-amylase minimized the absorption of glucose into blood by the delay of the digestion of carbohydrates. The in vitro α-amylase inhibitory activities of the *E. hirta* powder were assayed. The result of this study showed that spray powder has a good α-amylase inhibitory activity (Figure 2) with the IC50 value as evidence (Table 3). The *E. hirta* powder and acarbose—a standard drug in diabetes management (at concentration 10 mg/mL) showed a 51.19% and 94.69% inhibitory effect on α-amylase, respectively. The IC50 value of the *E. hirta* powder was 5.725 mg/L, meanwhile for acarbose it was 2.511 mg/mL. Although the α-amylase inhibitory activity of *E. hirta* powder is not strong enough as that of acarbose is, it also proved that *E. hirta* can inhibit the carbohydrate-hydrolyzing enzyme and has potential antidiabetic activity.

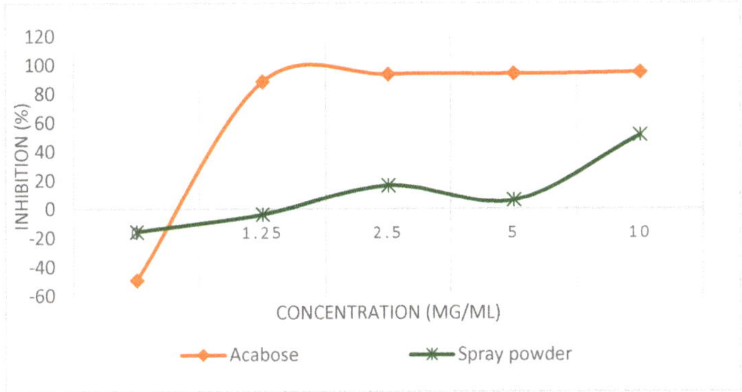

Figure 2. The inhibitory potency of the *E. hirta* powder against α-amylase activity. Each value is the average of three analyses ± standard deviation (*n* = 3).

Table 3. The IC50 of the *E. hirta* powder on α-amylase inhibition.

Sample	α-Amylase Equation (y = ax + b)	α-Amylase IC50 (mg/mL)
Acarbose	y = 29.246x − 23.446	2.511
Spray powder	y = 14.355x − 32.175	5.725

3.5. Acute Toxicity

The mice orally administered 10, 30, 100, 300, 1000, 3000, 5000 mg kg^{-1} doses of *E. hirta* powder were kept under observation for two weeks. After two weeks, all the mice were alive and did not show any toxic symptoms such as body weight loss. Therefore, it was found that 5000 mg kg^{-1} dose of the powder showed a confidence dose and was considered as safe. Therefore, it was concluded that the median lethal dose (LD50) was more than 5000 mg kg^{-1} when administered orally (Table 4). Therefore, the extracts used in this study are safe for long administration.

Table 4. Pyramiding dose.

Dosage	Mortality
10 mg kg^{-1}	0/2
30 mg kg^{-1}	0/2
100 mg kg^{-1}	0/2
300 mg kg^{-1}	0/2
1000 mg kg^{-1}	0/2
3000 mg kg^{-1}	0/2
5000 mg kg^{-1}	0/2
Both minimal lethal dosage and LD50 > 5000 mg kg^{-1}	

3.6. Anti-Hyperglycemic Test

This experiment was undertaken to evaluate the hypoglycemic activity of the spray powder containing *Euphorbia hirta* L. in the STZ-induced diabetic mice. Diabetic mice usually have fast blood glucose levels above 200 mg/dL. Acarbose is commonly used as a standard antidiabetic drug in STZ-induced diabetes to compare the anti-hyperglycemic extent of bioactive compounds. In normoglycemic rats, the test showed a significant reduction of blood glucose level till the end of 4 h.

The fast blood glucose levels in 4 h showed that after 4 h of treatment with the *E. hirta* powder (Figure 3), there was a 51% reduction in the fast blood glucose level, whereas the treatment with acarbose and distilled water at the same time produced a 44% and 35% reduction. In a short time, the *E. hirta* powder reduced fast blood glucose levels better than distilled water and acarbose.

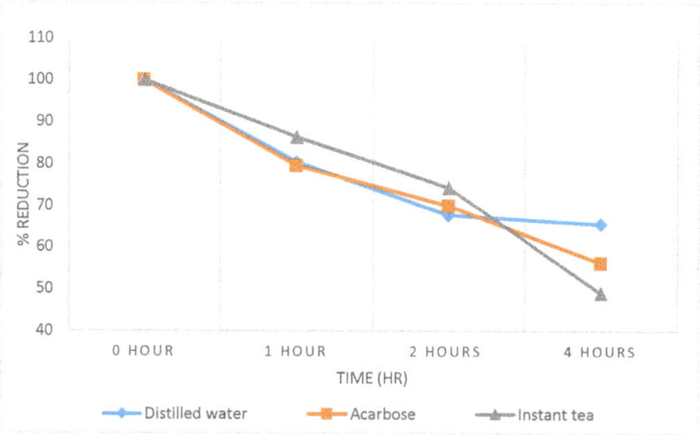

Figure 3. Effect of the *E. hirta* powder on the streptozotocin-induced diabetic mice in 4 h. Values are presented as the means ± standard deviation (*n* = 6).

A significant reduction of 23.32% in fast blood glucose was observed after 15 days of treatment with the *E. hirta* powder, compared to the 30.87% and 16.89% reduction when treated with acarbose and distilled water, respectively (Figure 4). In the longer time treatment, acarbose was the competitive inhibitor in the reduction of fast blood glucose levels, however, the reduction of the fast blood glucose levels of the *E. hirta* powder was close to acarbose and better than distilled water. Therefore, the result of this test showed that the *E. hirta* powder exhibited a significant reduction in the fast blood glucose levels.

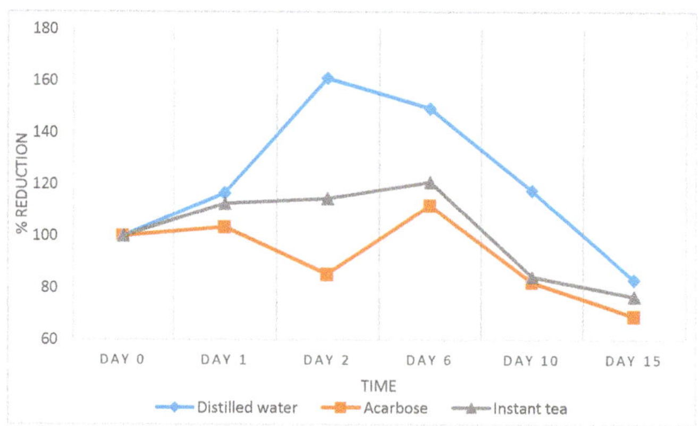

Figure 4. Effect of the *E. hirta* powder on the streptozotocin (STZ)-induced diabetic mice in 15 days. Values are presented as the means ± standard deviation (*n* = 6).

4. Discussion

Plants are a rich resource of natural bioactive compounds, such as phenolics, flavonoids and their derivatives. These compounds have received attention because of their bioactivities and physiological functions, including antioxidant, anti-allergic, anti-inflammatory, antimicrobial, and antidiabetic activities [35]. From the crude extract of *Euphorbia hirta* L. containing multiple different groups of bioactive compounds, liquid–liquid extractions can be employed to separate many different fractions. During this procedure, the chemical constituents of the extract are separated based on their polarity by specific solvents including petroleum ether, chloroform, and ethyl acetate. This procedure typically results in a total of four extracts. The qualitative phytochemical analysis in *E. hirta* was obtained from these extracts. In relation to the solvent used, high concentrations of the total phenolic content and total flavonoid content were found in the chloroform and ethyl acetate extracts. The value of the phenolic contents was 81.56 ± 1.72 and 214.21 ± 2.53 mg of GAE/g extract for the chloroform extract and the ethyl acetate extract, respectively. The concentration of flavonoids was 64.99 ± 2.60 and 91.69 ± 1.67 mg of RU/g extract for the chloroform extract and the ethyl acetate extract, respectively.

In the previous studies, the point of focus has been on elucidating the mechanism of action and the phytochemicals of plant extracts for traditional diabetes treatment. The study of the inhibitory activity of phytochemicals against α-amylase and α-glucosidase has been widely popular. Several phytochemicals including flavonoids and phenolics have been detected. The presence of these compounds, especially in chloroform and ethyl acetate extracts, demonstrated ability in the treatment of diabetes. The inhibition of the enzymes involved in the metabolism of saccharides such as α-amylase is an important therapeutic strategy for reducing hyperglycemia [14]. Polyphenolic compounds have popularly reported the inhibitory activity against α-amylase in both in vitro and in vivo experiments [11,12,36]. Phenolics and flavonoids found in *E. hirta* such as quercetin, quercitrin, and rutin were proved to be effective inhibitors of mammalian α-amylase. The *E. hirta* powder and acarbose (at concentration 10 mg/mL) showed 51.19% and 94.69% inhibitory effects on α-amylase, respectively. The IC50 value of the *E. hirta* powder was 5.725 mg/L, meanwhile acarbose was 2.511 mg/mL. Although the α-amylase inhibitory activity of *E. hirta* powder was not strong enough as acarbose, it also proved that *E. hirta* had a mild α-amylase inhibition and has potential antidiabetic activity. This result is consistent with the reported investigation [37,38].

The control of postprandial plasma glucose concentrations is the key to the the treatment of diabetes mellitus and its related complications. The results of this study showed that *Euphorbia hirta* L. powder from the highest number of bioactive compounds possessed antidiabetic activities. Phenolics and flavonoids are commonly distributed in plants and have been demonstrated to possess potential effects in the treatment of diabetes disease. The possible mechanism of antidiabetic action of *E. hirta* is the inhibition of α-amylase and the reduction of the fast blood glucose level on diabetic mice [39]. Therefore, this study supports the drug/functional food development from the *E. hirta* extract in the management of diabetes.

5. Conclusions

This study provided significant evidence for the antidiabetic activity of Euphorbia hirta L. One of the possible mechanisms of antidiabetic activity of this plant is related to the inhibitory action of the key enzymes involving the hydrolysis of carbohydrates, to control the blood glucose concentration in the body. Recent investigations have proved that phenolics and flavonoids from natural resources containing bioactive components have hypoglycemia in action. Therefore, in further studies, we will continue to investigate the stabilities in functional food application. Besides, the fragments of *E. hirta* extracts having antidiabetic activities will be carried out to isolate and identify the bioactive compounds for the in vitro and in vivo studies of their antidiabetic activity. Furthermore, it is necessary to elucidate the mechanisms of action of the extracts and phytochemicals of this plant at the cellular and molecular levels.

Author Contributions: N.T., M.T., H.T. and L.L. conceived and designed the experiments. N.T., M.T. and H.T. performed the experiments and analyzed the data. N.T. and M.T. wrote the paper. L.L. revised and approved the final version for submission. All authors have read and agreed to the published version of the manuscript.

Funding: This work was funded by Vietnam National University at Ho Chi Minh City under grant number GEN2019-28-01/HD-KHCN.

Acknowledgments: Support from School of Biochenology of International University at Ho Chi Minh City is gratefully acknowledged.

Conflicts of Interest: The authors declare no conflict of interest.

References

1. Heise, T.; Nosek, L.; Rønn, B.B.; Endahl, L.; Heinemann, L.; Kapitza, C.; Draeger, E. Lower within-subject variability of insulin detemir in comparison to NPH insulin and insulin glargine in people with type 1 diabetes. *Diabetes* **2004**, *53*, 1614–1620. [CrossRef] [PubMed]

2. American Diabetes Association. Classification and Diagnosis of Diabetes: Standards of Medical Care in Diabetes-2018. *Diabetes Care* **2018**, *41* (Suppl. 1), S13–S27. [CrossRef]

3. Faselis, C.; Katsimardou, A.; Imprialos, K.; Deligkaris, P.; Kallistratos, M.; Dimitriadis, K. Microvascular Complications of Type 2 Diabetes Mellitus. *Curr. Vasc. Pharmacol.* **2020**, *18*, 117–124. [CrossRef] [PubMed]

4. Mayfield, J. Diagnosis and classification of diabetes mellitus: New criteria. *Am. Fam. Physician* **1998**, *58*, 1355. [PubMed]

5. International Diabetes Federation. *IDF Diabetes Atlas*, 9th ed.; International Diabetes Federation: Brussels, Belgium, 2019.

6. Knight, K. A systematic review of diabetes disease management programs. *Am. J. Manag. Care* **2005**, *11*, 242–250. [PubMed]

7. Nathan, D.M.; Buse, J.B.; Davidson, M.B.; Ferrannini, E.; Holman, R.R.; Sherwin, R.; Zinman, B. Medical management of hyperglycemia in type 2 diabetes: A consensus algorithm for the initiation and adjustment of therapy. *Diabetes Care* **2009**, *32*, 193–203. [CrossRef]

8. Yakubu, O.E.; Imo, C.; Shaibu, C.; Akighir, J.; Ameh, D.S. Effects of Wthanolic Leaf and Sterm-bark Extracts of Adansonia digitate in Alloxan-induced Diabetic Wistar Rats. *J. Pharmacol. Toxicol.* **2020**, *15*, 1–7. [CrossRef]

9. James, P.; Sangeetha, P.; Davis, V.; Ravisankar, P.A. Nazeen and Deepu Mathew. Novel antidiabetic molecules from the medicinal plants of Western Ghats of India, Indentified Through wide-Spectrum in Silico Analyses. *J. Herbs Species Med. Plants* **2017**, *23*, 249–262. [CrossRef]

10. Asha, S. *Euphorbia hirta* linn—A review on traditional uses, phytochemistry and pharmacology. *World J. Pharm. Res.* **2014**, *3*, 180–205.

11. Widharna, R.M.; Soemardji, A.A.; Wirasutisna, K.R.; Kardono, L.B.S. Anti diabetes mellitus activity invivo of ethanolic extract and ethyl acetate fraction of *Euphorbia hirta* L. herb. *Int. J. Pharmacol.* **2010**, *6*, 231–240. [CrossRef]

12. Kumar, S.; Malhotra, R.; Kumar, D. Euphorbia hirta: Its chemistry, traditional and medicinal uses, and pharmacological activities. *Pharmacogn. Rev.* **2010**, *4*, 58–61. [CrossRef] [PubMed]

13. Gharsallaoui, A.; Roudaut, G.; Beney, L.; Chambin, O.; Voilley, A.; Saurel, R. Properties of spray-dried food flavours microencapsulated with two-layered membranes: Roles of interfacial interactions and water. *Food Chem.* **2012**, *132*, 1713–1720. [CrossRef]

14. Kwon, Y.I.; Apostolidis, E.; Kim, Y.C.; Shetty, K. Health benefits of traditional corn, beans and pumpkin: In vitro studies for hyperglycemia and hypertension management. *J. Med. Food.* **2007**, *10*, 266–275. [CrossRef] [PubMed]

15. Kazeem, M.I.; Ashafa, A.O.T. In-vitro antioxidant and antidiabetic potentials of Dianthus basuticus Burtt Davy whole plant extract. *J. Herb. Med.* **2015**, *5*, 158–164. [CrossRef]

16. Bhutkar, M.A.; Bhise, S.B. Bhise. In vitro assay of alpha amylase inhibitory activity of some indigenous plants. *Int. J. Chem. Sci.* **2012**, *10*, 457–462. [CrossRef]

17. Mahmood, N. A review of α-amylase inhibitors on weight loss and glycemic control in pathological state such as obesity and diabetes. *Comp. Clin. Pathol.* **2016**, *25*, 1253–1264. [CrossRef]

18. De Souza, P.M.; de Magalhães, P. Application of microbial α-amylase in industry—A review. *Braz. J. Microbiol.* **2010**, *41*, 850–861. [CrossRef]

19. Wu, P.; Deng, Q.; Ma, G.; Li, N.; Yin, Y.; Zhu, B.; Chen, M.; Huang, R. Spray Drying of Rhodomyrtus tomentosa (Ait.) Hassk. Flavonoids Extract: Optimization and Physicochemical, Morphological, and Antioxidant Properties. *Int. J. Food Sci.* **2014**, 420908. [CrossRef]

20. Gharsallaoui, A.; Roudaut, G.; Chambin, O.; Voilley, A.; Saurel, R. Application of spay-drying in microencapsulation of food ingredient: An overview. *Food Res. Int.* **2007**, *40*, 1107–1121. [CrossRef]

21. Susantikarn, P.; Donlao, N. Optimization of green tea extracts spray drying as affected by temperature and maltodextrin content. *Int. Food Res. J.* **2016**, *23*, 1327–1331.

22. Zhang, Q.; Lin, L.; Ye, W.-C. Techniques for extraction and isolation of natural products: A comprehensive review. *Chin. Med.* **2018**, *13*. [CrossRef] [PubMed]

23. Tambe, V.D.; Bhambar, R.S. Estimation of total phenol, tannin, alkaloid and flavonoid in *Hibiscus Tiliaceus* L. wood extracts. *Res. Rev. J. Pharmacogn. Phytochem.* **2014**, *2*, 41–47.

24. Bajalan, I.; Zand, M.; Goodarzi, M.; Darabi, M. Antioxidant activity and total phenolic and flavonoid content of the extract and chemical composition of the essential oil of *Eremostachys laciniata* collected from Zagros. *Asian Pac. J. Trop. Biomed.* **2017**, *7*, 144–146. [CrossRef]

25. Quettier, D.C.; Gressier, B.; Vasseur, J.; Dine, T.; Brunet, C.; Luyckx, M.C.; Cayin, J.C.; Bailleul, F.; Trotin, F. Phenolic compounds and antioxidant activities of buckwheat (Fagopyrum esculentum Moench) hulls and flour. *J. Ethnopharmacol.* **2000**, *72*, 35–42. [CrossRef]

26. Ahmed, M.; Akter, M.S.; Lee, J.; Eun, J. Encapsulation by spray drying of bioactive components, physicochemical and morphological properties from purple sweet potato. *LWT Food Sci Technol* **2010**, *43*, 1307–1312. [CrossRef]

27. Ré, M.I. Microencapsulation by spray drying. *Dry. Technol.* **1998**, *16*, 1195–1236. [CrossRef]

28. Silva, P.I.; Stringheta, P.C.; Teófilo, R.F.; de Oliveira, I.R.N. Parameter optimization for spray-drying microencapsulation of jaboticaba (*Myrciaria jaboticaba*) peel extracts using simultaneous analysis of response. *J. Food Eng.* **2013**, *117*, 538–544. [CrossRef]

29. Cano-Chauca, M.; Stringheta, P.C.; Ramos, A.M.; CalVidal, J. Effect of carriers in the microstructure of mango powder obtained by spray drying and its functional characteristic. *Innov. Food Sci. Emerg. Technol.* **2005**, *6*, 420–428. [CrossRef]

30. Kazeem, M.I.; Adamson, J.O.; Ogunwande, I.A. Modes of Inhibition of α-Amylase and α-Glucosidase by Aqueous Extract of Morinda lucida Benth Leaf. *Biomed Res. Int.* **2013**, 527570. [CrossRef]

31. Ali, H.; Houghton, P.J.; Soumyanath, A. α-Amylase inhibitory activity of some Malaysian plants used to treat diabetes; with particular reference to *Phyllanthus amarus*. *J. Ethnopharmacol.* **2006**, *107*, 449–455. [CrossRef]

32. Gad, S.C. Single-Dose (Acute) and Pilot (DRF) Toxicity Testing in Drug Safety Evaluation. In *Drug Safety Evaluation*, 2nd ed.; John Wiley & Son, Inc.: Hoboken, NJ, USA, 2010; Volume 6, pp. 1–49. [CrossRef]

33. Masiello, P.; Broca, C.; Gross, R.; Roye, M.; Manteghetti, M.; Hillaire-Buys, D.; Novelli, M.; Ribes, G. Experimental NIDDM: Development of a new model in adult rats administered streptozotocin and nicotinamide. *Diabetes* **1998**, *47*, 224–2299. [CrossRef] [PubMed]

34. Nasir, M.; Butt, M.S.; Anjum, F.M.; Sharif, K.; Minhas, R. Effect of Moisture on the Shelf Life of Wheat Flour. *Int. J. Agric. Biol.* **2003**, *5*, 458–459.

35. Wu, Y.; Qu, W.; Geng, D.; Liang, J.Y.; Luo, Y.L. Phenols and flavonoids from the aerial part of *Euphorbial hirta*. *Chin. J. Nat. Med.* **2012**, *10*, 0040–0042. [CrossRef]

36. Noreen, H.; Semmar, N.; Farman, M.; McCullagh, J.S.O. Measurement of total phenolic content and antioxidant activity of aerial parts of medicinal plant *Coronopus didymus*. *Asian Pac. J. Trop. Med.* **2017**, *10*, 792–801. [CrossRef] [PubMed]

37. Sheliya, M.A.; Begum, R.; Pillai, K.K.; Aeri, V.; Mir, S.R.; Ali, A.; Sharma, M. In vitro α-glucosidase and α-amylase inhibition by aqueous, hydroalcoholic, and alcoholic extract of *Euphorbia hirta* L. *Drug Dev. Ther.* **2016**, *7*, 26–30. [CrossRef]

38. Sheliya, M.A.; Rayhana, B.; Ali, A.; Pillai, K.K.; Aeri, V.; Sharma, M.; Mir, S.R. Inhibition of α-glucosidase by new prenylated flavonoids from *Euphorbia hirta* L. herb. *J. Ethnopharmacol.* **2015**, *176*, 1–8. [CrossRef]
39. Kumar, S.; Kumar, D. Evaluation of antidiabetic activity of Euphorbia hirta Linn in streptozotocin induced diabetic mice. *Indian J. Nat. Prod. Resour.* **2010**, *1*, 200–203.

 *foods*

Article

Assessment of the Antimicrobial, Antioxidant, and Antiproliferative Potential of *Sideritis raeseri* subps. *raeseri* Essential Oil

Gregoria Mitropoulou [1], Marianthi Sidira [2], Myria Skitsa [3], Ilias Tsochantaridis [3], Aglaia Pappa [3], Christos Dimtsoudis [2], Charalampos Proestos [4] and Yiannis Kourkoutas [1,*]

[1] Laboratory of Applied Microbiology and Biotechnology, Department of Molecular Biology & Genetics, Democritus University of Thrace, GR-68100 Alexandroupolis, Greece; gmitropo@mbg.duth.gr
[2] Research and Development Department, Macedonian-Thrace Brewery S.A., GR-69100 Komotini, Greece; sidiramania@yahoo.gr (M.S.); dimtsoudis@vergina.com.gr (C.D.)
[3] Cellular and Molecular Physiology Research Group, Department of Molecular Biology & Genetics, Democritus University of Thrace, GR-68100 Alexandroupolis, Greece; myriaskitsa@hotmail.com (M.S.); iliatsoc@gmail.com (I.T.); apappa@mbg.duth.gr (A.P.)
[4] Laboratory of Food Chemistry, Department of Chemistry, National and Kapodistrian University of Athens, GR-15771 Athens, Greece; harpro@chem.uoa.gr
*   Correspondence: ikourkou@mbg.duth.gr; Tel.: +30-25510-30633

Received: 8 May 2020; Accepted: 10 June 2020; Published: 1 July 2020

Abstract: The aim of the present study was to investigate the antimicrobial potential of *Sideritis raeseri* subps. *raeseri* essential oil (EO) against common food spoilage and pathogenic microorganisms and evaluate its antioxidant and antiproliferative activity. The EO was isolated by steam distillation and analyzed by GC/MS. The main constituents identified were geranyl-*p*-cymene (25.08%), geranyl-γ-terpinene (15.17%), and geranyl-linalool (14.04%). Initially, its activity against *Staphylococcus aureus*, *Staphylococcus epidermidis*, *Escherichia coli*, *Listeria monocytogenes*, *Salmonella* Enteritidis, *Salmonella* Typhimurium, *Pseudomonas fragi*, *Saccharomyces cerevisiae*, and *Aspergillus niger* was screened by the disk diffusion method. Subsequently, minimum inhibitory concentration (MIC), non-inhibitory concentration (NIC), and minimum lethal concentration (MLC) values were determined. Growth inhibition of all microorganisms tested was documented, although it was significantly lower compared to gentamycin, ciproxin, and voriconazole, which were used as positive controls. In a next step, its direct antioxidant properties were examined using 2,2-diphenyl-1-picrylhydrazyl (DPPH) and 2,2′-azino-bis(3-ethylbenzothiazoline-6-sulfonic acid) (ABTS) assays, and the IC_{50} values were determined. The potential cytoprotective activity of the oil against H_2O_2–induced oxidative stress and DNA damage was studied in human immortalized keratinocyte (HaCaT) cells using the comet assay. Finally, the antiproliferative activity of the oil was evaluated against a panel of cancer cell lines including A375, Caco2, PC3, and DU145 and the non-cancerous HaCaT cell line using the sulforhodamine B (SRB) assay, and the EC_{50} values were determined. The oil demonstrated weak radical scavenging activity, noteworthy cytoprotective activity against H_2O_2–induced oxidative stress and DNA damage in HaCaT cells, and antiproliferative activity against all cell lines tested, being more sensitive against the in vitro model of skin melanoma.

Keywords: *Sideritis raeseri* subsp. *raeseri*; essential oil; antimicrobial; antioxidant; antiproliferative activity

---

1. Introduction

The use of natural compounds isolated by plant origins with biological activity (antimicrobial, antioxidant, and antiproliferative) has always been a topic of great interest [1–5]. The rising trend for bioactive substances witnessed today can be mainly attributed to the growing number and severity of food poisoning outbreaks worldwide along with the recent negative consumer perception against artificial food preservatives and the demand for novel functional foods with a health potential.

Among the various herbs, *Sideritis*, known as the "mountain" tea, is a controversial botanic genus with a complex taxonomical classification due to the high number of hybridizations that occur between species. It comprises more than 150 perennial and annual vegetal species and several subspecies [6]. It belongs to the *Lamiaceae* family and is well known for its use as herbal medicine, commonly as an herbal tea. *Sideritis* is abundant in Mediterranean regions, the Balkans, the Iberian Peninsula, and Macaronesia, but it can also be found in Central Europe and in Asia. In Greece, the most commonly cultivated species is *Sideritis raeseri*, distinguished in three main subspecies, *Sideritis raeseri* subsp. *raeseri*, *Sideritis raeseri* subsp. *attica*, and *Sideritis raeseri* subsp. *florida* [7].

Essential oils (EOs) are naturally occurring volatile compound mixtures isolated by plant material by different methods, including solvent extraction, supercritical fluid extraction, hydro distillation, and steam distillation. Selection of the extraction method is crucial, since extraction yield and volatile composition depend greatly on the conditions applied. Distillation-based processes are considered advantageous due to flexibility, versatility, avoidance of volatile compounds decomposition, and ability for wide volume range operations, but the extraction yields may vary highly upon time, pressure, and temperature [8], as previously recorded for *Sideritis raeseri* subsp. *raeseri* EOs [9,10]. Likewise, remarkable differences on its chemical composition have been reported [7,9,10], which could be attributed to variations on extraction/isolation methods, plant chemotypes, harvesting periods, environment and climate, as well as improper taxonomical classification [11].

Despite the fact that a series of pharmacological activities such as antimicrobial, antioxidant, anti-inflammatory, and antiproliferative action of various extracts and oils isolated by several *Sideritis* spp. have been previously published [9,11–13], the biological activity of *S. raeseri* subsp. *raeseri* EO has been scarcely studied. In particular, no inhibitory action of *S. raeseri* subsp. *raeseri* EO against common food spoilage microorganisms, such as *Staphylococcus epidermidis*, *Pseudomonas fragi*, *Saccharomyces cerevisiae*, *Aspergillus niger*, etc. [14–16], and pathogens associated with food poisoning outbreaks worldwide, such as *Staphylococcus aureus*, *Escherichia coli*, *Listeria monocytogenes*, *Salmonella* spp., etc. [17], has been reported, although the antimicrobial activity of wide range of *Sideritis* spp. EOs and extracts has been tested [11]. There are also few reports on the antioxidant or other biological activities of various extracts or isolated compounds of plant preparations of *S. raeseri* species [13], but no information on EO preparations from *S. raeseri* subsp. *raeseri* exists.

Hence, the aim of the present study was to investigate the antimicrobial potential of *S. raeseri* subsp. *raeseri* EO against common food spoilage and pathogenic microorganisms and evaluate its antioxidant and antiproliferative activity, in order to assess potential commercial applications in food and pharmaceutical industries.

2. Materials and Methods

2.1. Standard Compounds

Standard compounds used for identification in GC/MS analysis were kindly provided by Professor L. Skaltsounis, Department of Pharmacy, National and Kapodistrian University of Athens, Athens, Greece. *n*-Pentane (Merck, Darmstadt, Germany) for spectroscopy (Uvasol®) was used for the dilution of EOs analyzed by GC–MS. The alkane C10–C40 analytical standard mixture was obtained from Sigma-Aldrich (Sigma-Aldrich, Darmstadt, Germany).

2.2. Plant Material

S. raeseri subsp. *raeseri* plant was cultivated in 2014 at Vrinena region of the Othrys mountain in Greece. The aerial parts of the plants were collected and air-dried naturally at temperature ranging 25–27 °C during the daytime and 16–22 °C during the night for 7–8 days.

2.3. Extraction of EO

EO was produced by steam distillation for 75 min using dried herb of *S. raeseri* subsp. *raeseri* (20 kg dry weight). The obtained EO (5 mL) was dried over anhydrous sodium sulfate and kept in a sealed vial at 4 °C. The yield (*w/w*) was 0.025% (on a dry weight basis).

2.4. Microbial Strains

Staphylococcus aureus ATCC 25923, *Staphylocccus epidermidis* FMCC B-202 C5M6 (kindly provided by Dr. A. Nisiotou, Athens Wine Institute, ELGO-DEMETER, Greece), *Escherichia coli* ATCC 25922, *Listeria monocytogenes* NCTC 10527 serotype 4b, *Salmonella enterica* subsp. *enterica* ser. Enteritidis FMCC B56 PT4 (kindly provided by Professor G.J.E. Nychas, Agricultural University of Athens, Greece), and *Salmonella enterica* subsp. *enterica* ser. Typhimurium DSMZ 554 were grown in Brain Heart Infusion (BHI) broth (LABM, Heywood, UK) at 37 °C for 24 h. *Pseudomonas fragi* 211 (kindly provided by Professor G.J.E. Nychas) was grown in BHI broth (LABM) at 25 °C for 24 h. *Saccharomyces cerevisiae* uvaferm NEM (Lallemand, Montreal, QC, Canada) was grown in YPD broth (yeast extract 10 g/L, peptone 20 g/L, and dextrose20 g/L) at 28 °C for 3 days. *Aspergillus niger* 19111 (kindly provided by Professor G.J.E. Nychas) was grown on Malt extract agar (LABM) for 7 days at 37 °C.

2.5. Analytical Procedures

2.5.1. GC/MS

EO was diluted with *n*-pentane at a concentration of 2 mg/mL, and 1 µL was injected to the GC/MS (in splitless mode). The analysis was performed in triplicate using Finnigan Trace GC Ultra 2000/Finnigan Trace DSQ MSD (Thermo Electron Corporation, Waltham, MA, USA) operating in EI mode. The separation was performed on a Trace TR-5MS (Thermo Scientific, Waltham, MA, USA) (30 m × 0.25 mm, 0.25 µm film thickness) capillary column. He at a flow rate of 0.8 mL/min was used as carrier gas. The initial temperature of the column was set at 60 °C, and then it was heated to 240 °C at a rate of 3 °C/min for 10 min. The injector temperature was kept at 200 °C. The detector voltage was 70 eV, and the temperature was 250 °C. Identification of the compounds was carried out by comparing the retention times and the mass spectra of volatiles to ADAMS, Wiley275, NIST, and *in-house* created libraries and by determining Kovats' retention indexes (KI) using *n*-alkanes (C10–C40) and comparing them with those reported in the literature.

2.5.2. Antimicrobial Assays

Screening of S. raeseri subp. raeseri EO Antimicrobial Activity by the Disc Diffusion Assay

The antimicrobial activity of the *Sideritis raeseri* subsp. *raeseri* EO was initially tested using the disk diffusion assay, as described previously by Mitropoulou et al. [18], using gentamycin (10 mg) (Oxoid Ltd., Basingstoke, UK) as positive control and sterile water as negative.

A similar procedure was also followed for screening the activity against yeasts and molds, using *S. cerevisiae* and *A. niger* as model microorganisms [18]. Voriconazole (1 mg) (BioRad Laboratories Inc., Hercules, CA, USA) was used as positive control and sterile water as negative.

Determination of Minimum Inhibitory), Non-Inhibitory, and Minimum Bactericidal Concentrations

Determination of minimum inhibitory concentration (MIC) and non-inhibitory concentrations (NIC) was carried out, as recently described by Mitropoulou et al. [18], by monitoring changes in optical density of bacterial suspensions in BHI broths containing multiple concentrations (ranging from 41–8786 mg/L) of the EO at 610 nm using a microplate reader (Molecular Devices, VERSAmax, San Jose, California, USA, Softmaxprov.5.0 software) during incubation at 37 °C for 24 h for all bacteria species, except *P. fragi*, which was incubated at 25 °C. Ciproxin and gentamycin were used as positive controls, and BHI broths with no inoculum and inoculated BHI broths with no essential oil were used as negative controls.

The calculation of MIC and NIC values was based on the Lambert–Pearson model (LPM) [19,20].

As the LPM model was not applicable to *S. cerevisiae* and *A. niger* due to yeast cell sedimentation and conidia flotation, the standard protocols described by The European Committee on Antimicrobial Susceptibility Testing (EUCAST) were applied for MIC determination [21,22], using voriconazole as positive control [23,24].

Minimum lethal concentration (MLC) was determined, as previously described by Mitropoulou et al. [18].

All experiments were carried out at least in four replicates.

2.6. Assessment of Cell-Free Antioxidant Activity by DPPH and ABTS Assays

The radical scavenging activity of *S. raeseri* subsp. *raeseri* EO was determined by the colorimetric 2,2-diphenyl-1-picrylhydrazyl (DPPH) and 2,2′-Azino-bis(3-ethylbenzothiazoline-6-sulfonic acid) (ABTS) assays [25,26], as previously described. Serial dimethylsulfoxide (DMSO) solutions of EO concentration ranging from 0.0042 to 42 mg/mL were prepared. Samples (10 μL) were mixed with DPPH or ABTS solution (200 μL total volume) in a 96-well plate and then incubated for 30 min in the dark (RT). Inhibition of formation of the radicals was monitored by measuring absorbance at 492 nm (DPPH assay) or 734 nm (ABTS assay) using a microplate reader (EnSpire Multimode Plate Reader, PerkinElmer, Waltham, MA, USA). The inhibition percentage of the radicals for each dilution was calculated as described previously [27]. Ascorbic acid was used as positive control. Based on the values derived from the inhibition percentage of each radical, reference curves were made for *S. raeseri* subsp. *raeseri* EO, from which the IC_{50} values (mg/mL) were calculated (EO concentration–inhibition percentage) by regression analysis using Sigma Plot Software v.10 (Systat Software Inc., San Jose, CA, USA). All determinations were performed in triplicates.

2.7. Sulforhodamine B Assay (SRB)

The antiproliferative effect of *S. raeseri* subsp. *raeseri* EO was evaluated by the SRB assay as previously described by Fitsiou et al. [27]. For this assay, 3×10^3 human immortalized keratinocyte (HaCaT) and A375 cells and 4×10^3 PC3, DU145, and Caco2 were cultured in 96-well microplates for 24 h and then treated with various concentrations of *S. raeseri* subsp. *raeseri* EO (0–0.84 mg/mL) for 72 h. Cell viability curves were plotted, and the EC_{50} values corresponding to efficient concentrations of *S. raeseri* subsp. *raeseri* EO required to cause 50% decrease in cell viability were determined by regression analysis using Sigma Plot Software v.10.

2.8. Single Cell Gel Electrophoresis (comet) Assay

Single cell gel electrophoresis (comet) assay was performed as described previously [28,29]. Briefly, HaCaT cells (2×10^4) were treated with *S. raeseri* subsp. *raeseri* EO (0.05 mg/mL and 0.5 mg/mL) for 20 min followed by incubation with H_2O_2 (50 μM) for 20 min in phosphate buffered saline (PBS) or left untreated or treated with EO/H_2O_2 alone for 20 min. Then, cells were embedded in low-melting agarose on microscope slides and processed for monitoring DNA damage levels by applying the alkaline version of the comet assay, which detects both single- and double-strand DNA breaks. Slides were observed by fluorescence microscopy (Zeiss Axio Scope.A1, Oberkochen, Germany). Image analysis and scoring of DNA damage in arbitrary units (AU) was performed as previously described by Panayiotidis et al. [30]. Results were expressed as fold-change relative to control.

2.9. Statistical Analysis

The mean values are presented, and standard deviation in MIC and NIC values determined by the Lambert–Pearson model (LPM) was calculated by Figure P.2.1 software (Fig.P Software Incorporated, Hamilton, ON, Canada).

The results were analyzed with analysis of variance (ANOVA) using Duncan's multiple range test to determine significant differences ($p < 0.05$) among results (coefficients, ANOVA tables, and significance ($p < 0.05$) were computed using Statistica software (v.10.0, StatSoft, Tulsa, USA).

3. Results and Discussion

3.1. GC/MS Analysis

A significantly low oil extraction yield (0.025%) compared to a previous study was recorded [9], probably due to the lower time allowed for steam distillation (75 min in contrast to 3 h usually applied) [8,9]. A short distillation process was decided to avoid decomposition and chemical changes of the major constituents, as time is a crucial factor affecting the quality of the distilled EO [8]. However, similar or even lower extraction yields have been previously reported for various *Sideritis* species [8,31,32].

The main constituents of *S. raeseri* subsp. *raeseri* EO were determined by GC/MS. The results are presented in Table 1, and a typical chromatogram is shown in Figure 1. In total, 19 compounds were identified. Geranyl-*p*-cymene (25.08%), geranyl-*γ*-terpinene (15.17%), and geranyl-linalool (14.04%) were the main compounds detected, accounting for approximately 54% of the total area. Of note, GC/MS analysis provided data about the content percentage of the volatile compounds and not their actual concentration.

Table 1. GC/MS analysis of *Sideritis raeseri* subsp. *raeseri* essential oil.

RT (min)	Component Name	MS Fragments	Area (%)
6.11	α-Pinene	136 (MW), 121, 105, 93, 92, 91, 77, 41, 39, 27	2.13
7.53	β-Pinene	136 (MW), 121, 93, 91, 79, 77, 69, 41, 39, 27	2.31
9.20	1,3,8-*p*-Menthatriene	134 (MW), 119, 91	2.67
18.78	1,1,6-Trimethyl-tetralin	174 (MW), 159, 131, 115	0.43
21.20	Carvacrol	150 (MW), 135, 91	0.93
23.91	β-Copaene	204 (MW), 161, 119, 105, 93	0.72
25.81	caryophyllene	204 (MW), 189, 175, 161, 147, 133, 120, 105, 93, 91, 79, 69, 41	2.88
28.96	γ-Elemene	204(MW), 161, 121, 107, 93	5.73
29.36	α-Bisabolene	204 (MW), 121, 119, 109, 93	1.56
29.80	(+)-δ-Cadinene	204 (MW), 189, 161, 134, 119, 105, 81, 41	0.60
30.03	Cadina-1,3,5-triene	202 (MW), 187, 159, 144, 129, 115, 105	3.25
32.31	(-)-Spathulenol	220 (MW), 205, 187, 159, 105, 91	2.82
32.49	Caryophyllene oxide	220 (MW), 205, 177, 161, 149, 135, 121, 109, 93, 79, 43, 41	1.24
36.32	α-Bisabolol	204 (MW), 189, 161, 139, 119, 109, 93, 69, 43, 41	2.90
41.58	(2E,6E)-Farnesyl acetate	264 (MW), 204, 161, 138, 123, 107, 93, 69, 43, 41	3.59
44.03	Geranyl-linalool	290 (MW), 272, 203, 161, 147, 135, 119, 107, 93, 81, 69, 41	14.04
44.32	(6E,10E)-7,11,15-Trimethyl-3-methylene-1,6,10,14-hexadecatetrene	272 (MW), 148, 132, 109, 93, 69, 41	1.54
45.37	Geranyl-*p*-cymene	242 (MW), 134, 119, 91	25.08
45.71	Geranyl-*γ*-terpinene	272 (MW), 136, 121, 93, 91, 77	15.17

MW: Molecular Weight.

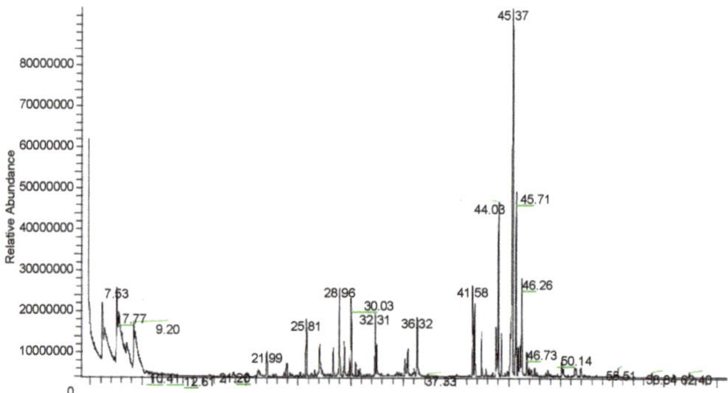

Figure 1. A typical GC/MS chromatogram of *Sideritis raeseri* subsp. *raeseri* essential oil.

Our results are in contradiction with previous studies [7,9,10]. Specifically, the *Sideritis raeseri* subsp. *raeseri* oil isolated by Alligiannis et al. [9] was characterized by the presence of *β*-pinene (9.06%), AR-curcumene (6.14%), *β*-phellandrene/limonene (6.06%), *δ*-cadinene (4.83%), *β*-caryophyllene (4.17%), and *α*-copaene (3.80%), while Koedam et al. [10] reported that the major constituents of the *S. raeseri* subsp. *raeseri* oil isolated by circulatory distillation were *β*-pinene (20.61%), *α*-pinene (16.50%), *α*-humulene (9.91%), limonene (6.73%), *β*-caryophyllene (6.52%), and D-germacrene (5.52%). However, compounds with similar structure to compounds identified in our study, such as 9-geranyl-*α*-terpinene and 9-geranyl-*p*-cymene, were found in *Sideritis dichotoma* oil [32]. Of note, 9-geranyl-*p*-cymene was also identified as a major constituent in *Sideritis trojana* oil [31]. These differences might be due to difficulties in proper taxonomical classification, as well as in differentiations in isolation/extraction methods, plant and location origin, climate and environmental conditions, time of harvesting, etc, that affect significantly the chemical composition [6,11,33]. In addition, the existence of more than one chemotype or ecotype of *S. raeseri* ssp. *raeseri* should not be excluded.

3.2. Antimicrobial Assays

The antimicrobial activity of *S. raeseri* subsp. *raeseri* EO was evaluated against a list of common food spoilage and pathogenic microorganisms [4,14–17], consisting of *Staphylococcus aureus*, *Staphylococcus epidermidis*, *Escherichia coli*, *Listeria monocytogenes*, *Salmonella* Enteritidis, *Salmonella* Typhimurium, *Pseudomonas fragi*, *Saccharomyces cerevisiae*, and *Aspergillus niger*. The presence of *Escherichia coli*, *Listeria*, and *Salmonella* spp. in foods is a primary concern due to their implication in a number of food poisoning outbreaks worldwide [17]. Similarly, food quality and safety is directly related to staphylococci counts, as its presence in high numbers constitutes a health hazard and results in spoilage [14]. Likewise, *Pseudomonas fragi* is a psychrotrophic bacteria associated mainly with food spoilage during storage at low temperatures [15]. *Saccharomyces cerevisiae* and *Aspergillus niger* are the usual cause of soft drinks and both alcoholic and non-alcoholic beverages [16] and served as model systems [34,35].

The effectiveness of the oil was initially confirmed using the disk diffusion method (data not shown). Subsequently, MIC and NIC values against were assessed, because their precise determination is crucial for food and pharmaceutical industries in order to regulate the optimum amount of the antimicrobial agent to secure microbial safety. MIC and NIC for bacteria were determined using a previously published model [19], combining the absorbance measurements with the common dilution method and non-linear regression analysis to fit the data, while the corresponding MIC values for *S. cerevisiae* and *A. niger* were estimated by standard protocols [21,22]. The effective growth inhibition of *S. raeseri* subsp. *raeseri* EO against all microorganisms tested (Table 2) was documented, although MIC, NIC, and MLC values were significantly ($p < 0.05$) higher compared to ciproxin, gentamycin, and voriconazole [18,27], which were used as positive controls.

Table 2. Minimum inhibitory concentration (MIC), non-inhibitory concentration (NIC), and minimum lethal concentration (MLC) (mg/L) of *Sideritis raeseri* subsp. *raeseri* essential oil (EO) against common food spoilage and pathogenic microorganisms. Lambert–Pearson (LPM) model was not applicable to *S. cerevisiae* and *A. niger* due to yeast cell sedimentation and conidia flotation. Thus, determination of MIC values was based on the standard protocols described by EUCAST [21,22].

Microbial Species	S. raeseri subsp. raeseri EO			Ciproxin (Data Reproduced by Fitsiou et al. (2016) [27]			Gentamycin (Data Reproduced by Mitropoulou et al. (2017) [18]			Voriconazole		
	MIC *	NIC *	MLC **	MIC *	NIC *	MLC **	MIC *	NIC *	MLC **	MIC *	NIC *	MLC **
Staphylococcus aureus	7116 ± 26	5887 ± 35	31630	0.982 ± 0.002	0.963 ± 0.003	4	3.332 ± 0.003	3.021 ± 0.001	16	-	-	-
Staphylococcus epidermidis	7732 ± 35	5535 ± 35	30751	0.979 + 0.002	0.957 + 0.002	4	3.421 ± 0.001	3.127 ± 0.001	16	-	-	-
Escherichia coli	6414 ± 26	5974 ± 26	26358	0.984 + 0.001	0.956 ± 0.002	4	3.952 ± 0.001	3.253 ± 0.002	16	-	-	-
Listeria monocytogenes	6853 ± 35	5799 ± 44	28115	0.979 ± 0.001	0.968 + 0.001	4	3.121 ± 0.002	3.001 ± 0.002	16	-	-	-
Salmonella Enteritidis	6326 ± 44	5799 ± 26	26358	0.976 ± 0.001	0.957 ± 0.001	8	4.942 ± 0.001	4.011 ± 0.001	18	-	-	-
Salmonella Typhimurium	5974 ± 26	5447 ± 53	26358	0.979 ± 0.001	0.964 ± 0.001	8	4.211 ± 0.002	4.026 ± 0.001	18	-	-	-
Pseudomonas fragi	5184 ± 35	4305 ± 26	21965	0.955 ± 0.001	0.940 ± 0.002	8	4.134 ± 0.002	4.009 ± 0.002	18	-	-	-
S. cerevisiae	7029	-	28115	-	-	-	-	-	-	0.25	-	1.00
A. niger	8786	-	35144	-	-	-	-	-	-	0.50	-	2.00

* Results are shown as mean ± SD when applicable. ** Standard deviation for MLC ranged in zero values.

Although similar results reporting remarkable antimicrobial activity of EOs or methanol and aqueous extracts isolated by various *Sideritis* spp. were previously reported [9,11,12], EO derived from *S. raeseri* subsp. *raeseri* had no inhibitory effect when tested against a series of spoilage and pathogenic microorganisms, such as *S. aureus*, *S. epidermidis*, *P. aeruginosa*, *E. cloacae*, *K. pneumonia*, *E. coli*, *C. albicans*, *C. tropicalis*, or *T. glabrata* [9], in contrast to our results. These findings highlight that chemical composition of the EOs and thus their biological activity may vary greatly depending on a high number of factors [11]. Hence, to the best of our knowledge, this is the first report estimating MIC and NIC values for *S. raeseri* subsp. *raeseri* EO, applying a reliable, rapid, and efficient method based on the LPM model [19,20]. Despite the fact that no antimicrobial action has been correlated to the main components of the *S. raeseri* subsp. *raeseri* EO separately, its activity could be attributed to all constituents, and possible synergistic effects should not be excluded.

3.3. Determination of Radical Scavenging Activity

To investigate the potential anti-oxidant potential of *S. raeseri* subsp. *raeseri* EO, we studied the in vitro radical scavenging activity by utilizing the DPPH• and the ABTS•$^+$ radical scavenging assays. The IC$_{50}$ value of the *S. raeseri* subsp. *raeseri* EO, corresponding to the sample concentration required to scavenge radicals by 50%, was estimated to be 24.77 ± 4.21 and 1.27 ± 0.59 mg/mL in the cases of DPPH• and ABTS•$^+$ assays, respectively (Table 3). Ascorbic acid (potent antioxidant agent) was used as a positive control. Low IC$_{50}$ values indicate strong antioxidant activities. Compared to ascorbic acid, the *S. raeseri* subsp. *raeseri* EO possesses weak in vitro antioxidant capacity. This is the first time reporting on the direct in vitro antioxidant activity of the *S. raeseri* subsp. *raeseri* EO preparation. To our best knowledge, there is no available information about the antioxidant potential of its major compounds (geranyl-*p*-cymene, geranyl-*γ*-terpinene, and geranyl-linalool).

Table 3. Antioxidant activity of *S. raeseri* subsp. *raeseri* essential oil in vitro.

	IC$_{50}$ (mg/mL) *	
	DPPH Assay	**ABTS Assay**
S. raesaeri	24.77 ± 4.21	1.27 ± 0.59
Ascorbic acid **	0.012 ± 0.004	0.0045 ± 0.0002

* Results are shown as mean ± SD. ** Ascorbic acid was used as positive control.

Additionally, limited literature information exists on the antioxidant activity of EO preparations of other *Sideritis* spp. Antioxidant, anticholinesterase, and anti-tyrosinase activities of the EOs of *Sideritis albiflora* and *Sideritis leptoclada* have been reported, and those preparations characterized by high concentrations of phenolic and flavonoid contents indicated the highest antioxidant and enzyme inhibitory activities [36]. In a follow-up study, the acetone extracts showed the highest activity in terms of antioxidant activity of both *Sideritis* species, while the hexane extracts exhibited superior urease inhibitory activity. Both species were found to be rich in rosmarinic and caffeic acids [37].

Previous studies on different parts of plant extracts of *Sideritis* spp. demonstrated antioxidant activity for methanolic/etnanolic or aqueous extracts measured similarly by DPPH• or ABTS•$^+$ assays, which was the weakest amongst 24 extracts from Greek domestic *Lamiaceae* species. Moderate antioxidant activity was reported for a methanolic extract of the aerial parts of the plant *Sideritis raeseri* subsp. *raeseri* estimated by Co(II) ethylenediaminetetraacetic acid (EDTA)-induced luminol chemiluminescence and DPPH• scavenging assay. The extract comprised nine 7-o-allosyl glucosides of 5,8-dihydroxy substituted flavones [38]. In most cases, the anti-oxidant effects reported for the samples studied were related to the total phenolic or flavonoid content of the extract preparations.

3.4. S. raeseri subsp. raeseri EO Protects Human Epidermal Keratinocytes (HaCaT) Cells from H_2O_2-Induced DNA Damage

Our previous results indicated that the *S. raeseri* subsp. *raeseri* EO possesses weak antioxidant activity using direct in vitro radical scavenging assays. However, the application of in vitro cell-based assays offers various advantages towards more accurately, accessing the antioxidant effects of tested compounds at a subcellular level. Cell-based assays can be more advantageous in the way that they may reveal more information about the antioxidant capacity of compounds that may trigger cell antioxidant mechanisms without direct radical scavenging action [39]. For this reason, we next explored the potential cytoprotective effect of *S. raeseri* subsp. *raeseri* EO against H_2O_2-induced DNA damage. Hydrogen peroxide is a non-radical derivative of oxygen with physiological significance as an oxidative agent; it is soluble, present in all biological systems, and capable of cell death induction [39]. Human keratinocytes (HaCaT) were pre-incubated in the presence or the absence of EO (0.05 and 0.5 mg/mL for 20 min, RT) and treated with H_2O_2 (100 μM) for 20 min to induce cellular oxidative DNA damage. DNA damage was monitored as single- and double-strand breaks in DNA by employing the alkaline version of single cell gel electrophoresis (comet) assay. The results are represented in Figure 2. Incubation of the HaCaT cells with the *S. raeseri* subsp. *raeseri* EO alone slightly caused a significant increase in DNA damage levels only in the case of the highest concentration of the EO used (0.5 mg/mL) (approximately 1.2-fold compared to untreated cells). Treatment of HaCaT cells with H_2O_2, significantly induced DNA damage levels (> 2.5-fold) compared to control. In the case of cell pretreatment with the two different concentrations of *S. raeseri* subsp. *raeseri* EO, the DNA damage observed was significantly lower and dose-dependent. More particularly, pre-treatment of HaCaT cells with 0.05 mg/mL or 0.5 mg/mL *S. raeseri* subsp. *raeseri* EO caused 34% and 44% decrease in the H_2O_2-induced DNA damage levels, respectively.

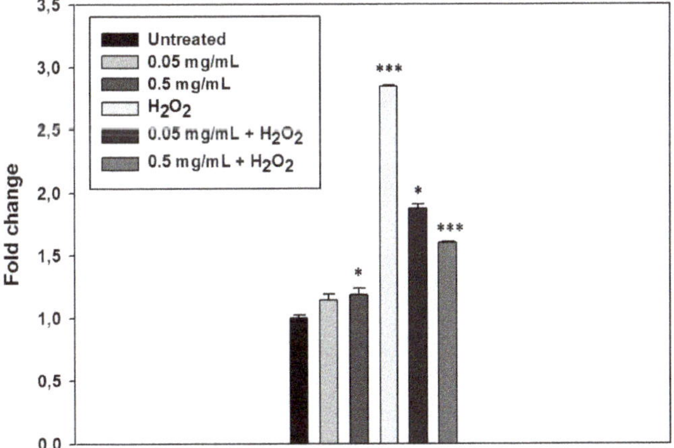

Figure 2. Cytoprotective activity of *S. raeseri* subsp. *raeseri* essential oil against H_2O_2-induced DNA damage on human immortalized keratinocyte (HaCaT) cells. HaCaT cells were incubated with *S. raeseri* subsp. *raeseri* essential oil (0.05 mg/mL and 0.5 mg/mL) for 20 min and then 20 min with H_2O_2 both in presence and in absence of the oil. Comet assay was performed to assess the H_2O_2-induced DNA damage. The data presented are the mean ± SD of three independent experiments performed in triplicates. * $p < 0.05$, *** $p < 0.001$ vs. untreated.

Previously, González-Burgos et al. [40] demonstrated the cytoprotective role of other species of *Sideritis* spp. on PC12 and U373-MG cells. They indicated that pre-treatment with isolated diterpenoids from *Sideritis* spp. prevented the H_2O_2-induced mitochondrial membrane disorder. Further studies are required to investigate the potential protective role of the EO against oxidative stress and the underlying mechanism(s) of action.

3.5. Determination of Antiproliferative Activity

Next, we evaluated the antiproliferative effect of the EO in various cancer cell lines. For this purpose, we utilized human melananoma A375, human colon adenosarcoma Caco2, and human prostate carcinoma cell lines PC3 and DU145. Moreover, a non-cancerous human cell line, the human keratinocytes HaCaT, was also employed in the study to discrete for preferential cancer-specific antiproliferative activity amongst the cell lines. Overall, cells were treated with increasing concentrations of *S. raeseri* subsp. *raeseri* EO for 72 h, and then cell viability was assessed as percent of control, and the corresponding EC_{50} value was also determined. As shown in Table 4, the observed patterns of antiproliferative activity were very similar between all the cell lines, and the EC_{50} value ranged approximately from 0.114–0.216 mg/mL. To our best knowledge, there is no available information about the antiproliferative/anticancer potential of its major compounds (geranyl-*p*-cymene, geranyl-γ-terpinene, and geranyl-linalool).

Table 4. Antiproliferative activity of *S. raeseri* subsp. *raeseri* essential oil against different cell lines.

Cell Line	EC_{50} (mg/mL) *
A375	0.151 ± 0.008
HaCaT	0.114 ± 0.015
Caco2	0.175 ± 0.080
PC3	0.216 ± 0.090
DU145	0.188 ± 0.060

* Results are shown as mean ± SD.

Few studies have investigated the antiproliferative potential of *Sideritis* spp. Tóth et al. [41] examined the antiproliferative effects of isolated diterpenoids derived from *Sideritis montana* on human cancer cell lines (HeLa, SiHa, and C33A) by using MTT assay. Moreover, Tadić et al. [42] investigated the cytotoxic effect of *Sideritis scardica* on PBMC, B16, and HL-60 cells, demonstrating also the potential anti-inflammatory effects of that extract.

4. Conclusions

The use of EOs as natural antimicrobial, antioxidant, and antiproliferative agents is less explored compared to their utilization as food flavorings and, thus, their application in food and pharmaceutical industries is limited. Our results revealed the growth inhibitory action of *S. raeseri* subsp. *raeseri* EO against food spoilage and pathogenic microoganisms, although its activity was significantly lower than gentamycin, ciproxin, and voriconazole that were used as positive controls, indicating that it represents a source of natural antimicrobial agent, which may be incorporated in food products to prevent spoilage and assure microbial safety. Furthermore, the oil exhibited low antioxidant activity by directly scavenging radicals. However, it indicated promising cytoprotective activity against H_2O_2-induced oxidative stress and DNA damage in HaCaT cells, suggesting the exertion of non-direct antioxidant mechanisms that require further exploration. Finally, the antiproliferative activity of the EO was evaluated against a panel of cell lines. It showed similar activity against all cell lines tested being more sensitive against the in vitro model of skin melanoma. However, similar activity was also observed against the non-carcinoma cell line HaCaT, indicating that it may not possess cancer-specific antiproliferative activity, although further studies are required. Although main components of the EO have been reported as main constituents in other plant extracts with similar biological activities [43–45], there are no available studies on the biological activities of the individual compounds, which is a valid area of further investigation. Overall, the results of our study indicated that EO form *S. raeseri* subsp. *raeseri* has favorable biological properties that may have potential applications in food and pharmaceutical industries.

Author Contributions: Conceptualization, A.P. and Y.K.; data curation, I.T., A.P., C.P., and Y.K.; funding acquisition, A.P.; investigation, G.M., M.S. (Marianthi Sidira), M.S. (Myria Skitsa), I.T., and C.P.; methodology, G.M., M.S. (Myria Skitsa), I.T., and C.P.; project administration, A.P. and Y.K.; resources, A.P., C.D., and Y.K.; supervision, A.P., C.D., and Y.K.; validation, A.P. and C.P.; visualization, A.P.; writing–original draft, G.M., M.S. (Marianthi Sidira), I.T., M.S. (Myria Skitsa), and C.P.; writing–review and editing, A.P. and Y.K. All authors have read and agreed to the published version of the manuscript.

Funding: This research was funded by the project "OPENSCREEN-GR: An Open-Access Research Infrastructure of Target-Based Screening Technologies and Chemical Biology for Human and Animal Health, Agriculture and Environment" (MIS 5002691), which is implemented under the Action "Reinforcement of the Research and Innovation Infrastructure", funded by the Operational Programme "Competitiveness, Entrepreneurship and Innovation" (NSRF 2014–2020) and co-financed by Greece and the European Union (European Regional Development Fund).

Acknowledgments: The authors thank C. Tassou for providing full access to the Microplate Reader, N. Chorianopoulos for his valuable technical support and scientific advice, L. Skaltsounis for providing standard compounds, and G.J.E. Nychas and A. Nisiotou for providing the microbial cultures.

Conflicts of Interest: The authors declare no conflict of interest. The funders had no role in the design of the study, in the collection, analyses, or interpretation of data, in the writing of the manuscript, or in the decision to publish the results.

References

1. Nakatsu, T.N.; Lupo, A.T.; Chinn, J.W.; Kang, R.K. Biological activity of essential oils and their constituents. *Stud. Nat. Prod. Chem.* **2000**, *21*, 571–631. [CrossRef]

2. Rios, J.; Recio, M. Medicinal plants and antimicrobial activity. *J. Ethnopharmacol.* **2005**, *100*, 80–84. [CrossRef]

3. Nychas, G.-J.; Skandamis, P.; Tassou, C. Antimicrobials from herbs and spices. In *Natural Antimicrobials for the Minimal Processing of Foods*; Elsevier BV: Amsterdam, The Netherlands, 2003; pp. 176–200.

4. Burt, S. Essential oils: Their antibacterial properties and potential applications in foods—A review. *Int. J. Food Microbiol.* **2004**, *94*, 223–253. [CrossRef] [PubMed]

5. Chorianopoulos, N.G.; Evergetis, E.T.; Aligiannis, N.; Mitakou, S.; Nychas, G.-J.E.; Haroutounian, S.A. Correlation between Chemical Composition of Greek Essential Oils and their Antibacterial Activity against Food-borne Pathogens. *Nat. Prod. Commun.* **2007**, *2*, 419–426. [CrossRef]

6. Patelou, E.; Chatzopoulou, P.; Polidoros, A.N.; Mylona, P.V. Genetic diversity and structure of *Sideritis raeseri* Boiss. & Heldr. (*Lamiaceae*) wild populations from Balkan Peninsula. *J. Appl. Res. Med. Aromat. Plants* **2020**, *16*, 100241. [CrossRef]

7. Papageorgiou, V.P.; Kokkini, S.; Argyriadou, N. Chemotaxonomy of the Greek Species of *Sideritis* I. Components of the volatile fraction of *Sideritis raeseri* ssp. *raeseri*. In *Aromatic Plants*; Springer Science and Business Media LLC: Berlin/Heidelberg, Germany, 1982; pp. 211–220.

8. Garzoli, S.; Navarra, A.; Garzoli, S.; Pepi, F.; Ragno, R. Esential oils extraction: A 24-hour steam distillation systematic methodology. *Nat. Prod. Res.* **2017**, *31*, 2387–2396. [CrossRef]

9. Aligiannis, N.; Kalpoutzakis, E.; Chinou, I.B.; Mitakou, S.; Gikas, E.; Tsarbopoulos, A. Composition and Antimicrobial Activity of the Essential Oils of Five Taxa of *Sideritis* from Greece. *J. Agric. Food Chem.* **2001**, *49*, 811–815. [CrossRef] [PubMed]

10. Koedam, A. Volatile oil composition of greek mountain tea (*Sideritis* spp.). *J. Sci. Food Agric.* **1986**, *37*, 681–684. [CrossRef]

11. González-Burgos, E.; Carretero, M.; Gómez-Serranillos, M.P. *Sideritis* spp.: Uses, chemical composition and pharmacological activities—A review. *J. Ethnopharmacol.* **2011**, *135*, 209–225. [CrossRef]

12. Stagos, D.; Portesis, N.; Spanou, C.; Mossialos, D.; Aligiannis, N.; Chaita, E.; Panagoulis, C.; Reri, E.; Skaltsounis, L.; Tsatsakis, A.; et al. Correlation of total polyphenolic content with antioxidant and antibacterial activity of 24 extracts from Greek domestic *Lamiaceae* species. *Food Chem. Toxicol.* **2012**, *50*, 4115–4124. [CrossRef]

13. Romanucci, V.; Di Fabio, G.; D'Alonzo, D.; Guaragna, A.; Scapagnini, G.; Zarrelli, A. Traditional uses, chemical composition and biological activities of *Sideritis raeseri* Boiss. & Heldr. *J. Sci. Food Agric.* **2016**, *97*, 373–383. [CrossRef]

14. Iqbal, M.N.; Anjum, A.A.; Ali, M.A.; Hussain, F.; Ali, S.; Muhammad, A.; Ahmad, A.; Irfan, M.; Shabbir, A. Assessment of Microbial Load of Un-pasteurized Fruit Juices and in vitro Antibacterial Potential of Honey Against Bacterial Isolates. *Open Microbiol. J.* **2015**, *9*, 26–32. [CrossRef]

15. Jay, J.M.; Loessner, M.J.; Golden, D.A. *Modern Food Microbiology*, 7th ed.; Springer Science Business Media, Inc.: New York, NJ, USA, 2005.

16. Juvonen, R.; Virkajarvi, V.; Priha, O.; Laitila, A. *Microbiological Spoilage and Safety Risks in Non-Beer Beverages Produced in a Brewery Environment*; VTT Tiedotteita-Research: Espoo, Finland, 2011.

17. Aneja, K.R.; Dhiman, R.; Aggarwal, N.K.; Kumar, V.; Kaur, M. Microbes Associated with Freshly Prepared Juices of *Citrus* and Carrots. *Int. J. Food Sci.* **2014**, *2014*, 1–7. [CrossRef] [PubMed]

18. Mitropoulou, G.; Fitsiou, E.; Spyridopoulou, K.; Tiptiri-Kourpeti, A.; Bardouki, H.; Vamvakias, M.; Panas, P.; Chlichlia, K.; Pappa, A.; Kourkoutas, Y. *Citrus medica* essential oil exhibits significant antimicrobial and antiproliferative activity. *LWT* **2017**, *84*, 344–352. [CrossRef]

19. Chorianopoulos, N.; Lambert, R.; Skandamis, P.; Evergetis, E.; Haroutounian, S.; Nychas, G.-J.E. A newly developed assay to study the minimum inhibitory concentration of *Satureja spinosa* essential oil. *J. Appl. Microbiol.* **2006**, *100*, 778–786. [CrossRef] [PubMed]

20. Lambert, R.; Lambert, R. A model for the efficacy of combined inhibitors. *J. Appl. Microbiol.* **2003**, *95*, 734–743. [CrossRef] [PubMed]

21. Arendrup, M.C.; Meletiadis, J.; Mouton, J.W.; Lagrou, K.; Hamal, P.; Guinea, J.; Subcommittee on Antifungal Susceptibility Testing (AFST) of the ESCMID European Committee for Antimicrobial Susceptibility Testing. EUCAST Definitive Document EDEF 9.3.2 Method for the Determination of Broth Dilution Minimum Inhibitory Concentrations of Antifungal Agents for Conidia Forming Moulds. 2020. Available online: https://www.eucast.org/fileadmin/src/media/PDFs/EUCAST_files/AFST/Files/EUCAST_E_Def_9.3.2_Mould_testing_definitive_revised_2020 (accessed on 7 April 2020).

22. Arendrup, M.C.; Meletiadis, J.; Mouton, J.W.; Lagrou, K.; Hamal, P.; Guinea, J.; Subcommittee on Antifungal Susceptibility Testing (AFST) of the ESCMID European Committee for Antimicrobial Susceptibility Testing. EUCAST Definitive Document EDEF 7.3.2 Method for the Determination of Broth Dilution Minimum Inhibitory Concentrations of Antifungal Agents for Yeasts. 2020. Available online: https://www.eucast.org/fileadmin/src/media/PDFs/EUCAST_files/AFST/Files/EUCAST_E_Def_7.3.2_Yeast_testing_definitive_revised_2020 (accessed on 8 April 2020).

23. Balkan, C.; Ercan, I.; Isik, E.; Akdeniz, E.S.; Balcioglu, O.; Kodedová, M.; Zimmermannová, O.; Dundar, M.; Sychrová, H.; Koc, A. Genomewide Elucidation of Drug Resistance Mechanisms for Systemically Used Antifungal Drugs Amphotericin B, Caspofungin, and Voriconazole in the Budding Yeast. *Antimicrob. Agents Chemother.* **2019**, *63*. [CrossRef]

24. Mandras, N.; Roana, J.; Scalas, D.; Fucale, G.; Allizond, V.; Banche, G.; Barbui, A.; Vigni, N.L.; Newell, V.A.; Cuffini, A.M.; et al. In vitro antifungal activity of fluconazole and voriconazole against non-*Candida* yeasts and yeast-like fungi clinical isolates. *New Microbiol.* **2015**, *38*, 583–587.

25. Lee, S.K.; Mbwambo, Z.H.; Chung, H.; Luyengi, L.; Gamez, E.J.; Mehta, R.G.; Kinghorn, A.D.; Pezzuto, J.M. Evaluation of the antioxidant potential of natural products. *Comb. Chem. High Throughput Screen.* **1998**, *1*, 35.

26. Re, R.; Pellegrini, N.; Proteggente, A.; Pannala, A.; Yang, M.; Rice-Evans, C. Antioxidant activity applying an improved ABTS radical cation decolorization assay. *Free. Radic. Boil. Med.* **1999**, *26*, 1231–1237. [CrossRef]

27. Fitsiou, E.; Mitropoulou, G.; Spyridopoulou, K.; Tiptiri-Kourpeti, A.; Vamvakias, M.; Bardouki, H.; Panayiotidis, M.; Galanis, A.; Kourkoutas, Y.; Chlichlia, K.; et al. Phytochemical Profile and Evaluation of the Biological Activities of Essential Oils Derived from the Greek Aromatic Plant Species *Ocimum basilicum*, *Mentha spicata*, *Pimpinella anisum* and *Fortunella margarita*. *Molecules* **2016**, *21*, 1069. [CrossRef]

28. Olive, P.L.; Banáth, J.P. The comet assay: A method to measure DNA damage in individual cells. *Nat. Protoc.* **2006**, *1*, 23–29. [CrossRef]

29. Fitsiou, E.; Mitropoulou, G.; Spyridopoulou, K.; Vamvakias, M.; Bardouki, H.; Galanis, A.; Chlichlia, K.; Kourkoutas, Y.; Panayiotidis, M.; Pappa, A. Chemical Composition and Evaluation of the Biological Properties of the Essential Oil of the Dietary Phytochemical *Lippia citriodora*. *Molecules* **2018**, *23*, 123. [CrossRef] [PubMed]

30. Panagiotidis, M.; Tsolas, O.; Galaris, D. Glucose oxidase-produced H_2O_2 induces Ca^{2+}-dependent DNA damage in human peripheral blood lymphocytes. *Free Radic. Boil. Med.* **1999**, *26*, 548–556. [CrossRef]

31. Kırmızıbekmez, H.; Karaca, N.; Demirci, F. Characterization of *Sideritis trojana* Bornm. essential oil and its antimicrobial activity. *Marmara Pharm. J.* **2017**, *21*, 860–865. [CrossRef]

32. Kirimer, N.; Baser, K.H.C.; Demirci, B.; Duman, H. Essential Oils of *Sideritis* Species of Turkey Belonging to the Section Empedoclia. *Chem. Nat. Compd.* **2004**, *40*, 19–23. [CrossRef]

33. Chorianopoulos, N.; Kalpoutzakis, E.; Aligiannis, N.; Mitaku, S.; Nychas, G.-J.E.; Haroutounian, S.A. Essential Oils of *Satureja, Origanum*, and *Thymus* Species: Chemical Composition and Antibacterial Activities Against Foodborne Pathogens. *J. Agric. Food Chem.* **2004**, *52*, 8261–8267. [CrossRef]

34. Battey, A.S.; Duffy, S.; Schaffner, D. Modeling Yeast Spoilage in Cold-Filled Ready-To-Drink Beverages with *Saccharomyces cerevisiae, Zygosaccharomyces bailii*, and *Candida lipolytica*. *Appl. Environ. Microbiol.* **2002**, *68*, 1901–1906. [CrossRef]

35. Garcia, D.; Ramos, A.J.; Sanchis, V.; Marín, S. Predicting mycotoxins in foods: A review. *Food Microbiol.* **2009**, *26*, 757–769. [CrossRef]

36. Deveci, E.; Tel-Çayan, G.; Usluer, Ö.; Duru, M.E. Chemical Composition, Antioxidant, Anticholinesterase and Anti-Tyrosinase Activities of Essential Oils of Two *Sideritis* Species from Turkey. *Iran J. Pharm. Res.* **2019**, *18*, 903–913.

37. Deveci, E.; Tel-Çayan, G.; Duru, M.E.; Öztürk, M. Phytochemical contents, antioxidant effects, and inhibitory activities of key enzymes associated with Alzheimer's disease, ulcer, and skin disorders of *Sideritis albiflora* and *Sideritis leptoclada*. *J. Food Biochem.* **2019**, *43*, e13078. [CrossRef]

38. Gabrieli, C.; Kefalas, P.; Kokkalou, E. Antioxidant activity of flavonoids from *Sideritis raeseri*. *J. Ethnopharmacol.* **2005**, *96*, 423–428. [CrossRef] [PubMed]

39. Da Silva, L.C.N.; Filho, C.M.B.; De Paula, R.A.; Silva, C.S.S.E.; De Souza, L.I.O.; Da Silva, L.C.N.; Correia, M.T.D.S.; De Figueiredo, R.C.B.Q. In vitro cell-based assays for evaluation of antioxidant potential of plant-derived products. *Free. Radic. Res.* **2016**, *50*, 801–812. [CrossRef] [PubMed]

40. González-Burgos, E.; Duarte, A.I.; Carretero, M.E.; Moreira, P.; Gómez-Serranillos, M.P. Mitochondrial-Targeted Protective Properties of Isolated Diterpenoids from *Sideritis* spp. in Response to the Deleterious Changes Induced by H_2O_2. *J. Nat. Prod.* **2013**, *76*, 933–938. [CrossRef] [PubMed]

41. Tóth, B.; Kúsz, N.; Forgo, P.; Bózsity, N.; Zupkó, I.; Pinke, G.; Hohmann, J.; Vasas, A. Abietane diterpenoids from *Sideritis montana* L. and their antiproliferative activity. *Fitoterapia* **2017**, *122*, 90–94. [CrossRef]

42. Tadić, V.; Jeremic, I.; Dobrić, S.; Isakovic, A.; Marković, I.; Trajkovic, V.; Bojović, D.; Arsić, I. Anti-inflammatory, Gastroprotective, and Cytotoxic Effects of *Sideritis scardica* Extracts. *Planta Med.* **2012**, *78*, 415–427. [CrossRef]

43. Jorge, L.; Meniqueti, A.; Silva, R.; Santos, K.; Da Silva, E.A.; Gonçalves, J.E.; De Rezende, C.; Colauto, N.B.; Gazim, Z.; Linde, G. Antioxidant activity and chemical composition of oleoresin from leaves and flowers of *Brunfelsia uniflora*. *Genet. Mol. Res.* **2017**, *16*. [CrossRef]

44. Tian, M.; Liu, T.; Wu, X.; Hong, Y.; Liu, X.; Lin, B.; Zhou, Y. Chemical composition, antioxidant, antimicrobial and anticancer activities of the essential oil from the rhizomes of *Zingiber striolatum* Diels. *Nat. Prod. Res.* **2019**, *25*, 1–5. [CrossRef] [PubMed]

45. Demiray, H.; Tabanca, N.; Estep, A.S.; Becnel, J.J.; Demirci, B. Chemical composition of the essential oil and *n*-hexane extract of *Stachys tmolea* subsp. Tmolea Boiss., an endemic species of Turkey, and their mosquitocidal activity against dengue vector *Aesdes aegypti*. *Saudi Pharm. J.* **2019**, *27*, 877–881. [CrossRef] [PubMed]

Article

Lemon Grass Essential Oil does not Modulate Cancer Cells Multidrug Resistance by Citral—Its Dominant and Strongly Antimicrobial Compound

Jitka Viktorová [1], Michal Stupák [2], Kateřina Řehořová [1], Simona Dobiasová [1], Lan Hoang [1], Jana Hajšlová [2], Tran Van Thanh [3], Le Van Tri [4], Nguyen Van Tuan [4] and Tomáš Ruml [1,*]

[1] Department of Biochemistry and Microbiology UCT Prague, Faculty of Food and Biochemical Technology, Technicka 3, Prague 166 28, Czech Republic; prokesoj@vscht.cz (J.V.); rehorova@vscht.cz (K.Ř.); dobiasos@vscht.cz (S.D.); hoangl@vscht.cz (L.H.)

[2] Department of Food Analysis and Nutrition UCT Prague, Faculty of Food and Biochemical Technology, Technicka 3, Prague 166 28, Czech Republic; stupakm@vscht.cz (M.S.); hajslovj@vscht.cz (J.H.)

[3] Vietnam University of Traditional Medicine, No. 2—Tran Phu, Ha Dong, Hanoi 100000, Vietnam; thanhtv63@gmail.com

[4] Vietnam Essential Oil Joint Stock Company, No. 814/3-Lang street, Dong Da, Hanoi 100000, Vietnam; biotech.jsc@gmail.com (L.V.T.); tuanbiotech.jsc@gmail.com (N.V.T.)

* Correspondence: rumlt@vscht.cz

Received: 21 February 2020; Accepted: 21 April 2020; Published: 5 May 2020

Abstract: With strong antimicrobial properties, citral has been repeatedly reported to be the dominant component of lemongrass essential oil. Here, we report on a comparison of the antimicrobial and anticancer activity of citral and lemongrass essential oil. The lemongrass essential oil was prepared by the vacuum distillation of fresh *Cymbopogon* leaves, with a yield of 0.5% (*w/w*). Citral content was measured by gas chromatography/high-resolution mass spectrometry (GC-HRMS) and determined to be 63%. Antimicrobial activity was tested by the broth dilution method, showing strong activity against all tested bacteria and fungi. Citral was up to 100 times more active than the lemongrass essential oil. Similarly, both citral and essential oils inhibited bacterial communication and adhesion during *P. aeruginosa* and *S. aureus* biofilm formation; however, the biofilm prevention activity of citral was significantly higher. Both the essential oil and citral disrupted the maturated *P. aeruginosa* biofilm with the IC$_{50}$ 7.3 ± 0.4 and 0.1 ± 0.01 mL/L, respectively. Although it may seem that the citral is the main biologically active compound of lemongrass essential oil and the accompanying components have instead antagonistic effects, we determined that the lemongrass essential oil-sensitized methicillin-resistant *S. aureus* (MRSA) and doxorubicin-resistant ovarian carcinoma cells and that this activity was not caused by citral. A 1 mL/L dose of oil-sensitized MRSA to methicillin up to 9.6 times and a dose of 10 µL/L-sensitized ovarian carcinoma to doxorubicin up to 1.8 times. The mode of multidrug resistance modulation could be due to P-glycoprotein efflux pump inhibition. Therefore, the natural mixture of compounds present in the lemongrass essential oil provides beneficial effects and its direct use may be preferred to its use as a template for citral isolation.

Keywords: multidrug resistance; doxorubicin; MRSA; quorum sensing; biofilm

---

1. Introduction

Cymbopogon citratus (monocotyledonous plant belonging to Poaceae family), known as lemongrass, is widely used for its characteristic lemon odor and flavor in the culinary industry [1] as a component of spices. Moreover, it is used for its curative effects in traditional and alternative medicine in Asia, Africa and Latin America. Lemongrass is usually used in the form of a concoction of the aboveground parts of the plant or as an essential oil. In addition to vitamins (esp. A, C, folate, niacin, β-carotene, source:

USDA National Nutrient data base), *C. citratus* contains flavonoids, alkaloids, tannins, phenols and saponins as beneficial compounds [2]. The chemical composition of the essential oil varies depending on the geographical origin of the plant. However, the main component is always citral (65%–85%), a mixture of isomers of geranial (citral a) and neral (citral b). It also contains citronellol, citronellal, limonene, linalool, nerol, etc. [3]. Lemongrass in traditional medicine is used for the treatment of colds, influenza, cough, diabetes, malaria [4], high blood pressure, high cholesterol, fever, inflammation, hypertension [5,6], dental hygiene [7], colorectal cancer [8], nervousness, toothache and sore throat [9].

The main demonstrated bioactivities are antimicrobial, anti-inflammatory, anticancer, antimutagenic and antidiabetic activities [2]. The antimicrobial activity is the most explained. Lemongrass essential oil has a nonselective activity against both Gram-negative (*Escherichia coli, Klebsiella pneumoniae, Pseudomonas aeruginosa, Proteus vulgaris*) and Gram-positive bacteria (*Bacillus subtilis, Staphylococcus aureus*), yeasts and fungi [2]. The oil was active against *Cronobacter sakazakii*, a food-borne pathogen associated with neonatal necrotizing enterocolitis. Its action is ascribed to decreasing quorum sensing, biofilm formation and endotoxin production [10]. The inhibition of sulfate-reducing bacteria, causing several issues in the petroleum industry, was demonstrated as well [11,12]. Recently, the inhibition of *Candida albicans* and *Cryptococcus* sp. biofilm formation and its disruption by the essential oil were published [13–15]. The activity of lemongrass oil against *Aspergillus flavus* was even higher than that of some synthetic fungicides, e.g., benzimidazole and diphenylamine [16]. Besides antimicrobial activity, an effect was also demonstrated against protozoa (*Plasmodium* sp., *Leishmania* sp.) [17,18] and insects (*Anopheles* sp., *Musca domestica, Aedes aegypti*) [19–21]. In addition, the essential oil of lemongrass is an effective repellent [22].

Despite showing strong antimicrobial activity and potency to treat diseases of human civilization, neither the ability of lemongrass essential oil nor citral to modulate drug resistance has ever been tested. Drug resistance is a major challenge for contemporary medicine. Antibiotic resistance is currently a frequent and typical complication of the treatment of diseases or injury. The first antibiotic, penicillin, was only introduced a few years before bacteria resistant to it were isolated. Similarly, antineoplastic resistance is a major challenge for treatment and for patient survival overall. Numerous generations of compounds overcoming drug resistance in therapy have been described; however, the usage of natural compounds still looks to be the most promising [23].

Although the effects of *C. citratus* extracts and especially its essential oil are studied relatively often, there are many areas that are controversial or deserve deeper exploration. In this paper, we report on the large spectrum of antimicrobial activities resulting from the chemical composition of lemongrass essential oil. Some of the activities have been already previously identified, however, we seek to provide deeper understanding and report some new explanations and context especially in the field of drug resistance modulation.

2. Materials and Methods

2.1. Chemicals

Mueller Hinton Broth 2 (MH broth, Sigma-Aldrich, St. Louis, MO, USA), Malt extract broth (ME broth, Oxoid, Hampshire, UK), Brain Heart Infusion Broth (BHI, Sigma-Aldrich), resazurin (Sigma-Aldrich), methicillin (Sigma-Aldrich), vancomycin (Sigma-Aldrich), citral (Sigma-Aldrich), casamino acids (Sigma-Aldrich), L-arginine (Sigma-Aldrich), Pgp-Glo Assay System (Promega, Madison, WI, USA), Na$_3$VO$_4$ (Promega), Dulbecco's Modified Eagle's Medium (DMEM, Sigma-Aldrich), doxorubicin (Sigma-Aldrich).

2.2. Plant Material and Essential Oil Extraction

Cymbopogon citratus (DC.) Stapf (family: Poaceae) was identified by the Viet Nam Institute of Medicinal Plants under the Ministry of Health (3 Quang Trung Str., Hoan Kiem Distr, Ha Noi, Viet Nam). The planting location was the Phu Thanh, Hung Thi and An Binh Communes, Lac Thuy District, Hoa Binh province, Mountainous North of Viet Nam. The plants were harvested during October and November 2018. The water content of the material before oil extraction was 50%. The water content was determined by weighing before and after drying.

The essential oil was distilled from the above-ground part of the plant by steam distillation under vacuum (0.5 atm, Shanghai EVP Vacuum Technology, China) at a temperature of 80–90 °C for two hours. By this procedure, which was modified according to [24], 1 kg of essential oil was obtained from 200 kg of fresh leaves. The yield of oil extraction was calculated as the percentage of obtained oil weight to fresh leaves weight. Density was determined gravimetrically.

2.3. Essential Oil Composition Screening

For the analysis of volatile and semi–volatile compounds in the lemongrass essential oil, gas chromatography coupled to high-resolution mass spectrometry (GC–HRMS) was employed as previously described by [25].

2.3.1. Sample Preparation

Fifty grams of the lemongrass essential oil was weighted into a 15 mL polypropylene centrifuge tube and 10 mL of ethyl acetate was added. One milliliter of ethyl acetate extract was diluted in 9 mL of isooctane. Finally, 2 µl of the isooctane extract was injected into the GC-HRMS instrumentation (Agilent Technologies, Santa Clara, CA, USA).

2.3.2. GC-HRMS

An Agilent 7200b system consisting of an Agilent 7890B gas chromatograph equipped with a quadrupole–time of flight mass spectrometer (Q–TOF) (Agilent Technologies, Santa Clara, CA, USA) was used for the instrumental analysis.

Sample components were separated in a 30 m HP-5MS capillary column (0.25 mm id, film thickness: 0.25 µm; Agilent Technologies, Santa Clara, CA, USA). The sample was injected in split mode (1:5) at 250 °C and the oven temperature was as follows: 40 °C (1 min), 10 °C/min to 150 °C, 5 °C/min to 190 °C, 20 °C/min to 310 °C (hold 5 min).

The mass spectrometric detector was operated in electron ionization mode and the temperature of the ion source was 230 °C. The mass range was 40–550 *m/z* and the resolution of the mass analyzer was set to >12 500 (FWHM).

Volatile and semi–volatile compounds were identified and verified by NIST library 2017, isotopic pattern, exact mass (mass error <5 ppm) and Kovats retention index.

2.4. Antimicrobial Activity

The inhibitory activity of the essential oil and citral was determined against both Gram-positive and Gram-negative bacterial strains, yeasts and micromycetes. Unless otherwise stated, microorganisms were obtained from the Collection of the Department of Biochemistry and Microbiology (DBM, UCT Prague, Prague, Czech Republic) as follows: *Salmonella enterica* (CCM, 4420), *Proteus vulgaris* (DBM, 3022), *Mycobacterium smegmatis* (ATCC, 70084), *Pseudomonas aeruginosa* (CCM, 3955), *Staphylococcus aureus* (ATCC, 25923), *Candida famata* (DBM, 23), *Candida albicans* (DBM, 2186), *Cryptococcus albidus* (DBM, 4), methicillin-resistant *Staphylococcus aureus* (DBM, 12). The resistant strain was obtained from the Collection of the Laboratory of Medical Microbiology (Czech Laboratory, Inc., NEM 449) and was previously characterized for its multidrug resistance properties. Unless otherwise stated, all strains are sensitive to commercial drug strains according to EUCAST (The European Committee on

Antimicrobial Susceptibility Testing. Breakpoint tables for interpretation of MICs and zone diameters. Version 10.0, 2020).

Antibacterial and anti-yeast activity was evaluated by the standard broth-dilution method using 96-well plates and MH broth or ME broth. Overnight microbial culture was diluted to a turbidity equal to 0.5 McFarland. After that, the standard broth microdilution method recommended by EUCAST (The European Committee on Antimicrobial Susceptibility Testing. Breakpoint tables for interpretation of MICs and zone diameters. Version 10.0, 2020) was used. The essential oil and citral were diluted with the prepared microbial culture. Subsequent binary dilution provided the range of tested concentrations from 5 to 10,000 µL/L. The positive control was the suspension of microorganisms without essential oil. Plates were incubated for 24 h at 37 or 28 °C after that, absorbance (500 nm) was recorded using a SpectraMax i3x Multi-Mode Detection Platform (Molecular Devices, San Jose, CA, USA). The turbidity of each strain was measured in eight replicates. The turbidity of the methicillin-resistant *S. aureus* was measured in eight replicates, both with and without chloramphenicol (0.2–200 mg/L), and both with and without lemongrass essential oil.

2.5. Anti-Biofilm Activity

The effect of lemongrass essential oil and citral on the bacterial biofilm was tested on *Staphylococcus aureus* (ATCC, 25923) and *Pseudomonas aeruginosa* (CCM, 3955). A static anti-biofilm assay was performed in 96-well polystyrene plates, as described previously with some modifications [26]. Both samples were evaluated for their activity in two main stages (i) planktonic cell adhesion and (ii) mature biofilm disruption. The overnight culture of the tested organism was diluted with BHI broth to obtain a turbidity equal to 0.5 McFarland, and 100 µl of the suspension was pipetted into each well. In the cell adhesion assay, the viability of adhered cells was evaluated by resazurin assay immediately after 24 h of incubation in the presence of the tested compounds at 37 °C and triple washing with PBS (pH 7.4). In the disruption of mature biofilm assay, fresh BHI medium containing various concentrations of lemongrass or citral was added to wells with a pre-formed biofilm. After 24 h of incubation, the medium was removed, the wells were washed three times with PBS and 100 µL of resazurin in PBS (0.03 mg/L) was added. The viability was evaluated by measuring fluorescence (560/590 nm, ex./em.) using a SpectraMax i3x Multi-Mode Detection Platform (Molecular Devices, USA). Each experiment was done in 16 repetitions.

2.6. Anti-Quorum Sensing Activity

For the evaluation of anti- quorum sensing activity, two commercial (ATCC) strains of *Vibrio campbellii* were used - BAA1118 and BAA1119 [27]. The bacteria were cultivated in Autoinducer Bioassay (AB-A) medium as described previously by [28]. The overnight culture was diluted with AB-A medium to a cell density of approximately 1×10^4 CFU/mL. The binary dilution of citral and lemongrass oil (0.1 µL/L—6 mL/L) was tested using *Vibrio campbellii* in a 96-well plate and after 24 h, the cell viability was determined by resazurin assay. IC_{10} was chosen for the subsequent experiment. The lemongrass essential oil or citral were applied at the IC_{10} concentration and further binary diluted with diluted cell culture. After that, the luminescence was recorded for 24 h with a measurement step of 20 min using a microplate reader (SpectraMax i3 Multi-Mode Detection Platform, Molecular Devices, San Jose, CA, USA) set to 30 °C; integration time of 10,000 ms; shaking for 60 s prior to measurement. The sum of luminescence was calculated and used for the determination of EC_{50}.

2.7. Inhibition of Transmembrane Efflux Pump

The inhibition of P-glycoprotein (P-gp) was tested using the in vitro Pgp-Glo Assay System according to the standard procedure [29] with only slight modification of time incubation. The incubation was prolonged in our case to 60 min. The luminescence (ΔRLU samples) was calculated as the difference between the relative luminescence of Na_3VO_4 and that of the samples. Orthovanadate was used as a known inhibitor of P-gp, Verapamil was used as a known activator of P-gp.

2.8. Sensitization of Doxorubicin-Resistant Human Ovarian Carcinoma

A human ovarian carcinoma cell line (HOC, A2780) and its doxorubicin-resistant sub-line (HOC/DOX, A2780/ADR) were purchased from Sigma-Aldrich (USA). Both cell lines were cultivated as described previously [29]. For the experiment, 1×10^5 cells/mL were seeded into the 96-well plates. After 24 h, the cells were washed 3× with PBS and fresh DMEM supplemented with lemongrass essential oil (10 µL/L) was added. Doxorubicin in the concentration range of 0.3–80 µM was then applied. The cell viability was evaluated after 72 h by resazurin assay as described previously [30]. The fold change was calculated as the ratio of IC_{50} for doxorubicin and IC_{50} for the doxorubicin co-treated with the lemongrass essential oil.

2.9. Data Processing and Statistical Analysis

The experiment was done with the appropriate number (n) of repetitions. The relative activity was evaluated as a percentage according to the formula: 100*(slope of sample fluorescence—average slope of PC)/(average slope of NC—average slope of PC). As the positive control (PC), non-treated reaction was used. As the negative control (NC), the blank reaction was used. IC_{50} values were determined using the software GraphPad Prism 7 and its function of nonlinear regression (Y = Bottom + (Top-Bottom)/(1 + 10^((LogIC-X)*HillSlope)). MIC was determined as IC_{90} using an online tool freely provided by AAT BioQuest. The data are presented as the averages of the repetitions with the standard error of the mean (SEM). Statistical significance was checked with the Excel t-test function (two-tailed distribution, heteroscedastic type). One-way analysis of variance (ANOVA) was used followed by Duncan's post hoc test ($P < 0.05$) to show the differences between the groups. For ANOVA, the software Statistica version 12 was used (Tibco Software Inc., Palo Alto, CA, USA).

3. Results

The *Cymbopogon citratus* essential oil was obtained by vacuum steam distillation at 80–90 °C. The procedure yielded 0.5% (*w/w*) of the oil with the density 0.891 + 0.004 g/mL.

3.1. Essential Oil Composition Screening

The composition of the essential oil (volatile and semi–volatile compounds) was characterized by GC–HRMS. To characterize the relative abundance of volatile components in the sample, deconvolution of all compounds was performed, and the detected compounds were normalized by percent of total peaks area (Figure 1). The compounds occurring at a level higher than 1% were identified. The most dominant components in the sample were as follows: two stereoisomers of citral (geranial and neral) >> geraniol–myrcene > γ-selinene > terpineol-1–geranyl acetate > β-caryophyllene > linalool > α-humulene > nerol > eucalyptol. The repeatability of the method, expressed as a relative standard deviation (RSD), was determined by the analysis of the oil sample in six repetitions, and ranged from 1.5% to 5.5% in all analyses.

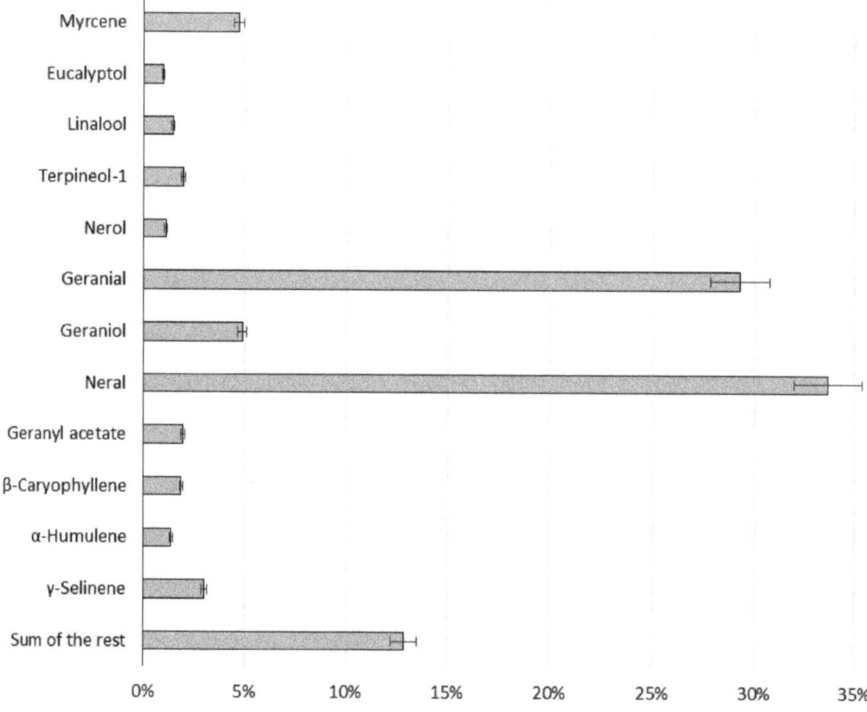

Figure 1. Chemical composition (%) of essential oil determined by GC–HRMS. Data are presented as an average of six repetitions (*n*) ± relative standard deviation (RSD).

3.2. Antimicrobial Activity

The antimicrobial activity of lemongrass essential oil was previously published with a special focus on its strong antifungal activity. In this paper, we summarize the spectrum of its antimicrobial activity, taking into account both Gram-positive and Gram-negative bacteria and yeasts (Table 1). In agreement with previous papers, its antifungal activity was higher than its antibacterial activity. The essential oil was the most active against *C. albidus*, which could cause ocular and systematic diseases in immunosuppressed patients and against other human/animal (*C. famata*, *C. albicans*) fungal pathogens, with an IC_{50} in the concentration range of 180–570 µL/L. This antibacterial activity was shown against both Gram-positive and Gram-negative bacteria. The essential oil was also active against *M. smegmatis*, the model organism for studies of *M. tuberculosis* and other mycobacterial pathogens. Among the tested organisms, the lowest activity was determined against the Gram-negative bacteria *P. aeruginosa* and *S. enterica* with an IC_{50} equal to 2.4–2.6 mL/L and an MIC of almost double that. In comparison to lemongrass essential oil, citral—the dominant oil component, exhibited an IC_{50} that was several times lower with no antimicrobial specificity. In contrast to the oil, it effectively inhibited the growth of both types of bacteria and yeast in the same concentration range. Similar to the oil, citral was the most active against *C. albidus*; however, its IC_{50} was 100 × lower than the IC_{50} of the oil. Similar results were observed for the other tested microorganisms, showing that citral is the main antimicrobial agent of oil and that its activity is antagonistically modulated by the accompanying components.

Table 1. Antimicrobial activity of lemongrass essential oil.

		Lemongrass Essential Oil		Citral	
species	Classification	IC$_{50}$ [μL/L]	MIC [μL/L]	IC$_{50}$ [μL/L]	MIC [μL/L]
Candida famata	Fungi, Ascomycota	177 ± 19 [a]	3684 ± 271 [c,d]	37 ± 7 [b]	142 ± 19 [b,c]
Cryptococcus albidus	Fungi, Basidiomycota	199 ± 25 [a]	265 ± 31 [a]	2 ± 0 [a]	20 ± 6 [a]
Candida albicans	Fungi, Ascomycota	571 ± 109 [a,b]	2734 ± 250 [b,c]	83 ± 8 [d,e]	110 ± 15 [b,c]
Mycobacterium smegmatis	Bacteria, Gram positive	860 ± 89 [b]	3409 ± 775 [c,d]	109 ± 12 [e]	137 ± 19 [b,c]
Proteus vulgaris	Bacteria, Gram negative	992 ± 37 [b]	1453 ± 40 [a,b]	97 ± 12 [e]	163 ± 34 [c]
Staphylococcus aureus	Bacteria, Gram positive	1841 ± 199 [c]	5830 ± 198 [e]	77 ± 2 [e]	92 ± 2 [b,c]
Pseudomonas aeruginosa	Bacteria, Gram negative	2385 ± 162 [d]	5308 ± 339 [e]	41 ± 2 [b]	93 ± 8 [b]
Salmonella enterica	Bacteria, Gram negative	2626 ± 301 [d]	4693 ± 634 [d,e]	66 ± 8 [c,d]	97 ± 3 [b,c]

The data are presented as an average of eight repetitions (n) ± standard error of the mean (SEM). The data were analyzed by one-way ANOVA with Duncan's post hoc test ($P > 0.05$) as indicated by the superscript letters. The letters indicate the differences between the groups within one assay. Statistically significant levels are denoted with different letters. The data are presented as the concentration (μM) that (i) halved the cell viability (IC$_{50}$) or (ii) reduced 90% of cell viability (MIC).

The antimicrobial activity of the essential oil was also tested against antibiotic-resistant bacteria. The clinical isolate of methicillin-resistant *S. aureus* (MRSA) used in this study was characterized as multidrug-resistant to a broad spectrum of antibiotics (data not shown). The IC$_{50}$ concentration of chloramphenicol for MRSA was 142 mg/L. To evaluate the MDR-modulating activity, MRSA was cultivated in the presence of lemongrass essential oil (1000 μL/L) or citral (400 μL/L) and chloramphenicol. Even though the citral content in the oil is 63%, this concentration (630 μL/L) was not applied because of the high toxicity of pure citral to the cells, as shown above for drug-sensitive *S. aureus* (IC$_{50}$ 77 μL/L). Therefore, we chose concentrations of lemongrass essential oil or citral equal to IC$_{10}$, which were applied simultaneously with chloramphenicol to determine the IC$_{50}$ of the mixture (Table 2). The addition of lemongrass oil reduced the IC$_{50}$ of MRSA almost tenfold (to 15 mg/L). The effect of citral on MRSA was almost three times lower, the sensitization of MRSA was only 3-fold.

Table 2. Sensitization of methicillin-resistant *Staphylococcus aureus* (MRSA) by lemongrass essential oil and citral.

	Chloramphenicol IC$_{50}$ [mg/L]	Chloramphenicol IC$_{50}$ [mg/L] Affected by Lemongrass Essential Oil [1000 μL/L]	Chloramphenicol IC$_{50}$ [mg/L] Affected by Citral [400 μL/L]
MRSA	142 ± 10	15 ± 2	42 ± 1
Fold		9.6 ± 1.9	3.4 ± 0.3

The data are presented as an average of eight repetitions (n) ± standard error of the mean (SEM).

3.3. Anti-Biofilm Activity

Biofilm formation plays a crucial role in many medical and industrial applications. Typically, the effect of anti-biofilm compounds can be seen in two main stages: a) adhesion of the planktonic cells to the surface and b) disruption of a maturated biofilm. Both citral and lemongrass essential oil inhibited the adhesion of Gram-positive (*S. aureus*) and Gram-negative (*P. aeruginosa*) bacteria in a dose-dependent manner. In contrast to direct antimicrobial activity, a higher effect was observed for the adhesion of *P. aeruginosa* than for *S. aureus* (Figure 2). In the eradication of mature biofilms, just only citral disrupted bacterial biofilms in concentration-dependent manner with significantly higher effect on gram negative bacteria (*P. aeruginosa*) while lemon grass essential oil was not considered significant (Figure 3). Both citral and lemongrass essential oil disrupted the biofilm in a concentration-dependent manner with a significantly higher effect on Gram-negative bacteria (*P. aeruginosa*).

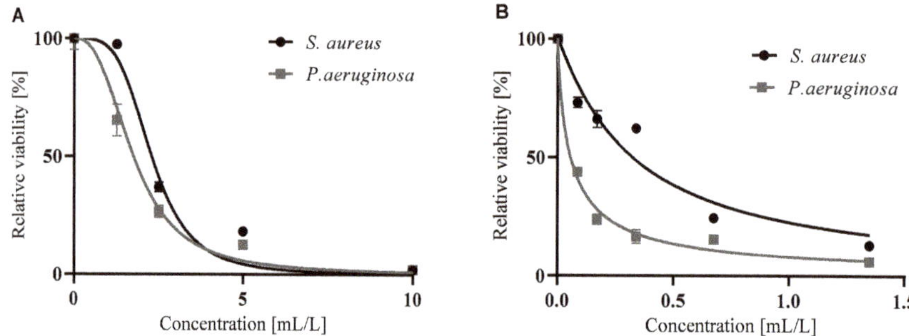

Figure 2. Inhibition of adhesion of bacteria forming a biofilm by lemongrass essential oil (**A**) and citral (**B**). The data are presented as an average of 16 repetitions with SEM.

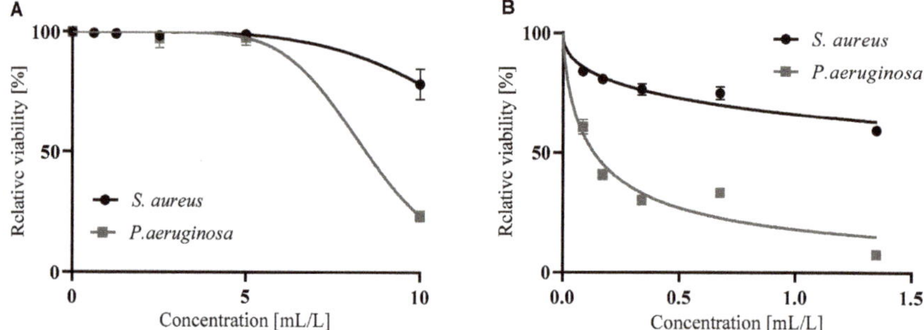

Figure 3. Disruption of maturated biofilm by lemongrass essential oil (**A**) and citral (**B**). The data are presented as an average of 16 repetitions with SEM.

As can be seen from Table 3, which compares the concentrations of citral and lemongrass essential oil that halve the respective activity, citral was more active against both phases of biofilm formation—adhering and maturated cells. As the IC_{50} values of citral were 7–70 × lower than the IC_{50} of lemongrass essential oil, it could be concluded that citral is responsible for the main activity of the essential oil against a biofilm. However, this activity is negated by the antagonistic effect of some other compounds that are also in the oil. Citral is up to 5 × more active against the adhesion phase of *P. aeruginosa* biofilm than against an *S. aureus* biofilm.

Table 3. Concentration of citral and lemongrass essential oil that halves the respective activity: (i) adhesion of bacteria forming biofilm and (ii) mature biofilm.

	Anti-adhesion IC_{50} [mL/L]		Anti-biofilm IC_{50} [mL/L]	
	S. aureus	*P. aeruginosa*	*S. aureus*	*P. aeruginosa*
Citral	0.32 ± 0.03	0.06 ± 0.01	>1.5	0.11 ± 0.01
Lemongrass essential oil	2.16 ± 0.11	1.90 ± 0.20	>10	7.34 ± 0.40

Data are presented as an average of 16 repetitions with SEM.

3.4. Inhibition of Bacterial Cell-To-Cell Communication

The inhibition of bacterial quorum sensing was measured by using the sensor system of *V. campbellii*. First, the direct toxicity of the oil and citral was determined in order to avoid false-positive results. Table 4 shows the concentrations of citral and lemongrass essential oil that halved the bacterial quorum sensing and viability of mutant sensor strains of *V. campbellii*, which responds either only to (i) AI-1 autoinducer (BAA1118) or (ii) AI-2 autoinducer (BAA1119). As can be seen from the table, the concentrations that halve viability [mL/L] and those that halved communication [µL/L] differ by a factor of 1000. The decrease in luminescence was therefore caused by an inhibition of communication rather than inhibition of cell growth (Figure 4). The lemongrass essential oil inhibited communication based on both autoinducer 1 (AI-1) and autoinducer 2 (AI-2) systems similarly to citral. However, in comparison to the oil, citral effectively inhibited both systems, with a slightly higher activity for the first one. Although the oil inhibited the AI-2-mediated communication, which is based on boron compounds and used by many Gram-negative and Gram-positive bacteria, its activity was significantly higher against AI-1, which is based on homoserine lactones used by Gram-negative bacteria.

Table 4. The effect of citral and lemongrass essential oil on the quorum sensing and viability of *Vibrio campbellii*.

	V. campbellii **BAA1118**			*V. campbellii* **BAA1119**		
	Viability [mL/L]		QS IC$_{50}$ [µL/L]	Viability [mL/L]		QS IC$_{50}$ [µL/L]
	IC$_{50}$	IC$_{10}$		IC$_{50}$	IC$_{10}$	
citral	0.24 ± 0.11	0.09 ± 0.05	0.26 ± 0.03	0.99 ± 0.04	0.68 ± 0.08	1.7 ± 0.8
lemongrass essential oil	0.17 ± 0.01	0.05 ± 0.01	1.4 ± 0.03	5.42 ± 0.09	4.07 ± 0.16	259 ± 88

The data are presented as an average of 3 repetitions with SEM.

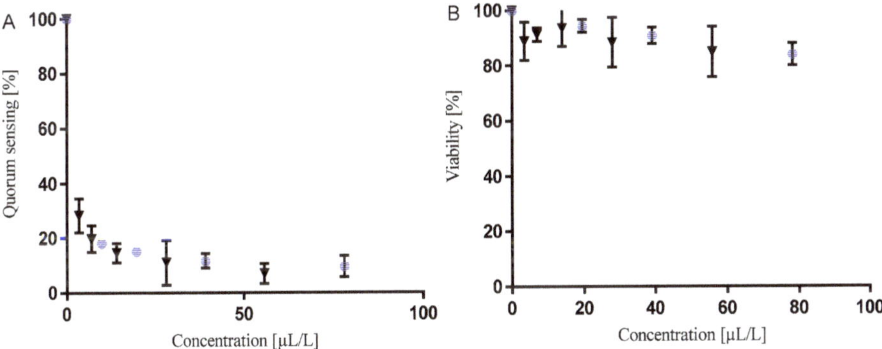

Figure 4. Dose-dependent effect of lemongrass essential oil (blue circle) and citral (black triangle) on *V. campbellii* BAA1118: (**A**) quorum sensing inhibition, (**B**) cell viability. The data are presented as an average of 3 repetitions with SEM.

3.5. Modulation of MDR in Cancer Cells

As was shown with the bacteria, lemongrass essential oil affected the multidrug resistance phenotype of MRSA. One of the main mechanisms involved in both bacteria and cancer drug-resistance is the overexpression of transmembrane efflux pumps, which transport the drug outside the cells and thus decrease its intracellular concentration [23]. Therefore, the activity of both citral and lemongrass essential oil was tested on a fraction of isolated membranes containing human P-glycoprotein (P-gp)—a transmembrane efflux pump responsible for most multidrug resistance phenotypes in tumors. As can be seen from Figure 5, lemongrass essential oil inhibited this pump. The function of P-gp is connected to the ATP consumption; therefore, the better the inhibitor, the lower the amount of consumed ATP. Compared with lemongrass essential oil, citral was unable to inhibit this pump even at higher concentration levels. Rather than being an inhibitor, citral was potentially the substrate of this pump, showing the typical trend, i.e., the higher the concentration, the higher the amount of consumed ATP.

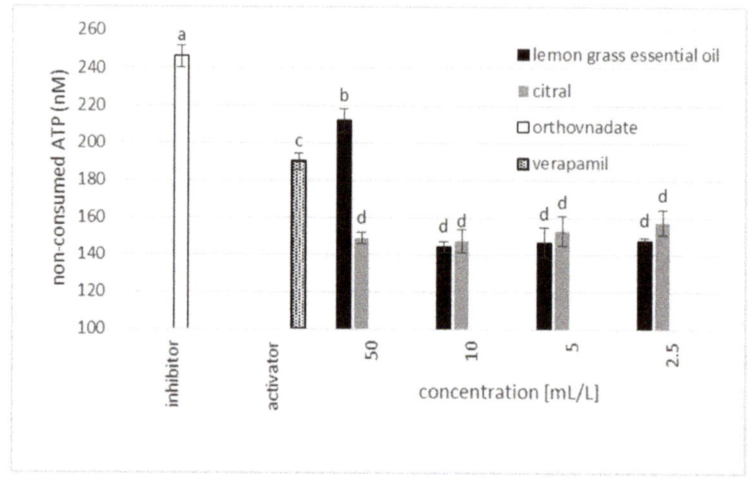

Figure 5. P-glycoprotein activity modulation. Orthovanadate was used as a known inhibitor of P-gp, Verapamil was used as a known activator of P-gp.

P-glycoprotein activity inhibition was determined as the amount of ATP which was not consumed during the efflux pump activation in comparison to the control. Orthovanadate (white columns), a known inhibitor of P-gp, was used as a positive control, verapamil (dotted columns) was used as known activator and lemongrass essential oil (black columns) and citral (gray columns) were tested in the concentration range 2.5–50 mL/L. Data are presented as an average of 3 repetitions (n) with SEM. Data were analyzed by one-way ANOVA with Duncan's post hoc test ($P > 0.05$) as indicated by the superscript letters. Statistically significant levels were denoted with different letters.

The activity of both the oil and citral were tested on an ovarian cancer cell line resistant to doxorubicin. As predicted based on the P-gp inhibition results, citral exhibited no potential to modulate the resistant phenotype of this cell line. In contrast, the co-administration of doxorubicin with the oil at a concentration far below the IC_{50} significantly decreased the cell viability (Figure 6).

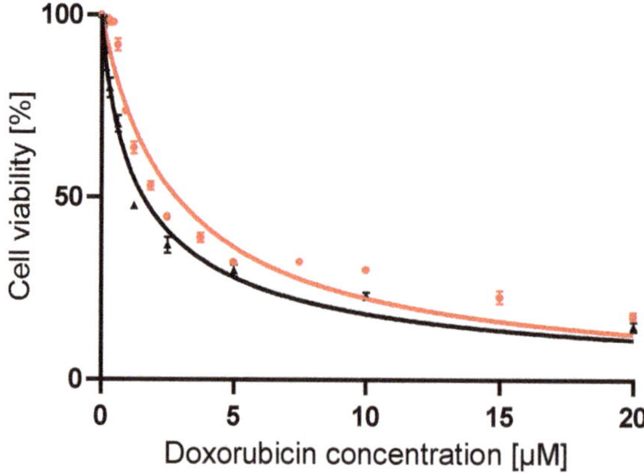

Figure 6. Inhibition of human ovarian carcinoma viability by doxorubicin (red line) and doxorubicin with lemongrass essential oil (10 µL/L, black line) after 72 h.

The sensitization of doxorubicin-resistant ovarian carcinoma was determined as the ratio of doxorubicin concentration needed to halve the cell viability with and without lemongrass essential oil

(Table 5). The resistant cell line treated with lemongrass essential oil was almost twice as sensitive as without oil, with the p value of the t-test being 0.022. As can be seen from the table, the resistant cell line is more than 100× more resistant than the sensitive line. The chosen dose of lemongrass was non-toxic even for the sensitive cancer cell line (note: viability 86% ± 4%); therefore, the sensitization effect was not caused by a direct cytotoxicity of the oil. Citral at the same concentration level (10 µL/L) did not modulate the resistant phenotype and its presence did not affect the values of doxorubicin's IC_{50}.

Table 5. Modulation of resistance to doxorubicin by lemongrass essential oil. The table shows the concentrations of doxorubicin that halved the viability of human ovarian carcinoma (HOC) cell lines in the presence of lemongrass essential oil and doxorubicin with a single-dose addition of lemongrass essential oil (10 µL/L). Both HOC and the same line resistant to doxorubicin (HOC/DOX) were tested.

	HOC	HOC/DOX
Doxorubicin, IC_{50} [µM]	0.022 ± 0.001	2.86 ± 0.18
Lemongrass, IC_{50} [µL/L]	55.2 ± 8.1	197.8 ± 5.7
Doxorubicin, IC_{50} [µM] with lemongrass essential oil [10 µL/L]		1.60 ± 0.14
Sensitization FOLD		1.78 ± 0.27

The sensitization factor was calculated as the ratio of IC_{50} [doxorubicin, µM] and IC_{50} [doxorubicin in the presence of lemongrass essential oil, µM]. The data are presented as an average of 3 repetitions with SE.

4. Discussion

Although the promising biologic activities of both lemongrass essential oil and citral have been known for decades, their modulation of the drug-resistant phenotype has never been tested in as much depth as by this paper. Here we report on the modulation of both bacterial and cancer drug resistance by lemongrass essential oil; however, based on our results, this activity is not caused by citral. The activity of *Cymbopogon flexuosus* essential oil against multi-drug resistant bacteria was previously studied using *Acinetobacter baumannii* strains [25]. The mean value of MIC was about 0.65% (*v/v*). Our IC_{50} value for MRSA (0.4%, *v/v*) agrees with published data. The sensitization factor of methicillin-resistant *S. aureus* (MRSA) by lemongrass essential oil was several times higher than the factor for citral. Therefore, the oil is more promising for modulating a bacterial drug-resistance phenotype. In addition, Berdejo et al. [31] reported on an enhanced resistance in *S. aureus* after sub-inhibitory doses of citral exposure. This was in agreement with the study of Chueca et al. [32], who published that the exposure of a hyper-resistant strain of *E. coli* to a sub-inhibitory concentration of citral did not increase the direct resistance globally; however, it led to the emergence of several mutants displaying an increased minimum inhibitory concentration of citral. Therefore, the application of oil as a natural component mixture is in agreement with the current trend of drug resistance treatment based on a combination of biologically active compounds with different targets.

Both lemongrass essential oil and citral possess strong antimicrobial activity, even against antibiotic-resistant strains; therefore, their activity against a biofilm was evaluated within this paper as well. The disruption of a matured biofilm was only observed for Gram-negative bacteria treated with either citral or with the oil. The application of 0.06 mL/L of citral eliminated half of the *P. aeruginosa* cells adhesion, which is in agreement with the results of Espina et al., who demonstrated that the biofilm formation of resistant *S. aureus* was significantly inhibited by 0.2 mL/L of citral [33]. Lemongrass essential oil was previously used for an oral spray preparation intended to inhibit *Streptococcus* sp. biofilm formation [34]. Similarly, both oil and citral prevented the biofilm formation of sulfate-reducing bacteria [12]. Both papers are in agreement with our results, which show that relatively low citral doses of 61 and 323 ppm inhibit the adhesion of both Gram-negative and positive bacteria by 50%, respectively. Lemongrass essential oil inhibited the adhesion of both cell types as well, however at several times higher concentrations. In another study, 0.2 µg/mL of citral significantly inhibited mixed biofilm (*S. aureus* and *S. enterica*) formation, which could be caused by decreasing AI-2-mediated quorum sensing communication [35]. The inhibition of AI-2 production by citral treatment was demonstrated by [36]. Our results also proved that citral significantly inhibited both biofilm formation and bacterial communication via AI-2; however, the inhibition of quorum sensing based on AI-1 was more effective and to our knowledge, this is the first report on such activity.

The potential of citral to interact with transmembrane efflux pumps has not been satisfactorily explained yet. It was previously shown that citral acts neither as a P-gp inhibitor (ABCB1) nor as a multidrug resistance protein 1 and 2 (MRP1, ABCC1; MRP2, ABCC2) inhibitor [37–39]. Moreover, citral should not affect ATPase activity in any of these pumps [38]. On the other hand, Queiroz et al. [40] observed that citral downmodulated the activity and inhibited the expression of multidrug resistance associated protein 1 (MRP1). In contrast to lemongrass essential oil, citral up to a concentration of 50 mL/L was unable to inhibit pure P-gp in the in vitro system and did not affect the DOX sensitivity of doxorubicin-resistant ovarian carcinoma over 72 h of incubation. The resistance of the cell line is caused by P-gp overexpression [29]. Therefore, the 10-μL/L dose of citral should not affect the expression level of P-gp, as the cells preserved the drug-resistant phenotype. As in bacteria, the lemongrass essential oil, being a natural mixture of biologically active compounds, is more promising in overcoming drug resistance in eukaryotic cells.

The antimicrobial activity of both lemongrass essential oil and citral have been reported many times before, but here we compared the activity of the same essential oil and pure citral against prokaryotic and eukaryotic organisms, i.e., Gram-positive, Gram-negative bacteria and yeasts, respectively. Despite the similarity of some of our results with published data, we found some differences, e.g., in the minimal oil concentration inhibiting *C. albicans* (MIC), which was published as 288 mg of oil per liter [41]. Our determination of the MIC for *C. albicans* gave 2734 μL/L, which corresponds to an MIC of 2435 mg of oil per liter (recalculated based on the gravimetrically defined density of our oil as 0.891 $\pm$ 0.004 g/mL). The difference could be explained by the different sensitivity of yeast strains. The same reasons may apply for differences between our MIC of *C. albicans*, which about 14 times higher than that obtained by Sacchetti et al. [42].

The antimicrobial activity of citral was previously measured for *E. coli*. The determined value ($\geq$0.01%) [43] corresponds to our observation for Gram-negative bacteria (41–97 μL/L). Similar to our results, Gupta et al. also demonstrated that citral has a stronger antimicrobial activity than lemongrass essential oil [44]. Citral was up to 10 times more active against some *Acinetobacter baumannii* strains than lemongrass essential oil in the study of Adukwu et al. [25].

The quality of the oil is generally assessed by its content of citral, high oxygenated monoterpenes, low monoterpene and sesquiterpene hydrocarbons, and finally low oxygenated sesquiterpenes. Our lemongrass essential oil had a standard composition comparable to previously published data. The yield of essential oil obtained by our distillation of fresh leaves was 0.5% (*w/w*), which corresponds to the data of other authors, who published yields of 0.6% [45] or 0.7% [46–48]. As expected, the dominant group of terpenes present in essential oil was comprised of oxygenated monoterpenes with the main components being neral (citral b, 33.7%) and geranial (citral a, 29.3%). The amounts of citral isomers differ in the literature, typically with a higher content of neral. The common contents are 36.2% [41], 39.5% [48], 41.3% [42], 41.8% [47], 42.2% [45], 50.5% [25], 52.9% [34], up to 56.8% [49] for citral a and 26.5% [41], 30.4% [47], 32.3% [42], 32.5% [45], 33.1 [49], 35.5% [48], 38.1% [34], up to 38.5% [25] for neral. The presence of 1.5% linalool is comparable with the data published by other authors—0.4% [49], 1.3% [42], 2.8% [25]. The level of geraniol was also similar, which we quantified as 4.9% and other research groups as 2.2%–8.2% [25,48].

5. Conclusions

Although the antimicrobial activity of citral is to some extent negated by the accompanying components of lemongrass essential oil, these components have promising activity against antibiotic-resistant bacteria and chemotherapeutic-resistant tumors. Citral preserves its strong antimicrobial activity, resulting in direct bacterial cytotoxicity, the inhibition of quorum sensing and influence on cell adhesion during biofilm formation as well as disruption of a matured biofilm. Further fractionation searching for bioactive compounds modulating the drug-resistance phenotype is a promising therapeutic approach.

Author Contributions: Conceptualization, J.V., M.S. and T.R.; methodology, J.V., M.S. and T.R.; software, J.V. and M.S.; validation, J.V., M.S., K.Ř., S.D. and L.H.; formal analysis, J.V., M.S. and T.R.; investigation, T.V.T., L.V.T., N.V.T., J.H. and T.R.; resources, T.V.T., L.V.T., N.V.T., J.H. and T.R.; data curation, J.V., M.S., K.Ř., S.D. and L.H.; writing—original draft preparation, J.V., T.R.; writing—review and editing, J.V., T.R., M.S. and J.H.; visualization, J.V. and M.S.; supervision, T.R. and J.H.; project administration, T.R. and J.H.; funding acquisition, T.R. and J.H. All authors have read and agreed to the published version of the manuscript.

Funding: This research was funded by the Czech Science Foundation, No 18-00150S; (ii) the European Union Horizon 2020 research and innovation Program, No 692195 ("MultiCoop"); (iii) mobility project from the Czech Ministry of Education, Youth and Sports INTER-COST LTC19007 (COST Action CA17104 STRATAGEM); (iv) the METROFOOD-CZ research infrastructure project (MEYS Grant No: LM2018100) including access to its facilities and (v) the Czech National Program of Sustainability, No LO1601.

Acknowledgments: The authors thank Ben Watson-Jones, MEng, for providing language corrections.

Conflicts of Interest: The authors declare no conflict of interest.

References

1. Ekpenyong, C.E.; Akpan, E.E. Use of *Cymbopogon citratus* essential oil in food preservation: Recent advances and future perspectives. *Crit. Rev. Food Sci. Nutr.* **2017**, *57*, 2541–2559. [CrossRef] [PubMed]

2. Ekpenyong, C.E.; Akpan, E.; Nyoh, A. Ethnopharmacology, phytochemistry, and biological activities of *Cymbopogon citratus* (dc.) stapf extracts. *Chin. J. Nat. Med.* **2015**, *13*, 321–337. [CrossRef]

3. Ganjewala, D. *Cymbopogon* essential oils: Chemical compositions and bioactivities. *Int. J. Essent. Oil Ther.* **2009**, *3*, 56–65.

4. Avoseh, O.; Oyedeji, O.; Rungqu, P.; Nkeh-Chungag, B.; Oyedeji, A. *Cymbopogon* species; ethnopharmacology, phytochemistry and the pharmacological importance. *Molecules* **2015**, *20*, 7438–7453. [CrossRef] [PubMed]

5. Bieski, I.G.; Leonti, M.; Arnason, J.T.; Ferrier, J.; Rapinski, M.; Violante, I.M.; Balogun, S.O.; Pereira, J.F.; Figueiredo Rde, C.; Lopes, C.R.; et al. Ethnobotanical study of medicinal plants by population of valley of juruena region, legal amazon, mato grosso, brazil. *J. Ethnopharmacol.* **2015**, *173*, 383–423. [CrossRef] [PubMed]

6. De Santana, B.F.; Voeks, R.A.; Funch, L.S. Ethnomedicinal survey of a maroon community in brazil's atlantic tropical forest. *J. Ethnopharmacol.* **2016**, *181*, 37–49. [CrossRef]

7. Fongnzossie, E.F.; Tize, Z.; Fogang Nde, P.J.; Nyangono Biyegue, C.F.; Bouelet Ntsama, I.S.; Dibong, S.D.; Nkongmeneck, B.A. Ethnobotany and pharmacognostic perspective of plant species used as traditional cosmetics and cosmeceuticals among the gbaya ethnic group in eastern cameroon. *S. Afr. J. Bot.* **2017**, *112*, 29–39. [CrossRef]

8. Ochwang'i, D.O.; Kimwele, C.N.; Oduma, J.A.; Gathumbi, P.K.; Mbaria, J.M.; Kiama, S.G. Medicinal plants used in treatment and management of cancer in kakamega county, kenya. *J. Ethnopharmacol.* **2014**, *151*, 1040–1055. [CrossRef]

9. Rehecho, S.; Uriarte-Pueyo, I.; Calvo, J.; Vivas, L.A.; Calvo, M.I. Ethnopharmacological survey of medicinal plants in nor-yauyos, a part of the landscape reserve nor-yauyos-cochas, peru. *J. Ethnopharmacol.* **2011**, *133*, 75–85. [CrossRef]

10. Shi, C.; Sun, Y.; Liu, Z.; Guo, D.; Sun, H.; Sun, Z.; Chen, S.; Zhang, W.; Wen, Q.; Peng, X.; et al. Inhibition of *Cronobacter sakazakii* virulence factors by citral. *Sci. Rep.* **2017**, *7*, 43243. [CrossRef]

11. Souza, P.M.; Goulart, F.R.V.; Marques, J.M.; Bizzo, H.R.; Blank, A.F.; Groposo, C.; Sousa, M.P.; Volaro, V.; Alviano, C.S.; Moreno, D.S.A.; et al. Growth inhibition of sulfate-reducing bacteria in produced water from the petroleum industry using essential oils. *Molecules* **2017**, *22*, 648. [CrossRef] [PubMed]

12. Korenblum, E.; Regina de Vasconcelos Goulart, F.; de Almeida Rodrigues, I.; Abreu, F.; Lins, U.; Alves, P.B.; Blank, A.F.; Valoni, E.; Sebastian, G.V.; Alviano, D.S.; et al. Antimicrobial action and anti-corrosion effect against sulfate reducing bacteria by lemongrass (*Cymbopogon citratus*) essential oil and its major component, the citral. *AMB Express* **2013**, *3*, 1–8. [CrossRef] [PubMed]

13. Madeira, P.L.B.; Carvalho, L.T.; Paschoal, M.A.B.; de Sousa, E.M.; Moffa, E.B.; da Silva, M.A.D.S.; Tavarez, R.D.J.R.; Gonçalves, L.M. In vitro effects of lemongrass extract on *Candida albicans* biofilms, human cells viability, and denture surface. *Front. Cell. Infect. Microbiol.* **2016**, *6*, 71. [CrossRef] [PubMed]

14. Khan, M.S.; Ahmad, I. Biofilm inhibition by cymbopogon citratus and syzygium aromaticum essential oils in the strains of candida albicans. *J. Ethnopharmacol.* **2012**, *140*, 416–423. [CrossRef]

15. Kumari, P.; Mishra, R.; Arora, N.; Chatrath, A.; Gangwar, R.; Roy, P.; Prasad, R. Antifungal and anti-biofilm activity of essential oil active components against *Cryptococcus neoformans* and *Cryptococcus laurentii*. *Front. Microbiol.* **2017**, *8*, 2161. [CrossRef]

16. Singh, P.; Shukla, R.; Kumar, A.; Prakash, B.; Singh, S.; Dubey, N.K. Effect of *Citrus reticulata* and *Cymbopogon citratus* essential oils on *Aspergillus flavus* growth and aflatoxin production on *Asparagus racemosus*. *Mycopathologia* **2010**, *170*, 195–202. [CrossRef]

17. Chukwuocha, U.M.; Fernandez-Rivera, O.; Legorreta-Herrera, M. Exploring the antimalarial potential of whole *Cymbopogon citratus* plant therapy. *J. Ethnopharmacol.* **2016**, *193*, 517–523. [CrossRef]

18. Santos Serafim Machado, M.; Ferreira Silva, H.B.; Rios, R.; Pires de Oliveira, A.; Vilany Queiroz Carneiro, N.; Santos Costa, R.; Santos Alves, W.; Meneses Souza, F.L.; da Silva Velozo, E.; Alves de Souza, S.; et al. The anti-allergic activity of *Cymbopogon citratus* is mediated via inhibition of nuclear factor kappa b (nf-kappab) activation. *BMC Complement. Altern. Med.* **2015**, *15*, 168. [CrossRef]

19. Akono Ntonga, P.; Baldovini, N.; Mouray, E.; Mambu, L.; Belong, P.; Grellier, P. Activity of *Ocimum basilicum*, *Ocimum canum*, and *Cymbopogon citratus* essential oils against *Plasmodium falciparum* and mature-stage larvae of anopheles funestus s.S. *Parasite* **2014**, *21*, 33. [CrossRef]

20. Chauhan, N.; Malik, A.; Sharma, S.; Dhiman, R.C. Larvicidal potential of essential oils against *Musca domestica* and *Anopheles stephensi*. *Parasitol. Res.* **2016**, *115*, 2223–2231. [CrossRef]

21. Soonwera, M.; Phasomkusolsil, S. Effect of *Cymbopogon citratus* (lemongrass) and *Syzygium aromaticum* (clove) oils on the morphology and mortality of *Aedes aegypti* and *Anopheles dirus* larvae. *Parasitol. Res.* **2016**, *115*, 1691–1703. [CrossRef] [PubMed]

22. Baldacchino, F.; Tramut, C.; Salem, A.; Liénard, E.; Delétré, E.; Franc, M.; Martin, T.; Duvallet, G.; Jay-Robert, P. The repellency of lemongrass oil against stable flies, tested using video tracking. *Parasite* **2013**, *20*, 21. [CrossRef] [PubMed]

23. Chambers, C.S.; Viktorova, J.; Rehorova, K.; Biedermann, D.; Turkova, L.; Macek, T.; Kren, V.; Valentova, K. Defying multidrug resistance! Modulation of related transporters by flavonoids and flavonolignans. *J. Agric. Food Chem.* **2020**, *68*, 1763–1779. [CrossRef] [PubMed]

24. Mirghani, M.; Liyana, Y.; Jamal, P. Bioactivity analysis of lemongrass (*Cymbopogan citratus*) essential oil. *Int. Food Res. J.* **2012**, *19*, 569–575.

25. Adukwu, E.C.; Bowles, M.; Edwards-Jones, V.; Bone, H. Antimicrobial activity, cytotoxicity and chemical analysis of lemongrass essential oil (*Cymbopogon flexuosus*) and pure citral. *Appl. Microbiol. Biotechnol.* **2016**, *100*, 9619–9627. [CrossRef]

26. Haney, E.F.; Trimble, M.J.; Cheng, J.T.; Valle, Q.; Hancock, R.E.W. Critical assessment of methods to quantify biofilm growth and evaluate antibiofilm activity of host defence peptides. *Biomolecules* **2018**, *8*. [CrossRef]

27. Bassler, B.; Greenberg, E.; Stevens, A. Cross-species induction of luminescence in the quorum-sensing bacterium *Vibrio harveyi*. *J. Bacteriol.* **1997**, *179*, 4043–4045. [CrossRef]

28. Bezek, K.; Kurinčič, M.; Knauder, E.; Klančnik, A.; Raspor, P.; Bucar, F.; Smole Možina, S. Attenuation of adhesion, biofilm formation and quorum sensing of *Campylobacter jejuni* by *Euodia ruticarpa*. *Phytother. Res.* **2016**, *30*, 1527–1532. [CrossRef]

29. Viktorova, J.; Dobiasova, S.; Rehorova, K.; Biedermann, D.; Kanova, K.; Seborova, K.; Vaclavikova, R.; Valentova, K.; Ruml, T.; Kren, V.; et al. Antioxidant, anti-inflammatory, and multidrug resistance modulation activity of silychristin derivatives. *Antioxidants* **2019**, *8*. [CrossRef]

30. Riss, T.L.; Moravec, R.A.; Niles, A.L.; Duellman, S.; Benink, H.A.; Worzella, T.J.; Minor, L. *Assay Guidance Manual*; Sittampalam, G.S., Brimacombe, K., Grossman, A., Arkin, M., Auld, D., Austin, C.P., Baell, J., Bejcek, B., Caaveiro, J.M.M., Chung, T.D.Y., et al., Eds.; Eli Lilly & Company and the National Center for Advancing Translational Sciences: Bethesda, MD, USA, 2013.

31. Berdejo, D.; Chueca, B.; Pagán, E.; Renzoni, A.; Kelley, W.L.; Pagán, R.; Garcia-Gonzalo, D. Sub-inhibitory doses of individual constituents of essential oils can select for *Staphylococcus aureus* resistant mutants. *Molecules* **2019**, *24*, 170. [CrossRef]

32. Chueca, B.; Berdejo, D.; Gomes-Neto, N.J.; Pagan, R.; Garcia-Gonzalo, D. Emergence of hyper-resistant *Escherichia coli* mg1655 derivative strains after applying sub-inhibitory doses of individual constituents of essential oils. *Front. Microbiol.* **2016**, *7*, 273. [CrossRef] [PubMed]

33. Espina, L.; Pagán, R.; López, D.; García-Gonzalo, D. Individual constituents from essential oils inhibit biofilm mass production by multi-drug resistant *Staphylococcus aureus*. *Molecules* **2015**, *20*, 11357–11372. [CrossRef] [PubMed]

34. Mitrakul, K.; Srisatjaluk, R.; Srisukh, V.; Lomarat, P.; Vongsawan, K.; Kosanwat, T. *Cymbopogon citratus* (lemongrass oil) oral sprays as inhibitors of mutans *Streptococci* biofilm formation. *J. Clin. Diagn. Res.* **2018**, *12*, 6–12. [CrossRef]

35. Zhang, H.M.; Zhou, W.Y.; Zhang, W.Y.; Yang, A.L.; Liu, Y.L.; Jiang, Y.; Huang, S.S.; Su, J.Y. Inhibitory effects of citral, cinnamaldehyde, and tea polyphenols on mixed biofilm formation by foodborne *Staphylococcus aureus* and *Salmonella enteritidis*. *J. Food Prot.* **2014**, *77*, 927–933. [CrossRef]

36. Sun, Y.; Guo, D.; Hua, Z.; Sun, H.; Zheng, Z.; Xia, X.; Shi, C. Attenuation of multiple *Vibrio parahaemolyticus* virulence factors by citral. *Front. Microbiol.* **2019**, *10*, 894. [CrossRef]

37. Zhang, W.; Lim, L.-Y. Effects of spice constituents on P-glycoprotein-mediated transport and cyp3a4-mediated metabolism *in vitro*. *Drug Metab. Dispos.* **2008**, *36*, 1283–1290. [CrossRef]

38. Nabekura, T.; Yamaki, T.; Kitagawa, S. Effects of chemopreventive citrus phytochemicals on human P-glycoprotein and multidrug resistance protein 1. *Eur. J. Pharmacol.* **2008**, *600*, 45–49. [CrossRef]

39. Wortelboer, H.M.; Usta, M.; van Zanden, J.J.; van Bladeren, P.J.; Rietjens, I.M.; Cnubben, N.H. Inhibition of multidrug resistance proteins mrp1 and mrp2 by a series of alpha,beta-unsaturated carbonyl compounds. *Biochem. Pharmacol.* **2005**, *69*, 1879–1890. [CrossRef] [PubMed]

40. Queiroz, R.M.; Takiya, C.M.; Guimaraes, L.P.; Rocha Gda, G.; Alviano, D.S.; Blank, A.F.; Alviano, C.S.; Gattass, C.R. Apoptosis-inducing effects of *Melissa officinalis* l. Essential oil in glioblastoma multiforme cells. *Cancer Investig.* **2014**, *32*, 226–235. [CrossRef]

41. Tyagi, A.K.; Malik, A. Liquid and vapour-phase antifungal activities of selected essential oils against *Candida albicans*: Microscopic observations and chemical characterization of *Cymbopogon citratus*. *BMC Complement. Altern. Med.* **2010**, *10*, 65. [CrossRef]

42. Sacchetti, G.; Maietti, S.; Muzzoli, M.; Scaglianti, M.; Manfredini, S.; Radice, M.; Bruni, R. Comparative evaluation of 11 essential oils of different origin as functional antioxidants, antiradicals and antimicrobials in foods. *Food Chem.* **2005**, *91*, 621–632. [CrossRef]

43. Onawunmi, G. Evaluation of the antimicrobial activity of citral. *Lett. Appl. Microbiol.* **2008**, *9*, 105–108. [CrossRef]

44. Gupta, A. A study on antimicrobial activities of essential oils of different cultivars of lemongrass (*Cymbopogon flexuosus*). *Pharm. Sci.* **2016**, *22*, 164–169. [CrossRef]

45. Boukhatem, M.N.; Ferhat, M.A.; Kameli, A.; Saidi, F.; Kebir, H.T. Lemon grass (*Cymbopogon citratus*) essential oil as a potent anti-inflammatory and antifungal drugs. *Libyan J. Med.* **2014**, *9*. [CrossRef] [PubMed]

46. Mohamed Hanaa, A.R.; Sallam, Y.I.; El-Leithy, A.S.; Aly, S.E. Lemongrass (*Cymbopogon citratus*) essential oil as affected by drying methods. *Ann. Agric. Sci.* **2012**, *57*, 113–116. [CrossRef]

47. Saraswathi, V.; Thara Saraswathi, K.J. Evaluation of polar and non-polar fractions of essential oil from *Cymbopogon citratus* (dc.) stapf. *Int. J. Green Herb. Chem.* **2013**, *2*, 923–929.

48. Kpoviessi, S.; Bero, J.; Agbani, P.; Gbaguidi, F.; Kpadonou-Kpoviessi, B.; Sinsin, B.; Accrombessi, G.; Frederich, M.; Moudachirou, M.; Quetin-Leclercq, J. Chemical composition, cytotoxicity and in vitro antitrypanosomal and antiplasmodial activity of the essential oils of four *Cymbopogon* species from benin. *J. Ethnopharmacol.* **2014**, *151*, 652–659. [CrossRef]

49. Shaikh, M.; Suryawanshi, Y.; Mokat, D. Volatile profiling and essential oil yield of *Cymbopogon citratus* (dc.) stapf treated with rhizosphere fungi and some important fertilizers. *J. Essent. Oil Bear. Plants* **2019**, *22*, 1–7. [CrossRef]

Article

The Antioxidant and Anti-Inflammatory Properties of Rice Bran Phenolic Extracts

Nancy Saji [1,2], Nidhish Francis [1,3], Lachlan J. Schwarz [1,4], Christopher L. Blanchard [1,2] and Abishek B. Santhakumar [1,2,*]

[1] Australian Research Council (ARC) Industrial Transformation Training Centre (ITTC) for Functional Grains, Graham Centre for Agricultural Innovation, Charles Sturt University, Wagga Wagga, NSW 2650, Australia; nsaji@csu.edu.au (N.S.); nfrancis@csu.edu.au (N.F.); lschwarz@csu.edu.au (L.J.S.); CBlanchard@csu.edu.au (C.L.B.)
[2] School of Biomedical Sciences, Charles Sturt University, Locked Bag 588, Wagga Wagga, NSW 2678, Australia
[3] School of Animal and Veterinary Sciences, Charles Sturt University, Locked Bag 588, Wagga Wagga, NSW 2678, Australia
[4] School of Agricultural and Wine Sciences, Charles Sturt University, Locked Bag 588, Wagga Wagga, NSW 2678, Australia
* Correspondence: asanthakumar@csu.edu.au; Tel.: +61-2-6933-2678

Received: 22 May 2020; Accepted: 22 June 2020; Published: 24 June 2020

Abstract: Oxidative stress and inflammation are known to be linked to the development of chronic inflammatory conditions, such as type 2 diabetes and cardiovascular disease. Dietary polyphenols have been demonstrated to contain potent bioactivity against specific inflammatory pathways. Rice bran (RB), a by-product generated during the rice milling process, is normally used in animal feed or discarded due to its rancidity. However, RB is known to be abundant in bioactive polyphenols including phenolic acids. This study investigates the antioxidant and anti-inflammatory effects of RB phenolic extracts (25, 50, 100, and 250 µg/mL) on RAW264.7 mouse macrophage cells stimulated with hydrogen peroxide and lipopolysaccharide. Biomarkers of oxidative stress and inflammation such as malondialdehyde (MDA), intracellular reactive oxygen species, nitric oxide and pro-inflammatory cytokines such as interleukin-6 (IL-6), monocyte chemoattractant protein 1 (MCP-1), interleukin-10 (IL-10), tumor necrosis factor-α (TNF-α), interleukin-12, p70 (IL-12p70), and interferon-γ (IFN-γ) were measured in vitro. Treatment with RB extracts significantly decreased the production of MDA, intracellular reactive oxygen species, nitric oxide and pro-inflammatory cytokines (IL-6, IL-12p70, and IFN-γ) when compared to the control. It is proposed that RB phenolic extracts, via their metal chelating properties and free radical scavenging activity, target pathways of oxidative stress and inflammation resulting in the alleviation of vascular inflammatory mediators.

Keywords: rice bran; polyphenols; oxidative stress; antioxidant; inflammation; anti-inflammatory

---

1. Introduction

Oxidative stress is generated due to an imbalance between the endogenous antioxidant systems in our body and free radicals such as reactive oxygen species [1]. This consequently results in chronic inflammation leading to cardiovascular complications [2]. The key role of macrophages in the innate immune system is to engulf foreign agents, eliminate apoptotic cells and resolve inflammation via inflammatory cytokines and other mediators [3]. During an oxidative stress environment or a chronic inflammatory state, malondialdehyde (MDA), intracellular reactive oxygen species, nitric oxide, pro-inflammatory cytokines such as interleukin-6 (IL-6), monocyte chemoattractant protein 1 (MCP-1), interleukin-10 (IL-10), tumor necrosis factor-α (TNF-α), interleukin-12, p70 (IL-12p70), and interferon-γ (IFN-γ) are known to be secreted by activated macrophages [4,5].

MDA is generated as a result of the peroxidation between polyunsaturated fatty acids [6], as lipids are susceptible to oxidation due to their molecular structure which is abundant in reactive double bonds [7]. MDA is known to interlink with proteins (e.g., lysine to generate lysine–lysine cross-links) and subsequently result in interactions between oxidized low-density lipoprotein and macrophages, to promote the pathogenesis of cardiovascular disease [8]. Generally, cellular antioxidants such as catalase, superoxide dismutase and glutathione peroxidase are known to maintain the homeostatic levels of reactive oxygen species in the vasculature. However, the excess production of reactive oxygen species as a result of oxidative stress is recognized to result in the impairment of cell membrane, proteins, and nucleic acids leading to a series of vascular restructures resulting in atherosclerosis [1]. Nitric oxide, a gaseous lipophilic free radical cellular messenger, is generally known to protect against cardiovascular diseases by regulating blood pressure, impeding platelet aggregation and inhibiting smooth muscle cell proliferation. However, under oxidative stress conditions, nitric oxide is recognized to change from being a protective agent to being an infectious agent [9]. Moreover, during inflammation, activated macrophages are recognized to secrete several pro-inflammatory cytokines to further up-regulate the inflammatory responses. For example, IL-6 is a key cytokine involved in the synthesis and secretion of C-reactive protein which is one of the risk factors for the development of cardiovascular disease [10]. MCP-1 is involved in atherosclerotic plaque formation as it allows monocytes to pass from the lumen to the sub-endothelial space where they become foam cells, initiating fatty streak formation [11]. IL-10 is known to have a cardioprotective role as it is recognized to inhibit the adhesion of monocytes to endothelial cells and result in the destabilization of atherosclerotic plaque [12]. Similarly, TNF is normally recognized as a modulator of both cardiac contractility and peripheral resistance, but increased levels of TNF are known to result in cardiovascular complications [13]. In stimulated mouse macrophages, the increased production of IL-12p70 has been previously observed as a result of protein kinase C activation, which is known to control multiple physiological processes in the heart [14,15]. IFN-γ is a macrophage-activating factor, vital for both innate and adaptive immunity. It is produced by T cells and is known to initiate the generation of reactive oxygen species, and is therefore expressed at high levels during cardiovascular events such as atherosclerosis [16]. Therefore, the regulation of these inflammatory responses, by modulating the production of pro-inflammatory and oxidant markers has been the key metabolic target towards the treatment of vascular disorders.

Several synthetic drugs are currently available in the treatment of acute inflammatory states, however, their long-term use has been associated with gastrointestinal, cardiac and renal complications [17]. Moreover, a rise in health-conscious consumers has resulted in the search for functional foods with nutritional and disease-preventive properties [18]. Natural plant-derived phytochemicals, such as polyphenols and phenolic acids, have been demonstrated to target pathways of inflammation and oxidative stress [18]. Furthermore, it is believed that polyphenols, via their free radical scavenging, metal-chelating properties and blunting cellular signalling pathways, modulate the risk factors associated with chronic ailments including cardiovascular disease and cancer [19]. Phenolic compounds are known to exert anti-inflammatory activity through the regulation of cellular activities in inflammatory cells, modulating enzymes associated with the arachidonic acid metabolism and by blunting the release of pro-inflammatory molecules [19].

Rice is the seed of the grass species *Oryza sativa*, and has been cultivated around the globe for centuries and is considered a staple food. During rice milling, the rice bran (RB) layer is removed due to its rancidity associated with storage and is primarily used as animal feed. However, RB is known to be abundant in macronutrients and bioactive compounds such as *p*-coumaric acid, ferulic acid, and caffeic acid [20]. Although studies have demonstrated that RB phenolic compounds have the potential to exhibit antioxidant and/or anti-inflammatory properties [2], their impact on specific pathways of inflammation in an oxidative stress-induced/inflammatory environment has not been studied. The current study aimed to determine the antioxidant and anti-inflammatory effect of RB phenolic extracts on RAW264.7 mouse macrophage cells stimulated with hydrogen peroxide (H_2O_2) and lipopolysaccharide (LPS). Biomarkers of oxidative stress such as MDA, intracellular reactive oxygen species, nitric oxide and pro-inflammatory cytokines such as IL-6, MCP-1, IL-10, TNF-α, IL-12p70, and IFN-γ were measured in vitro.

2. Materials and Methods

2.1. Chemicals and Reagents

Unless otherwise stated, all the chemicals and reagents used in this study were acquired from Sigma-Aldrich (St Louis, MI, USA) and BD Biosciences (Franklin Lakes, NJ, USA).

2.2. Rice Bran Extract Preparation

Stabilization of RB was previously conducted by SunRice, Leeton, Australia using a drum-drying method. Extraction of the RB phenolic compounds was performed using an acetone/water/acetic acid (70:29.5:0.5, *v/v*) mixture, and characterized using ultra-high-performance liquid chromatography coupled to an 2,2'-Azino-bis 3-ethylbenzothiazoline-6-sulfonic acid-online system and mass spectrometry (Agilent Technologies) [20]. Prior to the cell culture studies, the extract was reconstituted in 50% dimethyl sulfoxide (DMSO).

2.3. Cell Culture Conditions

Experiments were conducted on RAW264.7 cells purchased from Sigma-Aldrich (St Louis, MO, USA). RAW264.7 cells were cultured in high-glucose Dulbecco's modified Eagle's medium (DMEM) (Sigma-Aldrich: D6429) supplemented with 10% fetal bovine serum and 1% 10,000 U/mL penicillin–10 mg/mL streptomycin at 37 °C in 5% CO_2. Cultured RAW264.7 cells were used before reaching the 8th passage.

2.4. Cytotoxicity Assessment

A resazurin red cytotoxicity assay, as described by Saji, Francis [21], was utilized to examine the effect of RB phenolic extracts on RAW264.7 cells. Briefly, RAW264.7 cells were seeded into 96-well plates at a density of 50,000 cells/well and incubated overnight in the DMEM complete media. The cell count for the experimental seeding was achieved with a Muse® Cell Analyzer from Luminex Corporation (Austin, TX, USA). RAW264.7 cells were then treated with 200 µL of DMEM complete media containing RB phenolic extract (25 µg/mL, 50 µg/mL, 100 µg/mL, 250 µg/mL, 500 µg/mL, 750 µg/mL and 1000 µg/mL) for 6 h. For the positive control, 5 mM concentration of H_2O_2 was used, and 0.5% DMSO served as the negative control. Subsequently, all the treatment wells were emptied before incubation with 200 µL of resazurin red solution for an additional 4 h. The absorbance at 570 and 600 nm was measured on a microplate reader (FLUOstar Omega microplate reader, BMG Labtech, Offenburg, Germany) against a resazurin red blank. The percentage of cell viability was calculated. Each treatment was measured in quintuplicate.

2.5. Antioxidant Effect of RB Phenolic Extract

2.5.1. Malondialdehyde Determination

RAW264.7 cells (500,000 cells/well) were seeded into 6-well plates and incubated for 24 h. RB phenolic extracts at varying concentrations (25 µg/mL, 50 µg/mL, 100 µg/mL, 250 µg/mL) were then incubated with the cells for 6 h. To induce oxidative stress, RAW264.7 cells were subjected to 12 h incubation with a 500 µM concentration of H_2O_2. For MDA determination, a Lipid Peroxidation (MDA) Assay Kit (Sigma-Aldrich: MAK085) was used according to the manufacturer's instructions. Briefly, the cell lysate was centrifuged for 10 min at 13,000× g to remove any insoluble material. Subsequently, 600 µL of freshly prepared thiobarbituric acid (TBA) solution was combined with 200 µL of the sample supernatant and the previously prepared MDA standards (0 nmol, 4 nmol, 8 nmol, 12 nmol, 16 nmol, and 20 nmol) to form the MDA–TBA adduct. All the samples were incubated at

95 °C for 1 h and subsequently placed in an ice bath for 10 min. The sensitivity of each sample was enhanced by adding 300 μL of 1-butanol, followed by centrifugation for 3 min at 16,000× g to separate the layers. The 1-butanol layer (the top layer) was then transferred to another tube and evaporated at 55 °C. The remaining residue was dissolved in 200 μL of water and transferred to a 96-well plate for analysis. Absorbance was measured at 532 nm using a microplate reader (FLUOstar Omega microplate reader, BMG Labtech, Offenburg, Germany). Each treatment was measured in quintuplicate.

2.5.2. Intracellular Reactive Oxygen Species Generation

RAW264.7 cells were seeded at a density of 50,000 cells/well into a black, clear-bottom 96-well plate and incubated for 24 h. The cells were then incubated with RB extracts (25 μg/mL, 50 μg/mL, 100 μg/mL, 250 μg/mL) for 6 h. The DMSO-treated cells served as the control. Oxidative stress was induced by treatment with 500 μM concentration of H_2O_2 for 12 h. Formation of reactive oxygen species was determined by removing the old media and replenishing with 50 μM 2,7-dichlorofluorescein diacetate (Sigma-Aldrich: 35845) and incubating for an additional 30 min in the dark at 37 °C. The assessment of intracellular reactive oxygen species levels was conducted by measuring the fluorescence using a fluorescence microplate reader (FLUOstar Omega microplate reader, BMG Labtech, Offenburg, Germany) at 480 nm excitation and 530 nm emission at 37 °C. Each treatment was measured in octuplicate.

2.6. Anti-Inflammatory Effect of RB Phenolic Extract

2.6.1. Nitric Oxide Determination

RAW264.7 cells at a density of 500,000 cells/well were seeded into a 6-well plate and incubated for 24 h, followed by treatment with 25 μg/mL, 50 μg/mL, 100 μg/mL, 250 μg/mL RB phenolic extracts for 6 h. DMSO-treated cells served as the control. To induce inflammation, RAW264.7 cells were further subjected to 24 h incubation with 1 μg/mL of LPS. The supernatant was collected by centrifugation (3000× g, 10 min, 4 °C). To assess the nitric oxide levels, the cell supernatants and standards (400 μL) were combined with 400 μL of Griess reagent (Sigma-Aldrich: G4410) and incubated for 15 min in the dark at room temperature. Sodium nitrite was used to generate a standard curve (10 μM, 20 μM, 40 μM, 60 μM, 80 μM and 100 μM). Absorbance was measured at 540 nm using a microplate reader (FLUOstar Omega microplate reader, BMG Labtech, Offenburg, Germany). Each treatment was measured in quintuplicate.

2.6.2. Inflammatory Cytokine Determination

Seeding of RAW264.7 cells (500,000 cells/well) was conducted in a 6-well plate and incubated for 24 h. After confluency was reached, the RAW264.7 cells were treated with 25 μg/mL, 50 μg/mL, 100 μg/mL, 250 μg/mL RB phenolic extracts and DMSO (control) for 6 h. Inflammation was induced with the addition of LPS (1 μg/mL) and the cells were incubated for an additional 24 h, after which, the supernatant was collected by centrifugation (3000 × g, 10 min, 4 °C) and stored at −20 °C before analysis. The anti-inflammatory effect of RB phenolic extracts on inflammatory cytokines (IL-6, MCP-1, IL-10, TNF-α, IL-12p70, and IFN-γ) was determined using a BD™ Cytometric Bead Array Mouse Inflammation Kit (BD Biosciences: 552364). Briefly, 50 μL of the diluted samples (1:40) and standards were combined with 50 μL of the mixed capture beads and 50 μL of the mouse inflammation PE detection reagent. This was then incubated for 2 h in the dark at room temperature. After which, 1 mL of wash buffer was added to each assay tube and centrifuged (200 × g, 5 min). Following centrifugation, the supernatant was discarded, and the bead pellet was suspended in 300 μL of fresh wash buffer. Each sample was vortexed for 3–5 s and immediately inspected using a Gallios™ Flow Cytometer (Beckman Coulter, CA, USA). Kaluza Flow Cytometry Analysis Software (Beckman Coulter, CA, USA) was used for conducting the data analysis. Each treatment was measured in quintuplicate.

2.7. Statistical Analysis

GraphPad Prism 7 software (GraphPad Software Inc., San Diego, California, USA) was utilized for statistical analysis by one-way analysis of variance (ANOVA), followed by post-hoc Tukey's multiple comparisons test. Data were considered statistically significant when $p < 0.05$. Any significant statistical interactions were included in the analysis where applicable. The results are reported as mean ± standard deviation (SD).

3. Results

3.1. Cytotoxicity of RB Phenolic Extracts on RAW264.7 Cells

A time-course study was conducted on a range of incubation periods varying from 2–24 h (data not shown) from which a 6 h incubation period was selected as the optimal time point associated with cell viability. It was observed that RAW264.7 cells, post-exposure to RB phenolic extracts for 6 h, did not display any toxic effects at any of the doses tested (25–1000 µg/mL) when compared to the DMSO control. Therefore, the optimal and physiologically attainable concentrations of RB phenolic extract selected for further examinations were between 25 and 250 µg/mL (Figure 1).

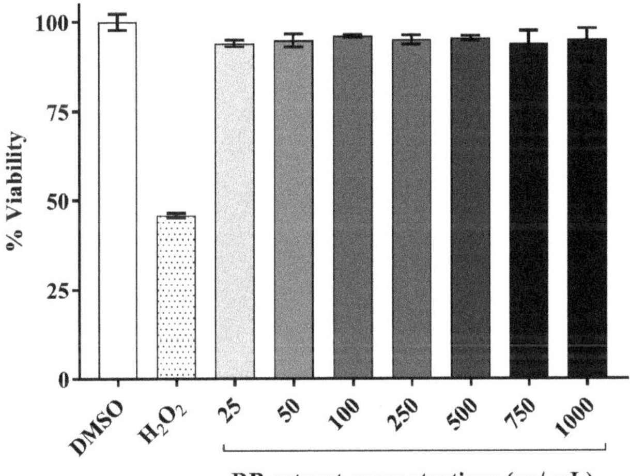

Figure 1. Cytotoxicity of RAW264.7 cells post-exposure to various doses of RB phenolic extracts (6 h treatment). The RB phenolic extracts did not display any cytotoxic effect on the RAW264.7 cells at the concentrations (25–1000 µg/mL) tested ($n = 5$). Data are presented as mean ± SD. Rice bran, RB.

3.2. Antioxidant Properties of RB Phenolic Extract

3.2.1. Malondialdehyde Concentration

MDA concentrations (nmol/mL) were significantly reduced in oxidative stress-induced RAW264.7 cells after treatment with 25 µg/mL ($p < 0.05$), 50 µg/mL ($p < 0.05$) and 100 µg/mL ($p < 0.01$) of RB phenolic extract compared to the H_2O_2 control (Figure 2). However, the treatment with 250 µg/mL RB phenolic extract did not reduce the MDA levels.

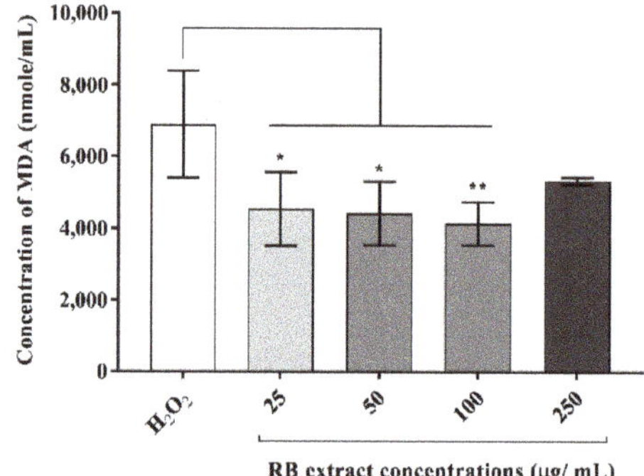

Figure 2. The impact of the RB phenolic extract treatment on oxidative stress-induced RAW264.7 macrophage cells. A significant reduction in the MDA concentration (nmol/mL) is observed after treatment with the different concentrations (25–100 µg/mL) of RB phenolic extracts compared to the control ($n = 5$). The level of significance is indicated by the asterisks, whereby * $p < 0.05$, ** $p < 0.01$. Data are presented as mean ± SD. Mouse macrophage cell, RAW264.7; hydrogen peroxide, H_2O_2; rice bran, RB.

3.2.2. Intracellular Reactive Oxygen Species Generation

RAW264.7 cells induced with H_2O_2 showed a significant reduction in the generation of intracellular reactive oxygen species after treatment with 25–250 µg/mL ($p < 0.0001$) of RB phenolic extract when compared to the control (Figure 3).

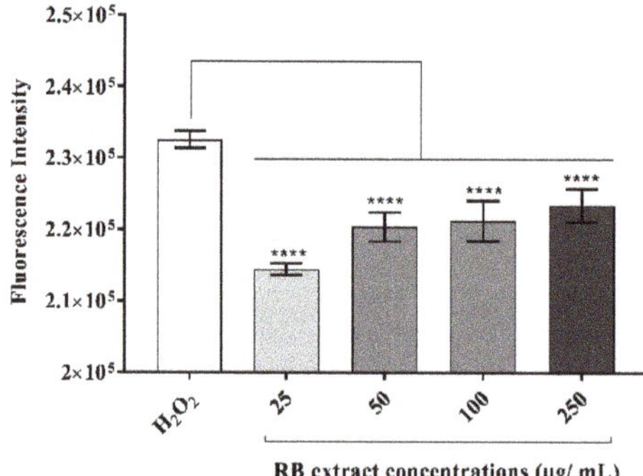

Figure 3. The effect of RB phenolic extracts on intracellular reactive oxygen species generation in oxidative stress-stimulated RAW264.7 cells. A significant reduction in intracellular reactive oxygen species generation, as indicated by the differences in the fluorescence intensity, is observed after treatment with the different concentrations (25–250 µg/mL) of RB phenolic extracts compared to the control ($n = 8$). The level of significance is indicated by the asterisks, whereby **** $p < 0.0001$. Data are presented as mean ± SD. Hydrogen peroxide, H_2O_2; rice bran, RB.

3.3. Anti-Inflammatory Effect of RB Phenolic Extract

3.3.1. Nitric Oxide Determination

RB phenolic extracts at all concentrations (25–250 µg/mL) reduced the expression of nitric oxide levels in LPS-stimulated RAW264.7 cells (Figure 4).

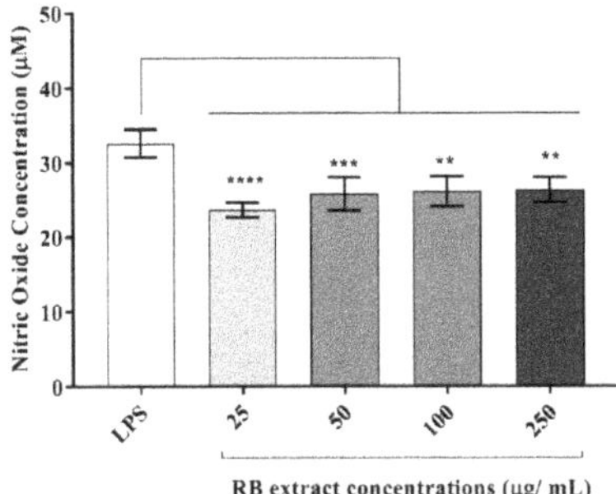

Figure 4. Effect of RB phenolic extract on nitric oxide levels in RAW264.7 cells induced with LPS. A significant reduction in the nitric oxide levels is observed after treatment with the different concentrations (25–250 µg/mL) of RB phenolic extracts compared to the control ($n = 5$). The level of significance is indicated by the asterisks, whereby ** $p < 0.01$, *** $p < 0.001$, **** $p < 0.0001$. Data are presented as mean ± SD. Lipopolysaccharide, LPS; rice bran, RB.

3.3.2. Inflammatory Cytokine Determination

It was observed that treatment with 250 µg/mL ($p < 0.01$) of RB phenolic extract reduced the expression of IL-12p70 and IL-6 under a pro-inflammatory environment. Furthermore, the RB phenolic extract at 100 µg/mL ($p < 0.05$) and 250 µg/mL ($p < 0.01$) concentrations reduced IFN-γ compared to the LPS control (Figure 5). No significant differences in the regulation of the other cytokines tested were observed (data not shown for IL-10, MCP-1, and TNF-α).

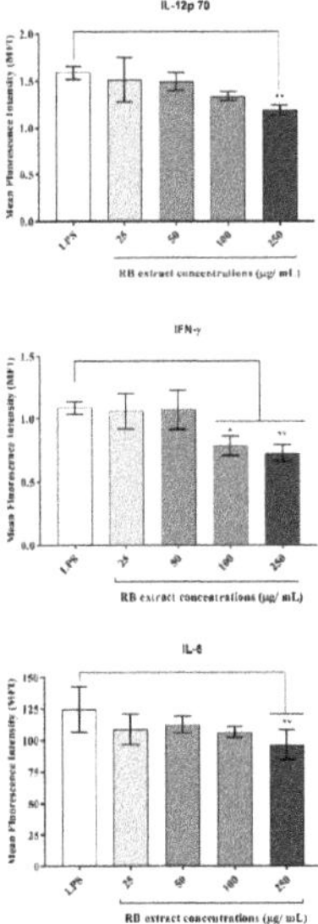

Figure 5. The effect of RB phenolic extracts in modulating pro-inflammatory cytokines. A significant reduction in IL-12p70 and IL-6 cytokines is observed after treatment with 250 µg/mL ($p < 0.01$). Furthermore, the pre-treatment with 100 µg/mL ($p < 0.05$) and 250 µg/mL ($p < 0.01$) of the RB phenolic extract is observed to have resulted in a significant reduction in the IFN-γ cytokine compared to the control ($n = 5$). The level of significance is indicated by the asterisks, whereby * $p < 0.05$, ** $p < 0.01$. Data are presented as mean ± SD. Lipopolysaccharide, LPS; rice bran, RB.

4. Discussion

Both oxidative stress and inflammation are known to be precursor risk factors that play a central role in the development of chronic conditions such as cardiovascular disease [2]. Studies have proposed that polyphenols possess anti-inflammatory properties by virtue of their free radical scavenging, enzyme and cell signalling pathway modulating effects [19]. The current study has demonstrated that phenolic extracts present in RB have significant antioxidant and anti-inflammatory properties under oxidative stress-induced/inflammatory environment in macrophage cells. The drum-dried RB sample used in this study was previously characterized and some of the bioactive compounds identified in the extract were caffeic acid, ethyl vanillate, ferulic acid, feruloyl glycoside, *p*-coumaric acid, shikimic acid, sinapic acid, syringic acid, tricin and vanillic acid. Among which, ferulic acid and its isomers were identified to be the predominant compound in the extract with tricin having the most antioxidant activity. Furthermore, the examination of the total phenolic content and antioxidant activity revealed that the drum-dried RB extract had a total free phenolic content of 362.17 ± 34.16 GAE/100 g of RB with an antioxidant activity of 975.33 ± 20.24 Fe^{2+}/100 g of RB and a total bound phenolic content of 160.65 ± 5.52 GAE/100 g of RB with an antioxidant activity of 551.91 ± 8.82 Fe^{2+}/100 g of RB. It is believed that the antioxidant and anti-inflammatory effects observed in this study could be due to the synergistic action of the antioxidant-rich phenolic compounds identified in the extract [20].

An increase in free radicals results in the overproduction of MDA and reactive oxygen species levels in the body. Hence, they serve as an accurate biomarker for the detection of oxidative stress and overall antioxidant status [4]. In this study, RAW264.7 cells stimulated with H_2O_2 resulted in a significant reduction in MDA concentration (nmol/mL) after treatment with 25, 50 and 100 μg/mL of RB phenolic extract (Figure 2). Furthermore, it was observed that all RB extract concentrations (25–250 μg/mL) reduced the intracellular reactive oxygen species levels under a H_2O_2-stimulated oxidative stress environment (Figure 3). Other studies have also demonstrated that plant-derived bioactive compounds can alleviate free radical production in vitro under oxidative stress conditions [22–24]. Curcumin has been shown to decrease MDA and reactive oxygen species levels in macrophage cells that were induced with oxidative stress conditions. The authors have highlighted curcumin to have increased the activity of antioxidant enzymes and activated the Nrf2-Keap1 pathway [22]. In addition, pigeon pea extracts rich in cyanidin-3-monoglucoside were also observed to prevent the reduction of antioxidant enzyme activity and decrease MDA production in oxidative stress-induced macrophage cells [23]. By reducing hydrogen peroxidase, reactive oxygen species generation has been shown to increase free metal ion production, consequently producing highly reactive hydroxyl radicals. Polyphenols, due to their thermodynamically lower redox potential, reduce these oxidizing free radicals via their metal chelating ability [19]. The results observed in this study are a clear demonstration of the ability of RB polyphenols to reduce the highly oxidizing free radicals produced in an oxidative stress environment. In summary, it is believed that the antioxidant potential of polyphenols is correlated to their capacity to suppress reactive oxygen species formation, the inhibition of free radical-producing enzymes and/or the upregulation of antioxidant defences. The structure of functional groups in polyphenols also plays an important role in their antioxidant capacity. We believe that the synergistic action of polyphenols, including the phenolic acids present in the RB extracts by virtue of the mechanisms listed above, is blunting MDA and reactive oxygen species generation in an oxidative stress environment in vitro.

During inflammation, the activation of macrophage occurs as a result of a cascade of events that are mediated by the nuclear factor kappa-light-chain-enhancer of activated B cells (NF-κB) and mitogen-activated protein kinase (MAPK) pathways [25,26]. Macrophages stimulated with LPS result in the release of NF-κB protein, which acts as a transcription factor that promotes inflammation [25]. Similarly, activated MAPKs contribute to the phosphorylation of downstream targets, including protein kinases and transcription factors, which subsequently facilitates the transcription of MAPK-regulated genes. This process results in the production of mediators such as pro-inflammatory cytokines and nitric oxide in stimulated macrophage cells [26]. Since pro-inflammatory cytokines are recognized to play a vital role in cell signalling and systemic inflammation, the excessive production of pro-inflammatory cytokines may result in tissue destruction and the development of several pro-inflammatory processes [27].

In the present study, the stimulation of RAW264.7 cells with LPS was observed to up-regulate several inflammatory markers which were subsequently suppressed by pre-treatment with the RB phenolic extract. This included a significant decrease in the nitric oxide levels and pro-inflammatory cytokine (IL-6, IL-12p70, and IFN-γ) production in RAW264.7 cells stimulated with LPS (Figures 4 and 5). However, no significant differences were observed in the IL-10, MCP-1, and the TNF-α production post-treatment with RB phenolic extracts (data not shown). Previous studies have also revealed that nitric oxide levels in LPS-induced RAW264.7 macrophages can be suppressed via treatment with extracts that have a high polyphenol content [28]. Rheosmin, a naturally occurring phenolic compound, was noted to dose-dependently suppress nitric oxide production by inhibiting the activity of NF-κB [29]. Moreover, it is also believed that nitric oxide inhibition by phenolic compounds may be achieved as a result of the substitution of the hydroxyl functional group [28]. Polyphenols are known to favourably modulate the activity of arachidonic acid metabolizing enzymes such as cyclooxygenase (COX), lipoxygenase (LOX) and nitric oxide synthase (NOS) [19]. The observed effect on the alleviation of the nitric oxide levels could be a direct impact of the synergistic action of RB polyphenols in inhibiting the key inflammation-mediating enzymes.

The results observed in this study demonstrate that RB phenolic compounds effectively down-regulated the expression of IL-6, IL-12p70, and IFN-γ production in LPS-stimulated RAW264.7 cells. IL-12 is a heterodimeric interleukin that plays an important role in defence against intracellular pathogens [30]. It is characterized by the high-level secretion of both IFN-γ and TNF-α [30]. IL-12p70 is one of the first pro-inflammatory cytokines released after antigens from the foreign agents, apparent in the microenvironment, and the secretion of IL-12p70 cytokine is critical for the initiation and polarization of the appropriate immune response [31]. IFN-γ is a cytokine that is critical for both innate and adaptive immunity and is widely recognized for its pro-inflammatory capability against infectious agents [32]. Furthermore, it is predominantly produced by natural killer cells and is known to inhibit viral replication directly [33]. An essential mediator of the inflammatory signalling pathway induced by LPS is a transcription factor known as a signal transducer and an activator of transcription 3 (STAT3) [34]. Activated STAT3 is recognized to translocate to the nucleus and regulate the transcription of inflammation-related genes. Upon LPS stimulation, the activation of STAT3 through the IL-6 signalling pathway can be observed to result in increased IL-6 production [34]. Polyphenols have been known to modulate the aforementioned pro-inflammatory products by inhibiting the enzymes associated with pro-inflammatory effects (COX-2, LOX and NOS); MAPK, protein kinase-C and activated protein-1 activation; and NFkB inhibition [19]. It is believed that RB polyphenols, by virtue of their structure-activity relationship and free radical scavenging attributes, might potentially target the mechanisms detailed above. Further studies are warranted to understand the role of RB phenolic extracts on specific molecular mechanisms associated with inflammation.

In the body, phytochemicals are processed as xenobiotics. This is because the human body is only capable of distinguishing between nutrients and constituents that are not nutrients, and is unable to differentiate between beneficial, neutral and toxic compounds. However, many of these phytochemicals are known to activate the adaptive cellular response pathways to oxidative stress by acting as low-dose stressors or pro-oxidants [35]. In the present study, treatment with lower doses of the RB phenolic extract (25 µg/mL) was observed to have resulted in a more pronounced inhibitory effect to certain markers compared to the higher doses (Figures 2–4). This occurs through a phenomenon known as hormesis, a biphasic dose-response with stimulation in low-doses and inhibition in higher doses (lesser activity) of the phytochemicals present in the RB extracts [36]. Furthermore, it is believed that the oversaturation of phenolic compounds, though excreted via the portal circulation in vivo, could result in the unfavourable activation of cellular signalling molecules . In addition, bioavailability is another factor that needs to be considered, as the effect of any dietary compound is mainly influenced by its bioavailability, or the proportion of the substance that actually enters the circulation when introduced into the body. Due to the remarkable diversity in the human population influenced by genetics and/or medications, the appropriate bioavailable dosage may be significantly different [35]. Furthermore, future studies that compare the in vitro effect of crude RB phenolic extracts against antioxidant/anti-inflammatory synthetic/therapeutic agents are warranted. This would also help provide novel insights into potential mechanisms associated with the mode of action of natural polyphenols in the pro-inflammatory process. Although the current study highlights the antioxidant and anti-inflammatory properties of RB at physiologically attainable concentrations in vitro, it is important to demonstrate similar effects in vivo due to the potential variation in the bioavailability of polyphenol sub-classes and individual predisposition. Well controlled human dietary intervention trials are warranted to confirm the thus observed in vitro impact of RB phenolic compounds.

5. Conclusions

This study highlights that RB phenolic extracts play a significant role in modulating the biomarkers of oxidative stress and inflammation in an oxidative stress/inflammatory environment in vitro. It is believed that the RB phenolic compounds, via their synergistic action, target pathways of inflammation and oxidative stress through their free radical scavenging potential, and modulate pro-inflammatory cytokine expression. The outcomes of this study suggest that RB phenolic extracts may potentially be used as a functional food alternative and could have preventive or therapeutic implications in chronic oxidative stress and inflammatory conditions. Further in vivo investigations that evaluate the bioactivity of RB phenolic extracts in pro-inflammatory populations are warranted.

Author Contributions: Conceptualization, N.S., N.F., L.J.S., C.L.B. and A.B.S; methodology, N.S., N.F., L.J.S., C.L.B. and A.B.S.; software, N.S., N.F. and A.B.S.; validation, N.S., N.F., L.J.S., C.L.B. and A.B.S.; formal analysis, N.S.; investigation, N.S.; resources, N.S., N.F., L.J.S., C.L.B. and A.B.S.; data curation, N.S.; writing–original draft, N.S.; writing–review and editing, N.S., N.F., L.J.S., C.L.B. and A.B.S.; supervision, N.S., N.F., L.J.S., C.L.B. and A.B.S.; funding acquisition, C.L.B.; project administration, N.F., L.J.S., C.L.B. and A.B.S. All authors have read and agreed to the published version of the manuscript.

Funding: The funding for this study was obtained from the AUSTRALIAN RESEARCH COUNCIL INDUSTRIAL TRANSFORMATION TRAINING CENTRE FOR FUNCTIONAL GRAINS (Project ID 100737) and from AGRIFUTURES, AUSTRALIA (PRJ-011503). We would like to acknowledge the Faculty of Science, Charles Sturt University, for providing the funding towards the publication cost of this article.

Acknowledgments: The authors acknowledge SunRice, Australia, for providing the rice bran sample used in this study.

Conflicts of Interest: The authors declare no conflict of interest.

Abbreviations

ANOVA	Analysis of variance
COX	Cyclooxygenase
DMSO	Dimethyl sulfoxide
DMEM	Dulbecco's modified Eagle's medium
H_2O_2	Hydrogen peroxide
IFN-γ	Interferon-γ
IL-10	Interleukin-10
IL-12p70	Interleukin-12, p70
IL-6	Interleukin-6
LPS	Lipopolysaccharide
LOX	Lipoxygenase
MDA	Malondialdehyde
MAPK	Mitogen-activated protein kinase
MCP-1	Monocyte chemoattractant protein 1
NOS	Nitric oxide synthase
NF-κB	Nuclear factor kappa-light-chain-enhancer of activated B cells
RB	Rice bran

STAT3	Signal transducer and activator of transcription 3
SD	Standard deviation
TBA	Thiobarbituric acid
TNF-α	Tumor necrosis factor-α

References

1. He, F.; Zuo, L. Redox roles of reactive oxygen species in cardiovascular diseases. *Int. J. Mol. Sci.* **2015**, *16*, 27770–27780. [CrossRef] [PubMed]
2. Saji, N.; Francis, N.; Schwarz, L.J.; Blanchard, C.L.; Santhakumar, A.B. Rice bran derived bioactive compounds modulate risk factors of cardiovascular disease and type 2 diabetes mellitus: An updated review. *Nutrients* **2019**, *11*, 2736. [CrossRef] [PubMed]
3. Korns, D.R.; Frasch, S.C.; Fernandez-Boyanapalli, R.; Henson, P.M.; Bratton, D.L. Modulation of macrophage efferocytosis in inflammation. *Front. Immunol.* **2011**, *2*, 57. [CrossRef]
4. Ellulu, M.S. Obesity, cardiovascular disease, and role of vitamin C on inflammation: A review of facts and underlying mechanisms. *Inflammopharmacology* **2017**, *25*, 313–328. [CrossRef] [PubMed]
5. Poffenberger, M.C.; Straka, N.; El Warry, N.; Fang, D.; Shanina, I.; Horwitz, M.S. Lack of IL-6 during coxsackievirus infection heightens the early immune response resulting in increased severity of chronic autoimmune myocarditis. *PLoS ONE* **2009**, *4*, e6207. [CrossRef] [PubMed]
6. Slatter, D.; Bolton, C.; Bailey, A. The importance of lipid-derived malondialdehyde in diabetes mellitus. *Diabetologia* **2000**, *43*, 550–557. [CrossRef]
7. Porter, N.A.; Caldwell, S.E.; Mills, K.A. Mechanisms of free radical oxidation of unsaturated lipids. *Lipids* **1995**, *30*, 277–290. [CrossRef]
8. Uchida, K. Role of reactive aldehyde in cardiovascular diseases. *Free Radic. Boil. Med.* **2000**, *28*, 1685–1696. [CrossRef]
9. Naseem, K.M. The role of nitric oxide in cardiovascular diseases. *Mol. Asp. Med.* **2005**, *26*, 33–65. [CrossRef]
10. Abeywardena, M.Y.; Leifert, W.R.; Warnes, K.E.; Varghese, J.N.; Head, R.J. Cardiovascular Boilogy of interleukin-6. *Curr. Pharm. Des.* **2009**, *15*, 1809–1821. [CrossRef]
11. Niu, J.; Kolattukudy, P.E. Role of MCP-1 in cardiovascular disease: Molecular mechanisms and clinical implications. *Clin. Sci.* **2009**, *117*, 95–109. [CrossRef]
12. Girndt, M.; Köhler, H. Interleukin-10 (IL-10): An update on its relevance for cardiovascular risk. *Nephrol. Dial. Transplant.* **2003**, *18*, 1976–1979. [CrossRef] [PubMed]
13. Ferrari, R. The role of TNF in cardiovascular disease. *Pharmacol. Res.* **1999**, *40*, 97–105. [CrossRef] [PubMed]
14. Zheng, X.; Huo, X.; Zhang, Y.; Wang, Q.; Zhang, Y.; Xu, X. Cardiovascular endothelial inflammation by chronic coexposure to lead (Pb) and polycyclic aromatic hydrocarbons from preschool children in an e-waste recycling area. *Environ. Pollut.* **2019**, *246*, 587–596. [CrossRef] [PubMed]
15. Singh, R.M.; Cummings, E.; Pantos, C.; Singh, J. Protein kinase C and cardiac dysfunction: A review. *Heart Fail. Rev.* **2017**, *22*, 843–859. [CrossRef]
16. Voloshyna, I.; Littlefield, M.J.; Reiss, A.B. Atherosclerosis and interferon-γ: New insights and therapeutic targets. *Trends Cardiovasc. Med.* **2014**, *24*, 45–51. [CrossRef]
17. Shattat, G.F. A review article on hyperlipidemia: Types, treatments and new drug targets. *Biomed. Pharmacol. J.* **2015**, *7*, 399–409. [CrossRef]
18. Callcott, E.T.; Blanchard, C.L.; Oli, P.; Santhakumar, A.B. Pigmented Rice-Derived Phenolic Compounds Reduce Biomarkers of Oxidative Stress and Inflammation in Human Umbilical Vein Endothelial Cells. *Mol. Nutr. Food Res.* **2018**, *62*, 1800840. [CrossRef]
19. Hussain, T.; Tan, B.; Yin, Y.; Blachier, F.; Tossou, M.C.; Rahu, N. Oxidative Stress and Inflammation: What Polyphenols Can Do for Us? *Oxidative Med. Cell. Longev.* **2016**, *2016*, 7432797. [CrossRef]
20. Saji, N.; Schwarz, L.J.; Santhakumar, A.B.; Blanchard, C.L. Stabilization treatment of rice bran alters phenolic content and antioxidant activity. *Cereal Chem.* **2020**, *97*, 281–292. [CrossRef]
21. Saji, N.; Francis, N.; Blanchard, C.L.; Schwarz, L.J.; Santhakumar, A.B. Rice Bran Phenolic Compounds Regulate Genes Associated with Antioxidant and Anti-Inflammatory Activity in Human Umbilical Vein Endothelial Cells with Induced Oxidative Stress. *Int. J. Mol. Sci.* **2019**, *20*, 4715. [CrossRef]

22. Lin, X.; Bai, D.; Wei, Z.; Zhang, Y.; Huang, Y.; Deng, H.; Huang, X. Curcumin attenuates oxidative stress in RAW264. 7 cells by increasing the activity of antioxidant enzymes and activating the Nrf2-Keap1 pathway. *PLoS ONE* **2019**, *14*, e0216711. [CrossRef]

23. Lai, Y.-S.; Hsu, W.-H.; Huang, J.-J.; Wu, S.-C. Antioxidant and anti-inflammatory effects of pigeon pea (Cajanus cajan L.) extracts on hydrogen peroxide-and lipopolysaccharide-treated RAW264. 7 macrophages. *Food Funct.* **2012**, *3*, 1294–1301. [CrossRef]

24. Hwang, K.-A.; Hwang, Y.-J.; Song, J. Antioxidant activities and oxidative stress inhibitory effects of ethanol extracts from Cornus officinalis on raw 264.7 cells. *BMC Complementary Altern. Med.* **2016**, *16*, 196.

25. Zhang, Q.; Luna-Vital, D.; de Mejia, E.G. Anthocyanins from colored maize ameliorated the inflammatory paracrine interplay between macrophages and adipocytes through regulation of NF-κB and JNK-dependent MAPK pathways. *J. Funct. Foods* **2019**, *54*, 175–186. [CrossRef]

26. Yu, H.-S.; Lee, N.-K.; Choi, A.-J.; Choe, J.-S.; Bae, C.H.; Paik, H.-D. Anti-inflammatory potential of probiotic strain Weissella cibaria JW15 isolated from Kimchi through regulation of NF-κB and MAPKs pathways in LPS-induced RAW 264. 7 Cells. *J. Microboil. Biotechnol.* **2019**, *29*, 1022–1032. [CrossRef]

27. Ben Lagha, A.; Andrian, E.; Grenier, D. Resveratrol attenuates the pathogenic and inflammatory properties of Porphyromonas gingivalis. *Mol. Oral Microboil.* **2019**, *34*, 118–130. [CrossRef]

28. Taira, J.; Nanbu, H.; Ueda, K. Nitric oxide-scavenging compounds in Agrimonia pilosa Ledeb on LPS-induced RAW264. 7 macrophages. *Food Chem.* **2009**, *115*, 1221–1227. [CrossRef]

29. Jeong, J.B.; Jeong, H.J. Rheosmin, a naturally occurring phenolic compound inhibits LPS-induced iNOS and COX-2 expression in RAW264. 7 cells by blocking NF-κB activation pathway. *Food Chem. Toxicol.* **2010**, *48*, 2148–2153. [CrossRef] [PubMed]

30. Guo, Y.; Cao, W.; Zhu, Y. Immunoregulatory Functions of the IL-12 Family of Cytokines in Antiviral Systems. *Viruses* **2019**, *11*, 772. [CrossRef] [PubMed]

31. Zhang, S.; Wang, Q. Factors determining the formation and release of bioactive IL-12: Regulatory mechanisms for IL-12p70 synthesis and inhibition. *Biochem. Biophys. Res. Commun.* **2008**, *372*, 509–512. [CrossRef] [PubMed]

32. Harizi, H.; Gualde, N. The impact of eicosanoids on the crosstalk between innate and adaptive immunity: The key roles of dendritic cells. *Tissue Antigens* **2005**, *65*, 507–514. [CrossRef] [PubMed]

33. Schmidt, S.; Tramsen, L.; Lehrnbecher, T. Natural killer cells in antifungal immunity. *Front. Immunol.* **2017**, *8*, 1623. [CrossRef] [PubMed]

34. Greenhill, C.J.; Rose-John, S.; Lissilaa, R.; Ferlin, W.; Ernst, M.; Hertzog, P.J.; Mansell, A.; Jenkins, B.J. IL-6 trans-signaling modulates TLR4-dependent inflammatory responses via STAT3. *J. Immunol.* **2011**, *186*, 1199–1208. [CrossRef] [PubMed]

35. Holst, B.; Williamson, G. Nutrients and phytochemicals: From bioavailability to bioefficacy beyond antioxidants. *Curr. Opin. Biotechnol.* **2008**, *19*, 73–82. [CrossRef] [PubMed]

36. Mattson, M.P. Hormesis defined. *Ageing Res. Rev.* **2008**, *7*, 1–7. [CrossRef] [PubMed]

Article

Quantitative Determination of Andrographolide and Related Compounds in *Andrographis paniculata* Extracts and Biological Evaluation of Their Anti-Inflammatory Activity

Emmanuelle Villedieu-Percheron [1], Véronique Ferreira [1], Joana Filomena Campos [1], Emilie Destandau [1], Chantal Pichon [2] and Sabine Berteina-Raboin [1,*]

[1] Institut de Chimie Organique et Analytique, Université d'Orléans, UMR CNRS 7311, BP 6759, 45067 Orléans CEDEX 2, France; emmanuelle.percheron@univ-orleans.fr (E.V.-P.); veronique.ferreira@univ-orleans.fr (V.F.); joana-filomena.mimoso-silva-de-campos@univ-orleans.fr (J.F.C.); emilie.destandau@univ-orleans.fr (E.D.)

[2] Centre de Biophysique Moléculaire CNRS UPR 4301, Université d'Orléans, Rue Charles Sadron, F-45071 Orléans CEDEX 2, France; Chantal.PICHON@cnrs.fr

* Correspondence: sabine.berteina-raboin@univ-orleans.fr

Received: 17 November 2019; Accepted: 11 December 2019; Published: 14 December 2019

Abstract: Extraction, isolation and characterization of *Andrographis paniculata* (*A.p.*) products were developed. Three natural diterpenes compounds were obtained and one was used for chemical modifications. Evaluation of their inhibition of TNFα induced NFκB transcriptional activity. A rapid analytical method for the determination and quantitation of three diterpenoid lactones (andrographolide 1, didehydroandrographolide 2, neoandrographiside 3) found in *A. paniculata* extracts was investigated. After some optimizations on column type and injection solvent, the separation was achieved in 9 min on a monolithic Chromolith Performance RP18e column (100 mm × 4.6 mm ID, 2 μm), with a gradient solvent system of water and methanol, UV detection at 220 nm and ELSD detection. The method was proved to be suitable for the quantitation of these three diterpenes in four different commercial *Andrographis* dietary supplements. The anti-inflammatory activities of a mixture of known composition have been evaluated showing differences in activity depending on the relative ratio of various diterpenes and also a possible synergic activity for some of them.

Keywords: andrographolide; reversed-phase liquid chromatography; quantitative analysis; method validation; anti-inflammatory activity

---

1. Introduction

Andrographis paniculata (Acanthaceae) (*A.p.*) is widely used as medicinal herb in traditional medicine [1–3]. Diterpenoid lactones isolated from these extracts have been the subject of intensive investigations and are reported to exhibit a wide spectrum of biological activities, including antibacterial [4], anti-inflammatory [5–8], hepatoprotective [9–11], neuroprotective [12], cardioprotective [13,14] and anticancer properties [15,16]. Andrographolide 1 (Figure 1) has been reported to be the main active compound in the plant [17,18]. Its structure and stereochemistry have been determined over many years using chemical and spectroscopic techniques [19], and its crystal structure was characterized [20]. Other minor diterpenoid lactones also have been isolated from the plant, and their structures have been investigated [21] but their activities are not well studied. Studies about the pharmacological mechanism of its known anti-inflammatory actions have shown in 2004 that andrographolide attenuates inflammation by inhibition of NFκB activation through a covalent modification of reduced cysteine [22].

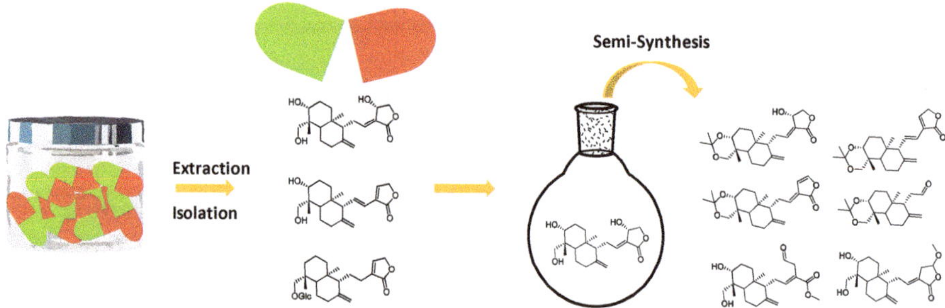

Figure 1. Structure of natural the compounds.

Considering the numerous biological applications of these compounds, commercial sources of preparations based on different parts of the plant have been available for a few years. However, the box does report if the given contents concern the major Compound **1** or the whole family of diterpenes, and the method of preparation. It is well-known that the composition could change depending on the origin and the season of gathering and on the part of plant studied (roots, leaves, stem, etc.). Nevertheless, a variation in the ratio of Compound 1 as well as related compounds had not been investigated and could lead to a change in therapeutic values. Considering that the composition of these dietary supplements is unclear and in order to control their quality, it appeared to be important for us to be able to quantify the composition of such commercial preparations, in order to connect their chemical composition with the anti-inflammatory activities generated by these commercially available preparations (Figure 2).

Figure 2. Composition of dietary supplements of natural compounds and pharmacomodulation.

HPLC is the most frequently used chromatographic technique in the laboratory setting. Some qualitative HPLC analyses have been reported in literature for these compounds [23,24]. HPTLC assays of *A.p.* extracts based on the quantification from spot UV visualizations have been described by Srivastava [25]; however, the extraction solvent influences the amount and nature of the extracted compounds and methanol leads to better extraction yields than chloroform or ethyl acetate. In addition, these methods often take long times and/or specific instrumentation. A few years ago, some groups developed a new rapid method for the determination of andrographolide in mixtures [24,26–28]. To complement these methods, we have developed a rapid and simple method for the separation and quantitation of andrographolide **1**, didehydroandrographolide **2** and neoandrographiside **3**. Based on the literature [23], using an Altima RP C18 column, an optimization of the elution gradient allowed us to obtain a better separation of the compounds present in the plant. Then we used a column dedicated to fast analysis of samples, the Chromolith Performance RP-18e column. This type of column allows to work at higher flow rates than with conventional columns of the same diameter without loss of efficiency or excessive increase in pressure. To assess to this work, the developed method was applied to four commercial preparations containing between 4% and

10% of andrographolides according to the labeling. Finally, the NFκB inhibition of these preparations was evaluated.

2. Materials and Methods

2.1. Chemicals and Reagents

Methanol from SDS Carlo Erba (Val de Reuil, France) and water purified (resistance < 18 MΩ) from ultra-pure water using an Elgastar UHQ II system (Elga, Antony, France) were used for the HPLC mobile phase and to prepare the stock solutions of analytes. Methanol from Sigma Aldrich (Steinheim, Germany) was used for the extraction and purification of crude extracts. Chloroform from SDS Carlo Erba (Val de Reuil, France) was used for purification of crude extracts.

2.2. Instrumentation and Conditions

An Agilent HP 1100 system (Waldbronn, Germany) with a 20 µL loop, coupled to a Kontron Ultra-Violet (UV) detector (Zurich, Switzerland) and piloted by EZchrome Elite workstation software was used. LC separation was performed on a Chromolith Performance RP-18e column (100 mm × 4.6 mm ID, 2 µm) provided by Merck (Darmstadt, Germany), with a gradient system containing water as solvent A and Methanol as solvent B, using solvent B from 40% to 51% over 9 min, at 3 mL·min^{-1}. To avoid any peak tailing and samples precipitation, they were injected in a mixture of water:methanol of 80:20. Detection was done by UV at 220 nm, and by ELSD giving the profile shown in Figure 3 where we can see the detection of the three main compounds present in the plant: andrographolide **1**, didehydroandrographolide **2** and neoandrographiside **3**.

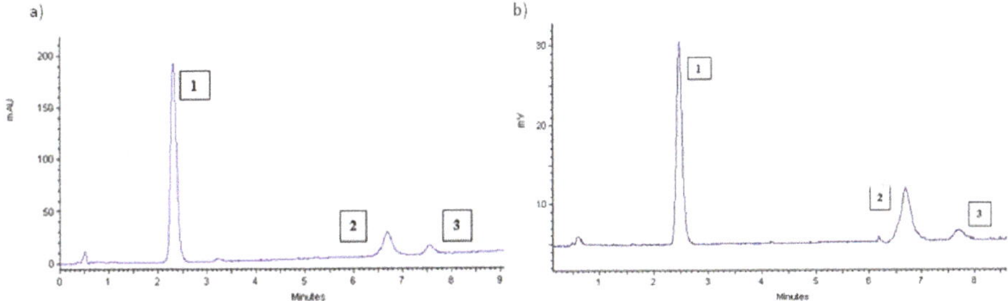

Figure 3. Chromatogram of the crude extract of *Andrographis* from sample A. Chromolith Performance RP-18e (100 × 4.6 mm ID, 2 µm); solvent A: water; solvent B: MeOH; gradient conditions: solvent B from 40% to 51% over 9 min, at 3 mL min, room temperature. (**a**) UV at λ = 220 nm, (**b**) ELSD.

2.3. Extraction Procedure

Commercial *Andrographis* products, all in the form of tablets, were obtained from online dietary supplements shops. To protect the manufacturer's identity, the samples were labeled with letters A–D. Extraction of diterpenoid lactones from tablets A–D was performed with a methanol ratio 15 mg·mL^{-1}, and the sample was sonicated at room temperature for 60 min. The mixture was filtered through filter paper (Whatman #1) and the residue was returned to the sample vial. The above extraction procedure was repeated two more times. The combined methanol extracts were evaporated under reduced pressure to give a green powder as the crude extract.

2.4. Preparation of Pure Compounds

External standards were obtained after purification of crude extracts from sample A by flash chromatography over silica gel (Geduran® Si 60, Merck), with an elution gradient of methanol from 3% to 15% in chloroform [29]. Maximal absorbance wavelengths were 250 nm and 224 nm for Compounds

1 and **2**, respectively, whereas Compound **3** absorbed up to 220 nm. The purity of the compounds was determined by HPLC with an Altima RP C18 column and gradient system containing water as solvent A and MeOH as solvent B, using solvent B from 20% to 100% over 45 min, at 1 mL·min^{-1}. Using UV and ELSD detection and washing the column for 10 min with MeOH, purity of each compound has been estimated to be superior to 99%. Further identification by NMR spectroscopy (^{1}H, ^{13}C, 1D, 2D) and LC-ESI-MS ([M + H]$^+$, [M − H]$^-$) were necessary to confirm the structures, and the results are in agreement with the literature and listed in the Results and Discussion section (Section 3.4). Semi-synthetic compounds from **4** to **7** (Figure 4) were obtained by treatment of **1** according to the procedure described in the literature [30] (Scheme 1).

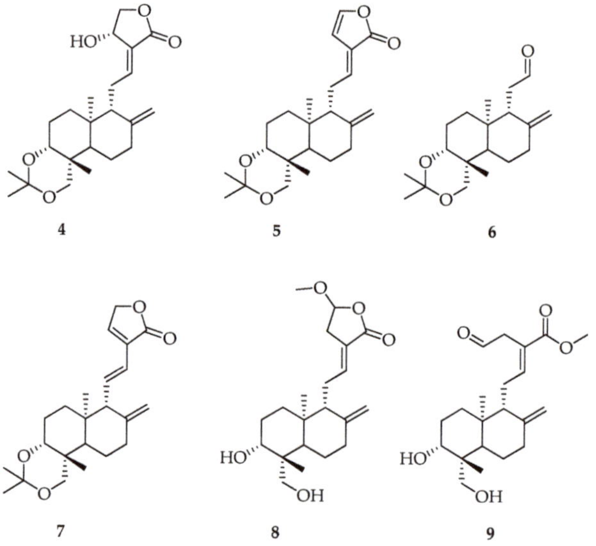

Figure 4. Structures of chemically modified derivatives.

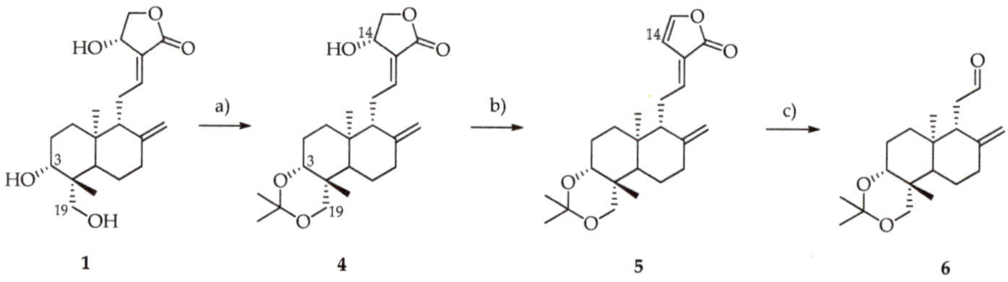

a) 2,2-dimethoxypropane, PPTS, DMF, room temperature, 5h, 90%
b) Ac$_2$O, pyridine, room temperature, 20h, 48%
c) KMnO$_4$, THF, 40°C, 2h, 30%

Scheme 1. Modification of andrographolide.

Compound **7** came from the standard isomerization of Compound **5** and the semi-synthetic Compounds **8** and **9** were obtained after treatment of **1** with MeONa in MeOH [31] at room temperature (Scheme 2) according to the mechanism proposed in Scheme 3.

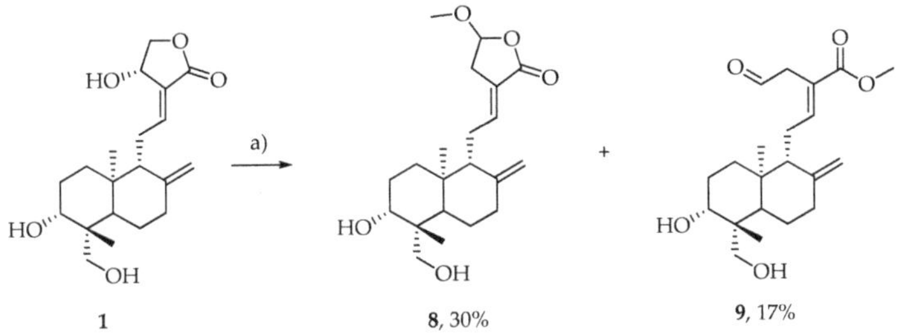

Scheme 2. Opening of the lactone.

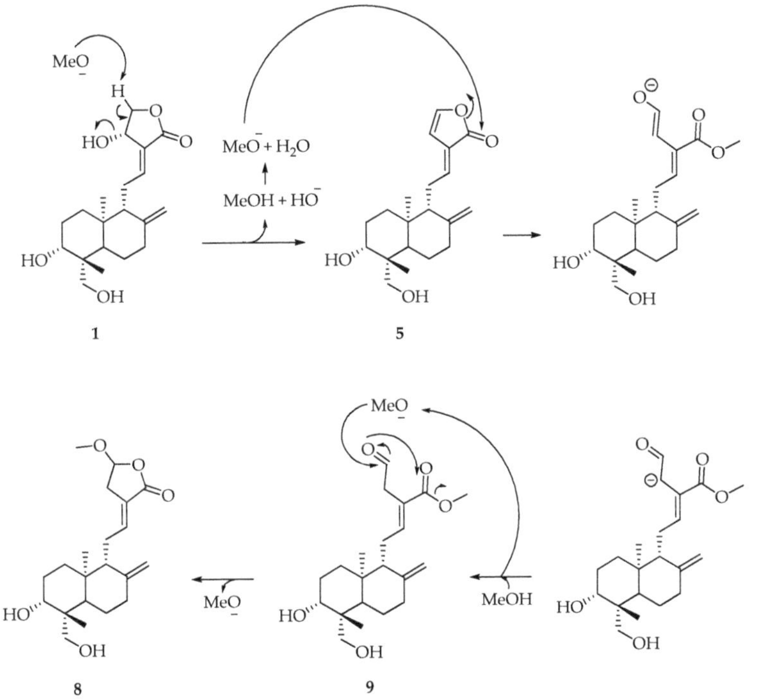

Scheme 3. Proposed mechanism for formation of Compounds **8** and **9**.

2.5. Standard and Sample Preparation

Stock solutions of each compound were prepared by dissolving them in methanol to obtain a 1 mg·mL^{-1} solution. Five calibration solutions were prepared by mixing the stock solution in a suitable proportion of water and methanol to have a final composition of injection solvent equal to a water:methanol of 20:80. The five concentrations were 50, 100, 125, 150 and 200 µg·mL^{-1} for Compound 1, 50, 75, 100, 150 and 200 µg·mL^{-1} for Compound 2, and 5, 10, 20, 40 and 50 µg·mL^{-1} for Compound 3. Sample preparations were made according to the following procedure: About 1 mg of crude extract was accurately weighted and methanol was added to obtain a 1 mg·mL^{-1} solution. This solution was diluted with a water:methanol ratio of 20:80 to a final concentration of 0.5 mg·mL^{-1}.

2.6. NFκB-Dependent Luciferase Activity Assay

A549/NFκB-luc cells were obtained by co-transfection of A549 cells (ATCC P/N CCL-185) with pNFκB-luc (Panomics P/N LR0051) and pHyg followed by hygromycin selection. These cells were obtained from Panomics and they maintain a chromosomal integration of a luciferase reporter gene regulated by multiple copies of NFκB response elements. They were cultured in DMEM medium supplemented with 10% FBS in presence of 100 µg/mL of hygromycin, 100 units/mL of penicillin and 100 µg/mL of penicillin and streptomycin, respectively. The NFκB-promoter luciferase reporter was used to assay the activity of NFκB activation. Stock solutions of samples were prepared in dimethyl sulfoxide (DMSO) with a concentration of 10 mg/mL. A549/NFκB-luc cells were incubated in with samples or its solvent and stimulated with TNFα (10 ng/mL, Sigma, Saint Quentin Fallavier Cedex, France) in 1 mL culture medium. After 24 h, the luciferase level was measured by luminescence. Cells were washed with PBS and centrifuged (7 min, 1500 rpm). After elimination of the supernatant, the homogenization buffer (0.2 mL of 8 mM $MgCl_2$, 1 mM dithiotreitol, 1 mM Ethylenediaminetetraacetic acid (EDTA), 15% glycetol, 1% Triton X-100, 25 mM Tris-phosphate buffer pH 7.8) was poured into each well. The tissue culture plates were shaken and kept at 20 °C for 10 min. The solution was recovered and centrifuged (10 min, 3000 rpm). ATP (95 µL of a 2 mM solution of homogenization buffer without Triton-X100) was added to 60 µL of supernatant and vigorously mixed. Luminescence was recorded for 4 s using a luminometer (LUMAT LB 9501, Berthold, Wildbach, Germany) upon addition of 0.15 mL of 167 mM luciferin solution in water. The relative luciferase activity can be given to evaluate the inhibition of NFκB.

2.7. Assay Validation, Linearity of Calibration and Limit of Quantification (LOQ)

Calibration standards were prepared at five concentration levels from 50 to 200 µg·mL^{-1} for **1** and **2**, and from 5 to 50 µg·mL^{-1} for **3**. Every calibration standard was injected in triples, on three different days. The nine replicates were then pooled. The linearity of the response was evaluated by plotting calculated against theoretical concentrations to represent the linearity of the method; slopes appeared to be near to 1, and the intercept near to 0; R^2 for **1**, **2** and **3** were evaluated at 0.9991, 0.9997 and 0.999, respectively. The calibration curves were estimated to be linear using this model, and the area of the peak corresponding to one compound was proportional to its concentration. The LOQ was defined as the lowest concentration on the calibration curve with a standard deviation in the tolerance domain of 5% for precision and accuracy. A LOQ of **1**, **2** and **3** was 10, 32 and 5 µg·mL^{-1}, respectively.

2.8. Assay Validation—Precision

Intra-day precision was determined by assaying three replications of samples at five different concentrations levels. The results were used to construct the calibration curves. Inter-day precisions were assessed by assaying three standard mixtures freshly prepared once a day over three days (J1, J2, J4). The precision of the method was calculated as the relative standard deviation (RSD) of the concentration determined in all replicates. Intra-day and inter-day precision (RSD) were 0.16% and 0.24% for andrographolide (25 µg/mL), 0.85% and 0.68% for didehydroandrographolide (100 µg/mL) and 0.76% and 0.68% for neoandrographiside (20 µg/mL).

2.9. Assay Validation—Accuracy

Accuracy was determined by analyzing three replicates at 125, 100 and 20 µg/mL for **1**, **2** and **3**, respectively, once a day over three days. A fresh mixture was prepared each day. The accuracies were assessed by comparing the determined concentrations with the nominal concentrations. We found that the repeatability of the method was below 2% RSD and the accuracy of the method was 99.9%–101.5%, reliable with a tolerant deviation of 5%.

3. Results and Discussion

3.1. Effects of Isolated and Modified Compounds on TNFα Induced NFκB Transcriptional Activation

Activation of A549 with TNFα increased NFκB transactivation luciferase activity and denoted as (T+). No activation by DMSO was reported as (T−). The samples were tested with a concentration of

25 µg/mL. The nine pure compounds and their inhibitory effects are shown in Figure 5. Inhibitory effects are in purple for the natural compounds (Figure 1), and in green for semisynthetic derivatives (Figure 4, Compounds **4** to **9**).

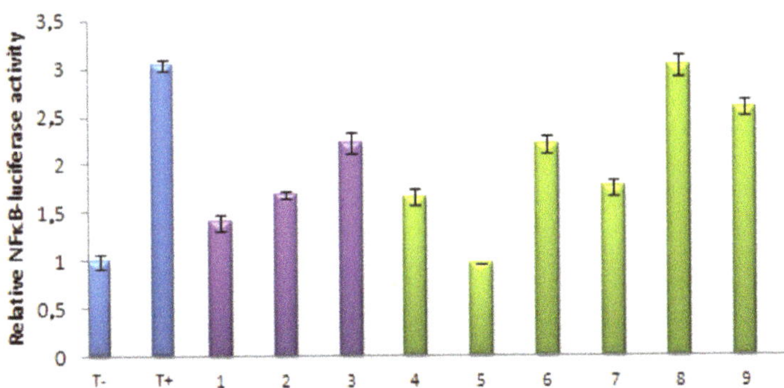

Figure 5. Relative NFκB-luciferase activity of pure compounds from **1** to **9**, 25 µg/mL.

The results obtained for Compounds **4** and **7** showed that a modification of the diol present on the starting andrographolide did not change the anti-inflammatory activity. While, for Compounds **5**, **6**, **8** and **9**, results showed that the modification of the lactone (5-membered ring) could lead to a radical change in the anti-inflammatory activity. Indeed, Compound **8** did not significantly inhibit NFκB transcriptional activity, aldehyde **6** showed the same activity than neoandrographiside and Compound **5** showed the best inhibition of these nine derivatives.

In fact, previous experiments have shown that modification of the compound at the lactone level (5-membered ring system) can modify the anti-inflammatory activity of andrographolide. In order to understand the mechanism of inhibition of NFκB by andrographolide, andrographolide **1** was treated with NHBoc-L-cysteine methyl ester. A Modification of NFκB by site-directed mutagenesis [22] has shown that a cysteine residue would be involved in the formation of a covalent bond with andrographolide. We therefore put andrographolide in the presence of a cysteine residue, in a buffered medium approaching the biological environment, and we were able to isolate Compound **10**, resulting from the addition of cysteine on andrographolide, at 40% yield.

Compound **10** as shown in Figure 6 was isolated and tested for its biological effect. The results showed that this compound did not inhibit NFκB transcriptional activity yet. This means that andrographolide acts as a Michael acceptor in the inhibition of NFκB and that in Compound **10** the lack of inhibition was due either to the absence of the acceptor, or to the steric hindrance of the compound due to the addition of NFκB.

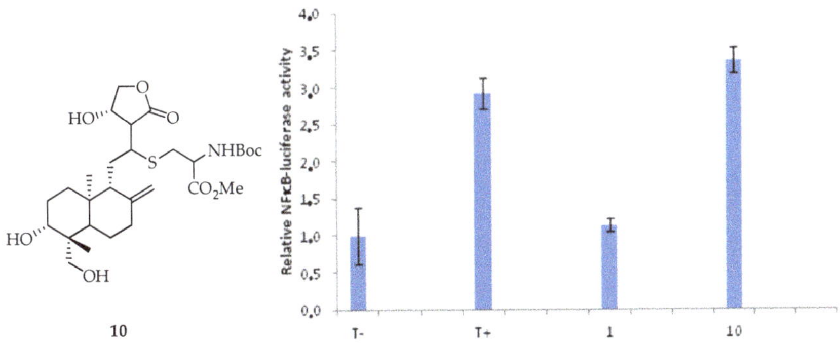

Figure 6. Structure of Compound **10** and relative NFκB-luciferase activity.

This last result could explain why Analogue **6** kept a significant inhibition. In this compound, addition of thiol is possible on carbonyl. Analogue **5** appeared to be the best Michael acceptor for the inhibition of NFκB transcriptional activity.

3.2. Quantitation of Commercial Preparations

As it has been shown that *Andrographis* extracts could have a better immunostimulant activity than andrographolide alone [32], we were interested in the difference of activity for mixtures with different ratios in andrographolide derivatives. Consequently, the quantification method was applied to four commercial preparations. The commercial samples were first extracted with methanol, as described in the Materials and Methods section. The extraction yields of dry substance were 44%, 40%, 17% and 33% for samples A to D, respectively. The four chromatograms were similar to those presented in Figure 3.

Concentrations determined by external standard quantitation method were presented in Table 1.

Table 1. Quantitation of three diterpenes in commercial preparations. Concentrations given in $\mu g \cdot mL^{-1}$ ± SD.

Diterpene	A	B	C	D
1	176.7 ± 12	138.8 ± 0.2	82.0 ± 0.1	127.3 ± 0.3
2	95.5 ± 0.7	136.9 ± 3.6	96.1 ± 2.1	51.3 ± 0.3
3	18.3 ± 0.7	13.4 ± 0.4	26.4 ± 0.4	9.8 ± 0.3

As expected, the analysis proved that the compositions of the commercial preparations are not the same. The dietary supplements B and C were reported to contain 10% *Andrographis* and it clearly appeared that both samples did not contain the same quantity neither in **1** nor in the rest of the compounds of the diterpenes family. Actually, B contains 10% andrographolide **1**, and C contains 10% of the three andrographolide derivatives **1**, **2** and **3**. Moreover, for the sample C, **1** is not the major compound. This analysis showed that there is a real difference in the labeling information of dietary supplements and the relative properties will not be the same.

3.3. Effects of the Extracts of Known Concentrations on TNFα Induced NFκB Transcriptional Activation

Using the quantitation of diterpenes in mixtures, we investigated the activity of herbal preparations incubating the same quantity of andrographolide, and we standardized the final concentration of andrographolide at 17 μg/mL. The results could show if there is a synergic or complementary effect of several compounds in the extracts for the inhibition of NFκB. The results are presented in Figure 7 and show that if the mixtures contain **1** as the major compound and only low amounts of **2** and **3** (as in A and D), the inhibitory activity follows the activity of Compound **1**. Nevertheless, for mixtures such as B and C where the quantity of **2** is not negligible, the inhibitory activity increases.

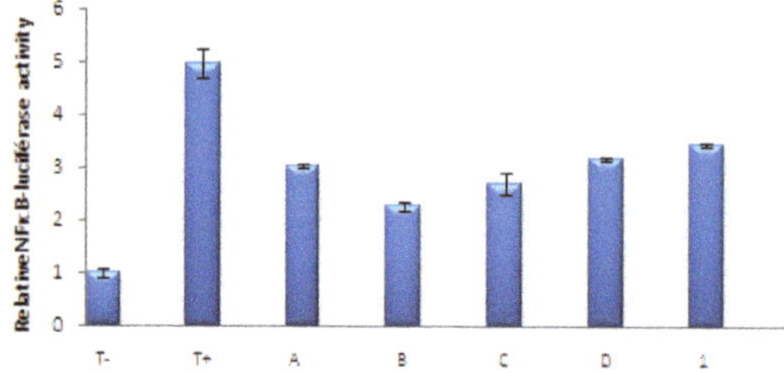

Figure 7. Relative NFκB-luciferase activity of A, B, C and D with 17 μg/mL andrographolide.

These results could explain those reported in the literature [31], showing that there were other compounds than andrographolide in *A. paniculata* extracts that act as an immunostimulant, and we proved herein that there was a complementary activity of **2** at the studied concentrations.

3.4. Characterization of Products 1, 2 and 3

Extracted compounds were characterized by standard spectral methods, NMR, mass Spectroscopy (ESIMS) and high resolution mass spectroscopy (HRMS), and conform to already-described data [33].

1: ^{1}H NMR (DMSO, 400 MHz) δ 6.62 (1H, dt, $J_{11,12}$ = 7.0 Hz, $J_{12,14}$ 1.5 Hz, H12), 5.71 (1H, d, $J_{14,OH}$ 6.0 Hz, OH14), 5.05 (1H, d, $J_{3,OH}$ 5.0 Hz, OH-3), 4.91 (1H, t, $J_{14,15}$ 6.0 Hz, H14), 4.81 (1H, s, H17A), 4.62 (1H, s, 17B), 4.39 (1H, dd, $J_{15A,15B}$ 10Hz, $J_{14,15A}$ 6.0 Hz, H15A), 4.13 (1H, dd, $J_{19B,OH}$ 7.5 Hz, $J_{19A,OH}$ 2.5 Hz, OH-19), 4.03 (1H, dd, $J_{15A,15B}$ 10.0 Hz, $J_{14,15B}$ 2.0 Hz, H15B), 3.84 (1H, dd, $J_{19A,19B}$ 11.0 Hz, $J_{19A,OH}$ 2.5 Hz, H19A), 3.29–3.20 (2H, m, H3, H19B), 2.46 (1H, m, H11A), 2.32 (1H, br d, $J_{7A,7B}$ 13.0 Hz, H7A), 1.6–1.84 (2H, m, H7B, H11B), 1.75–1.61 (5H, m, H1A, H2, H6A, H9), 1.35 (1H, dq, $J_{6A,6B}$ $J_{6B,7B}$ 13.0 Hz, $J_{6B,7A}$ 4.0 Hz, H6B), 1.23–1.17 (2H, H1B, H5), 1.08 (3H, s, H18), 0.66 (3H, s, H20). ^{13}C NMR (DMSO, 100 MHz) δ 170.8 (C16), 148.5 (C8), 147.2 (C12), 129.9 (C13), 109.2 (C17), 79.3 (C3), 75.2 (C15), 65.4 (C14), 63.6 (C19), 56.4 (C9), 55.3 (C5), 43.1 (C10), 39.5 (C4), 38.4 (C7), 38.1 (C11), 37.4 (C1), 28.8 (C2), 24.8 (C6), 24.0 (C18), 15.6 (C20). ESIMS (negative mode) m/z 331 [M-H$_2$O]$^{-}$, 349 [M-H]$^{-}$, 395 [M-H+HCOOH]$^{-}$, (positive mode) 351 [M+H]$^{+}$, 373 [M+Na]$^{+}$. HRMS: calcd for C$_{20}$H$_{31}$O$_5$ [M+H]$^{+}$ 351.2167, found 351.2166.

2: ^{1}H NMR (DMSO, 400 MHz) δ 7.65 (1H, brs, H14), 6.74 (1H, dd, $J_{11,12}$ 15.5 Hz, $J_{9,11}$ 10.0 Hz, H11), 6.12 (1H, d, $J_{11,12}$ 15.5 Hz, H12), 5.03 (1H, d, $J_{3,OH}$ 5.0 Hz, OH-3), 4,89 (2H, brs, H15), 4.73 (1H, s, H17A), 4.42 (1H, s, H17B), 4.12 (1H, t, $J_{OH,19B}$ 6.5 Hz, OH-19), 3.84 (1H, dd, $J_{19A,19B}$ 13.0 Hz, $J_{19A,OH}$ 5.0 Hz, H19A), 3.28–3.16 (2H, m, H3, 19B), 2.36–2.32 (2H, m, H9, H7A), 1.97 (1H, m, H7B), 1.72 (1H, m, H2A), 1.58–1.54 (2H, m, H2, H6A), 1.40 (1H, dd, $J_{6A,6B}$ 13.0 Hz, $J_{6B,7B}$ 4.0 Hz, H6B), 1.28–1.12 (2H, m, H1), 1.07 (3H, s, H18), 0.76 (3H, s, H20). ^{13}C NMR (DMSO, 100 MHz) δ 172.9 (C16), 149.5 (C8), 147.2 (C14), 134.7 (C11), 127.5 (C13), 121.6 (C12), 108.5 (C17), 79.1 (C3), 63.1 (C19), 61.0 (C9), 54.2 (C5), 42.8 (C4), 38.4 (C1), 36.7 (C7), 29.4 (C10), 28.1 (C2), 23.6 (C6), 23.4 (C18), 15.9 (C20). ESIMS (negative mode) m/z 331 [M-H]$^{-}$, 377 [M-H+HCOOH]$^{-}$, (positive mode) 355 [M+Na]$^{+}$. HRMS: calcd for C$_{20}$H$_{28}$O$_4$ [M+Na]$^{+}$ 355.2035, found 355.2036.

3: ^{1}H NMR (DMSO, 400 MHz) δ7.46 (1H, s, H14), 4.86–4.78 (6H, 3OH, H15, H17A), 4.59 (1H, s, H17B), 4.38 (1H, t, $J_{6',OH}$ 6.0 Hz, OH-6'), 4.02 (1H, d, $J_{1',2'}$ 8.0 Hz, 3.88 (1H, d, $J_{19A,19B}$ 10.0 Hz, H19A), 3.63 (1H, dd, $J_{6'A,6'B}$ 10.5 Hz, $J_{5',6'A}$ 6.0 Hz, H6A), 3.43 (1H, dd, $J_{6'A,6'B}$ 10.5 Hz, $J_{5',6'B}$ 5.0 Hz, H6B), 3.17–3.04 (3H, m, H3', H4', H5'), 2.92 (1H, m, H2'), 2.36–0.90 (16H, m, H1, H2, H3, H5, H6, H7, H9, H11, H12), 0.95 (3H, s, H18), 0.61 (3H, s, H20). ^{13}C NMR (DMSO, 100 MHz) δ 174.6 (C16), 148.1 (C8), 147.4 (C14), 132.6 (C13), 107.0 (C17), 104.0 (C1'), 77.3, 77.1, 70.6 (C3', C4', C5'), 74.0 (C2'), 71.2 (C15), 70.9 (C19), 61.5 (C6'), 56.1 (C9), 55.8 (C5), 38.8, 38.4, 36.1, 21.9, 19.0 (C1, C2, C3, C6, C7, C11, C12), 38.3 (C4), 28.0 (C18), 24.4 (C10), 15.6 (C20). ESIMS (negative mode) m/z 479 [M-H]$^{-}$, 525 [M-H+HCOOH]$^{-}$, (positive mode) 503 [M+Na]$^{+}$. HRMS: calcd for C$_{26}$H$_{41}$O$_8$ [M+H]$^{+}$ 481.2795, found 481.2796.

4. Conclusions

The isolation of natural andrographolide derivatives allowed to modify their structure and to study the inhibition of NFκB transcriptional activity of pure analogues. Compound **5** highlighted the best anti-inflammatory activity from the nine tested derivatives. The HPLC quantitation method has been developed in order to quantify three major diterpenes in *A. paniculata* extracts. This method was applied for the quantitation of four commercial herbal preparations. The composition was compared to the inhibitory activity of NFκB of such preparations and showed that even if andrographolide **1** is the major and more potent active compound in the plant, the amount of didehydroandrographolide **2** is not negligible.

Author Contributions: S.B.-R. conceived and designed the experiments; E.V.-P., performed the experiments; E.V.-P., E.D., C.P., V.F. and S.B.-R. analyzed the data; S.B.-R. supervised; E.V.-P., J.F.C. and S.B.-R. wrote the paper.

Conflicts of Interest: The authors declare no conflict of interest.

References

1. Acharya, R. A review of phytoconstituents and their pharmacological properties of Andrographis Paniculata (NEES). *Int. J. Pharma Biol. Sci.* **2017**, *8*, 77–83. [CrossRef]
2. Chowdhury, A.; Biswas, S.K.; Raihan, S.Z.; Das, J.; Paul, S. Pharmacological Potentials of Andrographis paniculata: An Overview. *Int. J. Pharmacol.* **2012**, *8*, 6–9. [CrossRef]
3. Hossain, M.S.; Urbi, Z.; Sule, A.; Rahman, K.M.H. *Andrographis paniculata* (Burm. f.) Wall. ex Nees: A Review of Ethnobotany, Phytochemistry, and Pharmacology. *Sci. World J.* **2014**, *2014*, 28. [CrossRef]
4. Zhang, L.; Bao, M.; Liu, B.; Zhao, H.; Zhang, Y.; Ji, X.; Zhao, N.; Zhang, C.; He, X.; Yi, J.; et al. Effect of Andrographolide and Its Analogs on Bacterial Infection: A Review. *Pharmacology* **2019**. [CrossRef]
5. Grainne, H.; Franco, S.; Antonio, G.; Diego, C. Herbal medicinal products for inflammatory bowel disease: A focus on those assessed in double-blind randomised controlled trials. *Phytother. Res.* **2019**. [CrossRef]
6. Rahman, H.; Kim, M.; Leung, G.; Green, J.A.; Katz, S. Drug-Herb Interactions in the Elderly Patient with IBD: A Growing Concern. *Curr. Treat. Options Gastroenterol.* **2017**, *15*, 618–636. [CrossRef]
7. Farzaei, M.H.; Shahpiri, Z.; Bahramsoltani, R.; Moghaddamnia, M.; Najafi, F.; Rahimi, R. Efficacy and Tolerability of Phytomedicines in Multiple Sclerosis Patients: A Review. *CNS Drugs* **2017**, *31*, 867–889. [CrossRef] [PubMed]
8. Tan, W.S.D.; Liao, W.; Zhou, S.; Wong, W.S.F. Is there a future for andrographolide to be an anti-inflammatory drug? Deciphering its major mechanisms of action. *Biochem. Pharmacol.* **2017**, *139*, 71–81. [CrossRef]
9. Majee, C.; Mazumder, R.; Choudhary, A.N. Medicinal plants with anti-ulcer and hepatoprotective activity: A review. *Int. J. Pharm. Sci. Res.* **2019**, *1*, 1–11. [CrossRef]
10. Chua, L.S. Review on Liver Inflammation and Antiinflammatory Activity of *Andrographis paniculata* for Hepatoprotection. *Phytother. Res.* **2014**, *28*, 1589–1598. [CrossRef] [PubMed]
11. Chao, W.-W.; Lin, B.-F. Hepatoprotective Diterpenoids Isolated from *Andrographis paniculata*. *Chin. Med.* **2012**, *3*, 136–143. [CrossRef]
12. Lu, J.; Ma, Y.; Wu, J.; Huang, H.; Wang, X.; Chen, Z.; Chen, J.; He, H.; Huang, C. A review for the neuroprotective effects of andrographolide in the central nervous system. *Biomed. Pharmacother.* **2019**, *117*, 109078. [CrossRef] [PubMed]
13. Vijaya, N.R.; Abinaya, R. An effect of cardioprotective activity in various medicinal plants-a review. *Int. J. Curr. Pharm. Res.* **2019**, *11*, 1–6. [CrossRef]
14. Rajendran, H.; Deepika, S.; Immanuel, S.C. An Overview of Medicinal plants for Potential Cardio-Protective Activity. *Res. J. Biotechnol.* **2017**, *4*, 104–113.
15. Shourie, A.; Chandwani, A.; Singh, S.; Rawal, A.; Vijayalakshmi, U. Anticancerous potential of medicinal plants—A review. *World J. Pharm. Res.* **2018**, *6*, 462–483. [CrossRef]
16. Kumar, M.S.; Swati, T.; Archana, S.; Hyun, O.S.; Mook, K.H. Andrographolide and analogues in cancer prevention. *Front. Biosci. (Elite Ed.)* **2015**, *7*, 255–266. Available online: www.bioscience.org/2015/v7e/af/732/fulltext.htm (accessed on 9 November 2019).
17. Gorter, M. The Bitter Constituent of Andrographis Paniculata Nees. *Recueil des Travaux Chimiques des Pays-Bas* **1911**, *30*, 151–160. [CrossRef]
18. Jayakumar, T.; Hsieh, C.-Y.; Lee, J.-J.; Sheu, J.-R. Experimental and Clinical Pharmacology of *Andrographis paniculata* and Its Major Bioactive Phytoconstituent Andrographolide. *Evid. Based Complementary Altern. Med.* **2013**, *2013*, 16. [CrossRef]
19. Cava, M.P.; Chan, W.R.; Stein, R.P.; Willis, C.R. Andrographolide: Further transformations and stereochemical evidence; the structure of isoandrographolide. *Tetrahedron* **1965**, *21*, 2617–2632. [CrossRef]
20. Smith, A.B.; Toder, B.H.; Carroll, P.; Donohue, J. Andrographolide: An X-ray crystallographic analysis. *J. Crystallogr. Spectrosc. Res.* **1982**, *12*, 309–319. [CrossRef]
21. Medforth, C.J.; Chang, R.S.; Chen, G.-Q.; Olmstead, M.M.; Smith, K.M. A conformational study of diterpenoid lactones isolated from the Chinese medicinal herb *andrographis paniculata*. *J. Chem. Soc. Perkin.* **1990**, *2*, 1011–1016. [CrossRef]

22. Xia, Y.F.; Ye, B.Q.; Li, Y.D.; He, X.J.; Lin, X.; Yao, X.; Ma, D.; Slungaard, A.; Hebbel, R.P.; Key, N.S.; et al. Andrographolide attenuates inflammation by inhibition of NF-kappa B activation through covalent modification of reduced cysteine 62 of p50. *J. Immunol.* **2004**, *173*, 4207–4217. [CrossRef] [PubMed]

23. Li, W.; Fitzloff, J. HPLC with Evaporative Light Scattering Detection as a Tool to Distinguish Asian Ginseng (Panax ginseng) and North American Ginseng (Panax quinquefolius). *J. Liq. Chromatogr. Relat. Technol.* **2002**, *25*, 1335–1343. [CrossRef]

24. Xu, L.; Xiao, D.-W.; Lou, S.; Zou, J.-J.; Zhu, Y.-B.; Fan, H.-W.; Wang, G.-J. A simple and sensitive HPLC–ESI-MS/MS method for the determination of andrographolide in human plasma. *J. Chromatogr. B* **2009**, *877*, 502–506. [CrossRef]

25. Srivastava, A.; Misra, H.; Verma, R.; Gupta, M. Chemical fingerprinting of *Andrographis paniculata* using HPLC, HPTLC and densitometry. *Phytochem. Anal.* **2004**, *15*, 280–285. [CrossRef]

26. Gu, Y.; Ma, J.; Liu, Y.; Chen, B.; Yao, S. Determination of andrographolide in human plasma by high-performance liquid chromatography/mass spectrometry. *J. Chromatogr. B* **2007**, *854*, 328–331. [CrossRef]

27. Chen, L.; Jin, H.; Ding, L.; Zhang, H.; Wang, X.; Wang, Z.; Li, J.; Qu, C.; Wang, Y.; Zhang, H. On-line coupling of dynamic microwave-assisted extraction with high-performance liquid chromatography for determination of andrographolide and dehydroandrographolide in *Andrographis paniculata* Nees. *J. Chromatogr. A* **2007**, *1140*, 71–77. [CrossRef]

28. Ding, L.; Luo, X.B.; Tang, F.; Yuan, J.B.; Guo, M.; Yao, S.Z. Quality control of medicinal herbs Fructus gardeniae, Common Andrographis Herb and their preparations for their active constituents by high-performance liquid chromatography–photodiode array detection–electrospray mass spectrometry. *Talanta* **2008**, *74*, 1344–1349. [CrossRef]

29. Du, Q.; Jerz, G.; Winterhalter, P. Separation of andrographolide and neoandrographolide from the leaves of *Andrographis paniculata* using high-speed counter-current chromatography. *J. Chromatogr. A* **2003**, *984*, 147–151. [CrossRef]

30. Nanduri, S.; Nyavanandi, V.K.; Thunuguntla, S.S.R.; Velisoju, M.; Kasu, S.; Rajagopal, S.; Kumar, R.A.; Rajagopalan, R.; Iqbal, J. Novel routes for the generation of structurally diverse labdane diterpenes from andrographolide. *Tetrahedron Lett.* **2004**, *45*, 4883–4886. [CrossRef]

31. Gomez-Bombarelli, R.; Calle, E.; Casado, J. Mechanisms of Lactone Hydrolysis in Neutral and Alkaline Conditions. *J. Org. Chem.* **2013**, *78*, 6868–6879. [CrossRef]

32. Puri, A.; Saxena, R.; Saxena, R.P.; Saxena, K.C.; Srivastava, V.; Tandon, J.S. Immunostimulant agents from *Androgravnis Paniculata*. *J. Nat. Prod.* **1993**, *56*, 995–999. [CrossRef]

33. Villedieu-Percheron, E. Study about an Anti-Inflammatory Natural Diterpenes Family. Ph.D. Thesis, University of Orléans, Orléans, France, 23 September 2011. Available online: https://tel.archives-ouvertes.fr/tel-00687029 (accessed on 9 November 2019).

 *foods*

Article

In Vitro Studies of Fermented Korean Chung-Yang Hot Pepper Phenolics as Inhibitors of Key Enzymes Relevant to Hypertension and Diabetes

Su-Jung Yeon [1], Ji-Han Kim [2], Won-Young Cho [3], Soo-Ki Kim [4], Han Geuk Seo [3] and Chi-Ho Lee [3,*]

1   Department of Agricultural, Life and Environmental Science, Tottori University, Tottori 680-8550, Japan; sujung0811@gmail.com
2   AgResearch (Grasslands Research Centre), Palmerston North 4442, New Zealand; aaddoo@nate.com
3   Department of Food Science and Biotechnology of Animal Resources, Konkuk University, Seoul 05029, Korea; ready1838@naver.com (W.-Y.C.); hgseo@konkuk.ac.kr (H.G.S.)
4   Department of Animal Science and Technology, Konkuk University, Seoul 05029, Korea; sookikim@konkuk.ac.kr
*   Correspondence: leech@konkuk.ac.kr; Tel.: +82-02-450-3681

Received: 21 August 2019; Accepted: 8 October 2019; Published: 14 October 2019

Abstract: This study was investigated to evaluate the antioxidant activity, the angiotensin I-converting enzyme (ACE) inhibition effect, and the α-amylase and α-glucosidase inhibition activities of hot pepper water extracts both before and after their fermentation. The fermented pepper water extract (FP) showed significantly higher total phenol content, 1,1-diphenyl-2-picrylhydrazyl (DPPH) radical inhibition effect, metal chelating activity and ACE inhibition activity compared to the non-fermented raw pepper water extract (RP) ($p < 0.05$). Meanwhile, the FP showed lower α-amylase and higher α-glucosidase inhibitory activities, but the RP showed similar levels of α-amylase and α-glucosidase inhibitory activities. Taken together, these results suggested that fermented pepper extract using water should be expected to have potentially inhibitory effects against both hyperglycemia and hypertension.

Keywords: pepper; fermentation; hyperglycemia; angiotensin I-converting enzyme (ACE) inhibition; antioxidant

1. Introduction

Hypertension is a condition of oxidative stress, therefore a decline in the antioxidant system can be fatal in hypertension patients, which can cause atherosclerosis and other hypertension-induced organ injuries [1]. Several researchers have reported the side effects of artificial antioxidants such as potential health hazards, for example, carcinogens [2,3]. Additionally, concerns about synthetic products have caused increasing interest in natural antioxidants made from food or other bioresources. Recently, there has been active interest in the research-based development of natural antioxidants such as vitamin E and ascorbic acid [3–5].

Meanwhile, angiotensin I is activated by renin secreted from the kidney; and angiotensin I-converting enzyme (ACE) changes angiotensin I to angiotensin II [6]. Angiotensin II leads to cardiovascular disease and vasoconstriction [7]. There are two kinds of prescription for treating hypertension—angiotensin converting enzyme inhibitor (ACEI) and angiotensin receptor blocker (ARB). Mostly, ACEI is used first with patients, because ACEI significantly reduces all-cause mortality at a rate greater than that of ARB [8,9]. Lee et al. [7] reported that general ACEI—such as captopril, quinapril, enalapril, lisinopril—has side effects such as cough, hypotension, and inflammatory response [10]. Therefore, the development of treatments having no side effect is needed, and much research along these lines has been reported [11,12].

α-amylase and α-glucosidase are related to the hydrolysis of dietary carbohydrates and the absorbance of glucose from the small intestine to the blood, respectively [13,14]. Control of these enzymes can play a key role in treating patients with diabetes mellitus [6,13]. Several inhibitors of these enzymes—such as acarbose, trestatin, and amylostatin—have recently shown adverse effects such as abdominal distention and meteorism [15]. This occurs because of the fermentation of undigested carbohydrates by anaerobic bacteria in the colon [13,16,17]. Thus, comparatively riskless inhibitors should be developed.

Pepper is currently being widely used for cooking in both traditionally and contemporary cuisine, and it contains a variety of phytochemicals such as capsaicin and ascorbate [18]. Capsaicin has an anti-obesity effect [19], can be used as an analgesic [20], and is an antioxidant [21]. However, due to its pungency, it is difficult to use it as a food additive in the food industry. In our previous studies, pepper fermented by *Bacillus licheniformis* showed significantly decreased capsaicin content, and pungency was decreased also, and this microorganism can utilize capsaicin as an energy source [22–24].

This study was performed to evaluate the antioxidant activity and the key enzyme inhibition effects related with hypertension and diabetes mellitus using fermented pepper, for the purpose of developing new functional food ingredients.

2. Materials and Methods

2.1. Materials

The pepper was obtained from a local market (Hwayang-dong, Seoul, Korea). All reagents were obtained from Sigma-Aldrich (St. Louis, MO, USA).

The pepper was blended and freeze dried. To prepare fermented pepper, 40 g of powdered pepper was mixed with 1 L of distilled water. After sterilization, the sample was inoculated with 5% of *Bacillus licheniformis* SK1230. It was then incubated for 11 days at 37 °C in a shaking incubator (Jisico, J-SIL-R, Seoul, Korea). After fermentation, the fermented pepper was freeze dried. For the raw pepper, the same amount of sterilized (inactive) Luria–Bertani (LB) broth (NaCl 10 g/L, tryptone 10 g/L, and yeast extract 5 g/L) was added (to substitute for the presence of microorganism), then freeze dried. Powdered raw pepper and fermented pepper were stored at −20 °C before use and were used as samples in this experiment. According to previously reported capsaicin analysis method [22], the amount of capsaicin in raw hot pepper was 1.31 mg/g and that of fermented pepper was significantly decreased to 1.20 mg/g ($p < 0.05$).

To prepare the ACE solution, 1 g of commercial rabbit lung acetone powder (Sigma-Aldrich, St. Louis, MO, USA) in 10 mL buffer (0.05 M $Na_2B_4O_7$: 0.2 M H_3BO_3 = 135 : 165, pH 8.3) was stirred at 4 °C for 24 h in dark room. Then, it was centrifuged at 12,000 rpm for 30 min. The supernatant was used as the ACE solution [25].

Dinitrosalicylic acid (DNS) reagent was prepared by mixing 1 g of DNS with 50 mL of distilled water. Then, 30 g of sodium potassium tartrate tetrahydrate was slowly added. 20 mL of 2 N NaOH was also added to the mixture, and distilled water was added until the mixture totaled 100 mL [26].

2.2. Extraction

The extract was prepared by means of a modified method published by Kwon et al. [13]. Of the sample 0.4 g was extracted with 10 mL of distilled water, over a period of 3 h. After centrifuging at 9300× *g* for 10 min, the supernatant was filtered using Whatman filter paper No. 2, and it was used for experimentation. (The "raw pepper water extract" and the "fermented pepper water extract" were expressed as RP and FP in this study, respectively.)

2.3. Total Phenol Compounds Content

The total content of the phenol compounds was determined according to literature [27]. Of the extract 30 μL was mixed with the same amount of 95% ethanol, 150 μL of distilled water, and 15 μL of 1 N Folin–Ciocalteu reagent. This mixture was allowed to react for 5 min, after which time 5% Na_2CO_3 30 µL was added. After 1 h, absorbance was measured at 725 nm. Gallic acid was used to establish a standard curve, and results were expressed as mg of gallic acid equivalents per gram of dried sample weight.

2.4. 1,1-Diphenyl-2-Picrylhydrazyl (DPPH) Radical Scavenging Activity

DPPH (1,1-diphenyl-2-picrylhydrazyl) radical scavenging activity by Blois [28] were performed to evaluate antioxidant activity of the extract. Of the extract, 20 µL was reacted with 180 µL of 0.1 mM DPPH in a 96 well plate. The absorbance was recorded at 517 nm after 30 min in a dark room. Ethanol, instead of extract, was used for the control. DPPH radical inhibition was calculated by:

$$\text{Inhibition (\%)} = (1 - A_{517}(\text{extract})/A_{517}(\text{control})) \times 100 \tag{1}$$

2.5. Metal Chelating Activity

Another antioxidant activity of extract was performed by metal chelating activity [29]. Of the extract, 100 µL was mixed with 50 µL of 2 mM ferrous chloride, and the reaction was started by adding 0.2 mL of 5 mM ferrozine. Then, absorbance was recorded at 562 nm after 10 min in a dark room. Distilled water, instead of extract, was used for the control. Metal chelating activity was calculated by:

$$\text{Inhibition (\%)} = (1 - A_{562}(\text{extract})/A_{562}(\text{control})) \times 100 \tag{2}$$

2.6. Angiotensin I-Converting Enzyme (ACE) Inhibition Effect

The ACE inhibition effect was performed by Chushman et al.'s method [25]. An extract of 40 µL, 12.5 mM of hippuryl-histidyl-leucine in an 80 µL buffer (0.05 M $Na_2B_4O_7$: 0.2 M H_3BO_3 = 135 : 165, 0.4 M NaCl, pH 8.3) and an ACE solution of 80 µL was reacted for 30 min in a 37 °C water bath. Then, 200 µL 1 N HCl was added to stop the reaction. Ethylacetate (1.2 mL) was added, and the mixture was vortexed for 5 min. After centrifugation at 3000 rpm for 15 min, the supernatant of 0.8 mL was allowed to stand at 90 °C for removal of ethylacetate. After that, the residue was mixed with 2.4 mL of distilled water, and the absorbance was recorded at 228 nm (OPTIZEN 2120UV, Mecasys Co. Ltd., Daejeon, Korea). The blank contained HCl from the beginning to prevent the reaction of substrate and enzyme. Distilled water, instead of extract, was used for the control. The ACE inhibition activity was calculated as below:

$$\text{Inhibition (\%)} = (1 - (A_{228}(\text{extract}) - A_{228}(\text{extract blank}))/(A_{228}(\text{control}) - A_{228}(\text{control blank}))) \times 100 \tag{3}$$

2.7. α-Amylase Inhibition Effect

The α-amylase inhibition effect was evaluated by a modification of the assay described by the Worthington Enzyme Manual [26]. Of the extract, 30 µL and 0.02 M of sodium phosphate buffer (pH 6.9) containing 400 U/mL of α-amylase was mixed and pre-incubated at 25 °C for 10 min. After adding 30 µL of 1% starch solution in 0.02 M sodium phosphate buffer (pH 6.9) to the mixture, it was stored at 25 °C for 10 min. To stop the reaction, 60 µL of DNS color reagent was added. Then, it was allowed to stand in a boiling water bath for 5 min, prior to cooling to room temperature. After dilution with distilled water in the amount of 0.4 mL, absorbance was read at 540 nm. Only 0.02 M sodium phosphate buffer (pH 6.9), instead of buffer containing α-amylase and extract, was used for the blank and the control, respectively. Percentage of inhibition was calculated as follows:

$$\text{Inhibition (\%)} = (1 - (A_{540}(\text{extract}) - A_{540}(\text{blank}))/A_{540}(\text{control})) \times 100 \tag{4}$$

2.8. α-Glucosidase Inhibitory Activity

The α-glucosidase inhibitory activity was measured by the modified method of Ranilla et al. [6]. Of the extract, 50 µL and 0.1 M potassium phosphate buffer (pH 6.9) containing 1 U/mL of α-glucosidase, was mixed and pre-incubated at 25 °C for 10 min. After adding 50 µL of 5 mM p-nitro-phenyl-α-D-glucopyranoside in a 0.1 M potassium phosphate buffer (pH 6.9) to the mixture, absorbance was measured at 405 nm. Then, after incubating the mixture at 25 °C for 5 min, absorbance was read at 405 nm. Potassium phosphate buffer (0.1 M, pH 6.9), in the place of extract, was used for control. Inhibitory activity was calculated as below:

$$\text{Inhibition (\%)} = (1 - (\Delta A_{405}(\text{extract}))/\Delta A_{405}(\text{control})) \times 100 \tag{5}$$

2.9. Statistical Analysis

All results were stated as mean ± standard error of mean (SEM) and, analyzed by *t*-test using SPSS 18.0 (SPSS Inc., Chicago, IL, USA). The experiments were performed in triplicates. Significant difference was evaluated by $p < 0.05$.

3. Results

3.1. Amount of Total Phenol Contents

Total phenol contents in extracts from raw and fermented peppers varied from 150 to 250 µg/mL (Table 1), and it was similar to the range reported by Kwon et al. [18]. The total phenol content of FP was 40% higher than that of the raw pepper extract significantly ($p < 0.05$).

Table 1. The amounts of total phenol compound in pepper extract, both before and after fermentation (gallic acid µg/mL).

	RP	FP
Total phenol compound (gallic acid µg/mL)	156.47 ± 2.12 [b]	224.55 ± 2.51 [a]

RP, raw pepper extract prepared using water. FP, fermented pepper extract prepared using water. [a,b] Superscripts with different letters indicate significant difference ($p < 0.05$).

3.2. Antioxidant Activity

Antioxidant activities were determined by both DPPH radical inhibition assay and metal chelating assay (Table 2). The FP showed significantly higher radical inhibition activity and chelating activity than did those of the RP ($p < 0.05$). Meanwhile, metal chelating activity was significantly higher than DPPH radical inhibition activity in both RP and FP ($p < 0.05$).

Table 2. Antioxidant activities of raw and fermented pepper extracts.

	RP	FP
DPPH radical inhibition (%)	52.00 ± 0.28 [b B]	66.12 ± 0.72 [a B]
Metal chelating activity (%)	67.32 ± 0.75 [b A]	71.99 ± 0.82 [a A]

RP, raw pepper extract prepared using water. FP, fermented pepper extract prepared using water. DPPH: 1,1-diphenyl-2-picrylhydrazyl. [a,b] Superscripts with different letters indicate significant difference between RP and FP ($p < 0.05$). [A,B] Superscripts with different letters indicate significant difference between DPPH radical inhibition and metal chelating activity ($p < 0.05$).

3.3. ACE Inhibition Effect

The inhibition rate of angiotensin I-converting enzyme is exhibited in Table 3. The water extract of fermented pepper showed a significantly higher inhibition rate as compared to that of raw pepper extract ($p < 0.05$).

Table 3. Angiotensin I-converting enzyme (ACE) inhibition effects of raw and fermented water-prepared pepper extracts.

	RP	FP
ACE inhibition (%)	54.29 ± 0.71 [b]	71.04 ± 0.58 [a]

RP, raw pepper extract prepared using water. FP, fermented pepper extract prepared using water. ACE: angiotensin I-converting enzyme. [a,b] Superscripts with different letters indicate significant difference ($p < 0.05$).

3.4. α-Amylase Inhibition and α-Glucosidase Inhibitory Activity

α-amylase and α-glucosidase inhibition activity were seen in the RP and FP (Table 4). Fermented pepper extract exhibited significantly lower α-amylase and α-glucosidase inhibition rates than the RP ($p < 0.05$). Meanwhile, α-amylase and α-glucosidase inhibition activity in RP were similar each other ($p > 0.05$), but in FP, low α-amylase and high α-glucosidase inhibition activity were shown significantly ($p < 0.05$).

Table 4. Inhibition rates of α-amylase in raw pepper water extract (RP) and fermented pepper water extract (FP).

	RP	FP
α-amylase inhibition (%)	25.15 ± 0.27 [a NS]	8.26 ± 0.16 [b B]
α-glucosidase inhibition (%)	24.32 ± 0.86 [a]	14.47 ± 0.67 [b A]

RP, raw pepper extract prepared using water. FP, fermented pepper extract prepared using water. [a,b] Superscripts with different letters indicate significant difference between RP and FP ($p < 0.05$). [A,B] Superscripts with different letters indicate significant difference between amylase and glucosidase inhibition ($p < 0.05$). [NS] Not significant between amylase and glucosidase inhibition ($p > 0.05$).

4. Discussion

Consistent with this study, Sim and Han [30] reported that the amount of phenolic compounds derived from red pepper seed on Kimchi significantly increased during fermentation ($p < 0.05$). Generally, phenolics exist as conjugated forms together with sugars or other moieties. Glucosidase is related to the hydrolysis of glycosidic linkages, and it leads to the release of free phenolics [31]. In this study, glucosidase inhibition activity of FP decreased significantly as compared to that of the RP (Table 4, $p < 0.05$). This means that the FP had higher glucosidase activity than the RP. Therefore, it suggested that it resulted in the release of free phenolics with increasing glucosidase activity during fermentation of the pepper.

Meanwhile, antioxidant activity generally correlates with the amount of phenolic compounds. Vattem et al. [31] reported that total phenolics and antioxidant activity correlated positively with glucosidase activity. Accordingly, significantly higher DPPH radical inhibition effect in the FP was suggested due to a higher total phenolics level and a lower glucosidase inhibition rate in comparison to the RP (Tables 1 and 4, $p < 0.05$). There were several ways to protect oxidation. Elimination of over produced free radicals which can react with biomolecules and cause chronic diseases [32] and heavy metals by chelating agent from fermented pepper extract would be the way for antioxidation.

There have been several reports about ACE inhibition activities by fermented milk peptides [33], flavanol-rich foods [34], and various fermented foods such as soy sauce, fish sauce, natto, and cheese [35]. Actis-Goretta et al. [34] referenced that the ACE inhibition effect was related to amounts of phenolics and flavanols. Meanwhile, Okamoto et al. [35] implied, on the basis of their research, that an ACE inhibitor would be produced during fermentation. For example, cottage cheese—which has no maturing step in the manufacturing process—did not show ACE inhibition activity. By contrast, red cheddar, blue, and camembert cheese—all of which have a maturing process—showed ACE inhibition effect. In this study, pepper exhibited a significantly higher amount of total phenolics after fermentation ($p < 0.05$). A relationship with higher ACE inhibition activity than before fermentation has been suggested. Higher ACE inhibition means lower conversion of angiotensin I to angiotensin II, a potent vasoconstrictor, and it seems to be a strategic treatment to protect against hypertension—which is a representative complication of diabetes [6].

Ranilla et al. [6] recently reported that high phenolic-linked plants showed a high α-glucosidase inhibition rate with a low activity of α-amylase. Consistent with this result, FP—having a higher total phenol contents than the RP—had higher α-glucosidase inhibition activity with lower α-amylase inhibition significantly ($p < 0.05$). Martin and Montgomery [36] referenced a finding that such a result could have functionality regarding the potential controlling of glucose absorption and of adverse effects from high α-amylase inhibition activity.

In this study, we fermented pepper for 11 days with the intention of "capsaicin decrease" not "no capsaicin" in pepper for adjustable pungency. If the effect of fermented pepper is higher than that of raw hot pepper, it is thought to be resulted from the degradation of the ingredient of capsaicin or other fermentation by-product even if capsaicin remains in the fermented pepper. Indeed, the chromatogram of fermented pepper showed a greater area in some peaks compared to that of raw hot pepper (blue square in Figure 1) and what this compound is and whether it has functionality will be studied in future.

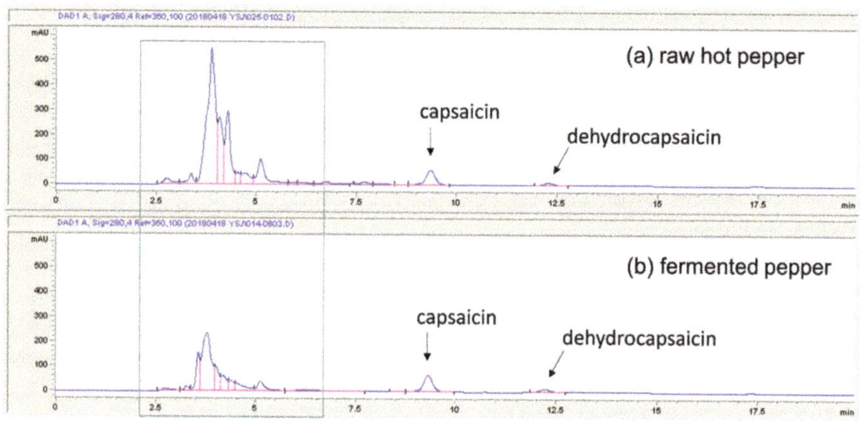

Figure 1. The chromatogram of raw hot pepper and fermented pepper.

5. Conclusions

Fermented pepper extract using water was evaluated about antioxidant activity, ACE and enzymes relevant for hyperglycemia inhibition effects. Results showed significantly higher total phenol contents, DPPH radical inhibition activity, metal chelating activity, and ACE inhibition effect than those of RP ($p < 0.05$). Furthermore, lower α-amylase and higher α-glucosidase inhibition effects in FP could have inhibitory potential in diabetes. Therefore, it is suggested that water extract of fermented pepper might be appropriate for a functional prevention ingredient of hypertension and hyperglycemia as a natural resource.

Author Contributions: Conceptualization, S.-J.Y.; methodology, S.-J.Y.; formal analysis, S.-J.Y.; resources, S.-K.K.; data curation, S.-J.Y.; writing—original draft preparation, S.-J.Y.; writing—review and editing, J.-H.K., W.-Y.C., H.G.S., and C.-H.L.; supervision, C.-H.L.; project administration, C.-H.L.

Funding: This work was supported by the National Research Foundation of Korea (NRF) grant funded by the Korean Government (Ministry of Science, ICT and Future Planning) (No. 2014K1A3A1A08045094).

Conflicts of Interest: The authors declare no conflict of interest.

References

1. Redón, J.; Oliva, M.R.; Tormos, C.; Giner, V.; Chaves, J.; Iradi, A.; Saez, G.T. Antioxidant activities and oxidative stress byproducts in human hypertension. *Hypertension* **2003**, *41*, 1096–1101. [CrossRef]
2. Kumar, K.S.; Ganesan, K.; Rao, P.V.S. Antioxidant potential of solvent extracts of *Kappaphycus alvarezii* (Dory) Doty–An edible seaweed. *Food Chem.* **2008**, *107*, 289–295. [CrossRef]
3. Ngo, D.H.; Vo, T.S.; Ngo, D.N.; Wijesekara, I.; Kim, S.K. Biological activities and potential health benefits of bioactive peptides derived from marine organisms. *Int. J. Biol. Macromol.* **2012**, *51*, 378–383. [CrossRef]
4. Li, J.; Lin, J.; Xiao, W.; Gong, Y.; Wang, M.; Zhou, P.; Liu, Z. Solvent extraction of antioxidants from steam exploded sugarcane bagasse and enzymatic convertibility of the solid fraction. *Bioresour. Technol.* **2013**, *130*, 8–15. [CrossRef]
5. Lorenzo, J.M.; Gonzalez-Rodriguez, R.M.; Sanchez, M.; Amado, I.R.; Franco, D. Effects of natural (grape seed and chestnut extract) and synthetic antioxidants (buthylatedhydroxytoluene, BHT) on the physical, chemical, microbiological and sensory characteristics of dry cured sausage "chorizo". *Food Res. Int.* **2013**, *54*, 611–620. [CrossRef]
6. Ranilla, L.G.; Kwon, Y.-I.; Apostolidis, E.; Shetty, K. Phenolic compounds, antioxidant activity and in vitro inhibitory potential against key enzymes relevant for hyperglycemia and hypertension of commonly used medicinal plants, herbs and spices in Latin America. *Bioresour. Technol.* **2010**, *101*, 4676–4689. [CrossRef] [PubMed]
7. Lee, S.E.; Song, J.; Seong, N.S. Analysis of ACE inhibitory activity. *Korean J. Crop. Sci.* **2004**, *49*, 203–207.
8. Cheng, J.; Zhang, W.; Zhang, X.; Han, F.; Li, X.; He, X.; Li, J. Effect of angiotensin-converting enzyme inhibitors and angiotensin II receptor blockers on all-cause mortality, cardiovascular deaths, and cardiovascular events in patients with diabetes mellitus. *JAMA Intern. Med.* **2014**, *174*, 773–785. [CrossRef] [PubMed]
9. Vark, L.C.V.; Bertrand, M.; Akkerhuis, K.M.; Brugts, J.J.; Fox, K.; Mourad, J.-J.; Boersma, E. Angiotensin-converting enzyme inhibitors reduce mortality in hypertension: A meta-analysis of randomized clinical trials of renin–angiotensin–aldosterone system inhibitors involving 158 998 patients. *Eur. Heart J.* **2012**, *33*, 2088–2097. [CrossRef]
10. Mark, K.S.; Davis, T.P. Stroke: Development, prevention and treatment with peptidase inhibitors. *Peptides* **2000**, *21*, 1965–1973. [CrossRef]
11. Lee, D.H.; Kim, J.H.; Park, J.S.; Choi, Y.J.; Lee, J.S. Isolation and characterization of a novel angiotensin I-converting enzyme inhibitory peptide derived from the edible mushroom *Tricholoma giganteum*. *Peptides* **2004**, *25*, 621–627.
12. Sheih, I.C.; Fang, T.J.; Wu, T.K. Isolation and characterization of a novel angiotensin I-converting enzyme (ACE) inhibitory peptide from the algae protein waste. *Food Chem.* **2009**, *115*, 279–284. [CrossRef]
13. Kwon, Y.-I.; Apostolidis, E.; Shetty, K. Evaluation of pepper (*Capsicum annuum*) for management of diabetes and hypertension. *J. Food Biochem.* **2007**, *31*, 370–385. [CrossRef]
14. Bischoff, H. Pharmacology of glucosidase inhibitor. *Eur. J. Clin. Investig.* **1994**, *24*, 3–10.
15. Puls, W.; Keup, U. Metabolic studies with an amylase. In *Recent Advances in Obesity Research*; Howard, A., Ed.; Newman Publisher: London, UK, 1975.
16. Bischoff, H.; Puls, W.; Krause, H.P.; Schutt, H.; Thomas, G. Pharmacological properties of the novel glucosidase inhibitors BAY m 1099 (miglitol) and BAY o 1248. *Diabetes Res. Clin. Pract.* **1985**, *1*, 53–62.
17. Horii, S.; Fukasse, K.; Matsuo, T.; Kameda, K.; Asano, N.; Masui, Y. Synthesis and α-D-glucosidase inhibitory activity of N-substituted valiolamine derivatives as potent oral antidiabetic agents. *J. Med. Chem.* **1987**, *29*, 1038–1046. [CrossRef]
18. Hwang, I.G.; Shin, Y.J.; Lee, S.; Lee, J.; Yoo, S.M. Effects of different cooking methods on the antioxidant properties of red pepper (*Capsicum annuum* L.). *Prev. Nutr. Food Sci.* **2012**, *17*, 286–292. [CrossRef]
19. Iwai, K.; Yazawa, A.; Watanabe, T. Roles as metabolic regulators of the non-nutrients, capsaicin and capsiate, supplemented to diets. *Proc. Jpn. Acad.* **2003**, *79*, 207–212. [CrossRef]
20. Cordell, G.A.; Araujo, O.E. Capsaicin: Identification, Nomenclature, and Pharmacotherapy. *Ann. Pharmacother.* **1993**, *27*, 330–336. [CrossRef]
21. Irena, P.; Malgorzata, M. Phenylalanine ammonia-lyase and antioxidant activities of lipophilic fraction of fresh pepper fruits *Capsicum annum* L. *Innov. Food Sci. Emerg. Technol.* **2001**, *2*, 189–192.

22. Yeon, S.J.; Kim, S.K.; Kim, J.M.; Lee, S.K.; Lee, C.H. Effects of fermented pepper powder on body fat accumulation in mice fed a high-fat diet. *Biosci. Biotechnol. Biochem.* **2013**, *77*, 2294–2297. [CrossRef] [PubMed]

23. Yeon, S.J.; Hong, G.E.; Kim, C.K.; Park, W.J.; Kim, S.K.; Lee, C.H. Effects of yogurt containing fermented pepper juice on the body fat and cholesterol level in high fat and high cholesterol diet fed rat. *Korean J. Food Sci. Anim. Resour.* **2015**, *35*, 479–485. [CrossRef] [PubMed]

24. Yeon, S.J.; Kim, J.H.; Hong, G.E.; Park, W.J.; Kim, S.K.; Seo, H.G.; Lee, C.H. Physical and sensory properties of ice cream containing fermented pepper powder. *Korean J. Food Sci. Anim. Resour.* **2017**, *37*, 36–41. [CrossRef] [PubMed]

25. Chushman, D.W.; Cheung, H.S. Spectrophotometric assay and properties of the angiotensin-converting enzyme of rabbit lung. *Biochem. Pharmacol.* **1971**, *20*, 1637–1648. [CrossRef]

26. Worthington, V. Alpha amylase. In *Worthington Enzyme Manual*; Worthington Biochemical Corporation, Ed.; Worthington Biochemical Corporation: Freehold, NJ, USA, 1993; pp. 36–41.

27. Ratnasari, N.; Walters, M.; Tsopmo, A. Antioxidant and lipoxygenase activities of polyphenol extracts from oat brans treated with polysaccharide degrading enzymes. *Heliyon* **2017**, *3*, e00351. [CrossRef] [PubMed]

28. Blois, M.S. Antioxidant determination by the use of a stable free radical. *Nature* **1958**, *181*, 1199–1200. [CrossRef]

29. Baakdah, M.M.; Tsopmo, A. Identification of peptides, metal binding and lipid peroxidation activities of HPLC fractions of hydrolyzed oat bran proteins. *J. Food Sci. Technol.* **2016**, *53*, 3593–3601. [CrossRef]

30. Sim, K.H.; Han, Y.S. Effect of red pepper seed on Kimchi antioxidant activity during fermentation. *Food Sci. Biotechnol.* **2008**, *17*, 295–301.

31. Vattem, D.A.; Shetty, K. Solid-state production of phenolic antioxidants from cranberry pomace by *Rhizopus Oligosporus*. *Food Biotechnol.* **2002**, *16*, 189–210. [CrossRef]

32. Wong, S.P.; Leong, L.P.; Koh, J.H.W. Antioxidant activities of aqueous extracts of selected plants. *Food Chem.* **2006**, *99*, 775–783. [CrossRef]

33. Hernández-Ledesma, B.; Miralles, B.; Amigo, L.; Ramos, M.; Recio, I. Identification of antioxidant and ACE-inhibitory peptides in fermented milk. *J. Sci. Food Agric.* **2005**, *85*, 1041–1048. [CrossRef]

34. Actis-Goretta, L.; Ottaviani, J.I.; Fraga, C.G. Inhibition of angiotensin converting enzyme activity by flavanol-rich foods. *J. Agric. Food Chem.* **2006**, *54*, 229–234. [CrossRef] [PubMed]

35. Okamoto, A.; Hanagata, H.; Matsumoto, E.; Kawamura, Y.; Koizumi, Y.; Yanagida, F. Angiotensin I converting enzyme inhibitory activities of various fermented foods. *Biosci. Biotechnol. Biochem.* **1995**, *59*, 1147–1149. [CrossRef] [PubMed]

36. Martin, A.; Mongtgomery, P. Acarbose: An alpha-glucosidase inhibitor. *Am. J. Health Syst. Pharm.* **1996**, *53*, 2277–2290. [CrossRef] [PubMed]

Review

Phenolic Acids of Plant Origin—A Review on Their Antioxidant Activity In Vitro (O/W Emulsion Systems) Along with Their In Vivo Health Biochemical Properties

Sotirios Kiokias [1], Charalampos Proestos [2] and Vassiliki Oreopoulou [3],*

[1] Research Executive Agency (REA), Place Charles Rogier 16, 1210 Bruxelles, Belgium; Sotirios.KIOKIAS@ec.europa.eu
[2] Laboratory of Food Chemistry, Department of Chemistry, National and Kapodistrian University of Athens, Panepistimiopolis Zografou, 15784 Athens, Greece; harpro@chem.uoa.gr
[3] Laboratory of Food Chemistry and Technology, School of Chemical Engineering, National Technical University of Athens, Iron Politechniou, 9, 15780 Athens, Greece
* Correspondence: vasor@chemeng.ntua.gr; Tel.: +30-210-772-3078

Received: 23 March 2020; Accepted: 22 April 2020; Published: 24 April 2020

Abstract: Nature has generously offered a wide range of herbs (e.g., thyme, oregano, rosemary, sage, mint, basil) rich in many polyphenols and other phenolic compounds with strong antioxidant and biochemical properties. This paper focuses on several natural occurring phenolic acids (caffeic, carnosic, ferulic, gallic, p-coumaric, rosmarinic, vanillic) and first gives an overview of their most common natural plant sources. A summary of the recently reported antioxidant activities of the phenolic acids in o/w emulsions is also provided as an in vitro lipid-based model system. Exploring the interfacial activity of phenolic acids could help to further elucidate their potential health properties against oxidative stress conditions of biological membranes (such as lipoproteins). Finally, this review reports on the latest literature evidence concerning specific biochemical properties of the examined phenolic acids.

Keywords: phenolic acids; emulsions; antioxidants; health properties

---

1. Introduction and Target of This Review

By "plant phenolics", we refer to a wide range of natural compounds (e.g., anthocyanins, flavonoids, phenolic acids etc.) with varying structural characteristics that modulate their antioxidant activity and their subsequent health and biological effects [1,2].

The phenolic acids, in particular, offer an important group of powerful natural compounds that having substantial lipid and water solubility can inhibit oxidative deterioration when added as functional ingredients in emulsion model systems [3]. Research studies in this field have focused on the free radical scavenging capacity of phenolic acids and more specifically on radical quenching that can be validly measured through kinetic parameters [4]. Over the last decade, the food industry has increasingly considered the use of natural phenolic antioxidants as an efficient strategy to retard oxidative deterioration in food-based systems and thereby maintain their sensory characteristics [5,6].

In emulsified foods, (e.g., dressings, sauces, soups, and desserts) lipid oxidation can occur rapidly due to their large surface area [7,8] with mechanisms that are more complex and not fully understood compared to bulk oils [9,10]. Furthermore, the antioxidant activity in interfaces can be crucial not only for developing novel food applications but also for further exploring their potential health properties against oxidative stress [11]. Emulsion systems generally mimic the amphiphilic nature and the basic structural characteristics of important biological membranes (e.g., lipoproteins) that are prone to oxidative degradation when attacked by singlet oxygen and free radicals [12]. A typical example of a "bio-interfacial system" is offered by plasma lipoproteins, which are complex aggregates of lipids and proteins that render the hydrophobic lipids accessible by the aqueous environment of body fluids and thereby by reactive oxygen species potentially present [13]. The initiation of these harmful biochemical processes leads to in vivo oxidative damage of biomolecules and ultimately the development of serious human health conditions such as aging, carcinogenesis, and cardiovascular diseases [14,15].

This paper first examines in Section 2 the most common naturally occurring phenolic acids that can be extracted from various plant sources, including edible herbs and well-known botanicals. In addition, the authors provide a summary of the available research findings concerning the antioxidant activity of each examined phenolic acid in o/w emulsion systems. Overall, in vitro research on the oxidative stability and antioxidant effects of phenolic acids in model emulsions could provide useful background knowledge and information of nutritional interest to support in vivo clinical trials.

In the last few years, an increasing body of clinical research has focused on the potential effects of phenolic acids against the development of cancer, cardiovascular diseases, and other health disorders (such as skin problems, inflammations, bacterial infections etc.). The main target of this publication is to provide in Section 3 an overview of the most recent literature evidence concerning the health beneficial effects of the examined phenolic acids.

2. Natural Sources and Antioxidant Activities of Phenolic Acids in Food Emulsions

This section provides a summary of the most important natural sources for a few common—in nature and food—phenolic acids that are listed along with their chemical structure in Table 1. In parallel, the authors have performed a literature search on the in vitro antioxidant activities of each examined phenolic acid against the oxidation of oil-based emulsion systems.

2.1. Caffeic Cid (CA)

CA is a hydroxycinnamic acid structurally composed of both phenolic and acrylic functional groups, the derivatives of which are *trans* in nature [16,17]. It is found at high levels in some herbs, especially in the South American herb yerba mate (1.5 g/kg) [18], and thyme (1.7 mg/kg), [19]. In fruits (such as berries, apples, and pears) CA was quantified in high amounts, representing together with p-coumaric acid 75–100% of the total hydroxycinnamic acids [20]. Also, CA can be found in the bark of *Eucalyptus globulus* [21] and was identified as the main phenolic constituent in coffee and coffee oil [22]. Boke et al. (2019) [23] analysed samples of *Cephalaria* species using high-performance liquid chromatography coupled with tandem mass spectrometry (HPLC-MS/MS) and determined CA as major phenolic acid.

A few researchers have reported a clear antioxidant effect of CA in various Tween-based emulsions prepared with linoleic acid [4] or other vegetable oils such as corn, flaxseed, and sunflower oils [24]. Sorensen et al. (2017) [25] observed that CA presents a clear antioxidant activity in Citrem-and Tween-stabilised emulsions in the presence of endogenous tocopherols but acted as a prooxidant in the absence of tocopherols. The authors suggested that the observed differences in antioxidant efficiency with different emulsifiers (and with or without endogenous tocopherols) were caused due to emulsifier–antioxidant interactions and antioxidant–antioxidant interactions in the emulsions.

2.2. Gallic Acid (GA)

GA (also known as 3,4,5-trihydroxybenzoic acid) is the main phenolic acid in tea [26] but also found in high amounts in chestnuts and several berries [19]. It is encountered in a number of land plants, such as the parasitic plant *Cynomorium coccineum*, the aquatic plant *Myriophyllum spicatum*, and the blue-green alga *Microcystis aeruginosa* [27,28]. Very recently, Souza et al. (2020) [29] has isolated gallic acid from black tea extract at a concentration around 0.8 mg/kg.

There is some contradictory evidence in the literature about the effect of GA against the oxidative deterioration of emulsions. Bou et al. (2011) [30] did not see any statistically remarkable effect of

GA following its addition in Tween-based sunflower o/w emulsions. Alavi Rafiee et al. (2018) [31] reported that GA exerted a high antioxidant action in the bulk oils but showed lower activity in o/w emulsions, highlighting the critical role of the carboxyl group and the effect of the degree of lipid unsaturation in GA antioxidant activity. Di Mattia et al. (2009) [32] reported that GA contributed to the colloidal stabilisation of the o/w emulsion systems, whilst exhibited a low activity towards secondary oxidation. Zhu et al. (2019) [33], however, observed clear antioxidant effects—in terms of peroxide values and hexanal content- of GA and its alkyl esters in o/w emulsions in the following order of activity: propyl gallate > lauryl gallate > octyl gallate > gallic acid > stearyl gallate. In a study by Wang et al. (2019) [34] GA or its alkyl esters were added in combination with α-tocopherol in o/w emulsions. The results showed that all the tested gallate esters (propyl, octyl and dodecyl gallate) exerted antioxidant activities combined with α-tocopherol, and propyl gallate, with the shortest alkyl chain length, possessed the highest synergistic action. Other researchers have also observed an enhancement of antioxidant activity of esterified GA derivatives in double emulsions by use of encapsulation [35].

2.3. Rosmarinic Acid (RA)

RA is an ester of caffeic acid, present as the main phenolic component in several members of the Lamiaceae family including among others: *Rosmarinus officinalis*, *Origanum* spp., *Perilla* spp., and *Salvia officinialis* [36,37]. A few researchers reported RA as the main phenolic acid of various culinary herbs (oregano, thyme sage, and rosemary) in concentrations varying between 0.05 and 26 g/kg dry weight [38,39]. Additionally, the results of Tsimogiannis et al. [40] indicate an amount of 19.5 g/kg in the leaves of pink savory (*Satureja thymbra* L.).

A body of research has reported antioxidant activities of RA (in terms of both hydroperoxides and volatiles formation) in o/w emulsions based on (i) corn oil and stabilised by various emulsifiers [41]; (ii) Tween-based emulsions prepared with linoleic acid [14] or soybean oil [42]. Bakota et al. (2015) [43] incorporated both pure RA and RA-rich extract (from *Salvia officinalis* leaves), at a concentration of ~30 mg/g, into o/w emulsions and observed that both treatments were effective in suppressing lipid oxidation.

2.4. Carnosic Acid (CarA)

CarA is a labdane-type diterpene present in plant species of the Lamiaceae family, such as rosemary and common *salvia* species [44]. CarA is commonly found in the dried leaves of sage in 1.5 to 2.5% concentration [45]. CarA is used as a preservative in food and non-food products, e.g., toothpaste, mouthwash, and chewing gum, since it is endowed with antioxidative and antimicrobial properties [46].

Through the pioneer work of Frankel et al. (1996) [47], CarA was reported to exhibit an antioxidant activity in emulsions that was enhanced at pH 4–5. The following years, several studies used extracts from rosemary and other Lamiaceae family herbs in emulsions and attributed the antioxidant activity mainly to CarA and carnosol [48,49]. Very recently, Lei et al. (2019) [50] have developed a α-tocopherol-based microemulsion aiming to improving the antioxidant stability of CarA and thereby set up a model system for potential applications in bioactive components.

Overall, although this lipid-soluble compound is recognised for its high antioxidative capacities, (which led to many industrial applications in foods and beverages), its mode of action against the oxidation of emulsions have not been yet fully elucidated [51].

2.5. Ferulic Acid (FA)

FA is a phenolic acid commonly found in the seeds of coffee, apple, artichoke, peanut, and orange [52]. Flaxseed has been reported as the richest natural source of FA glucoside (4.1 ± 0.2 g/kg) [53]. According to various researchers [54,55], black beans contain FA at an average concentration of 0.8 g/kg. In addition, FA can be found in *Brassica* vegetables and tomatoes [20].

There is a certain body of literature evidence claiming clear antioxidant effects of FA in various o/w systems including: (i) corn oil-based emulsions stabilised with the use of various emulsifiers [56]; (ii) Tween-linoleic acid-based emulsions [14] (iii) in Tween-menhaden o/w emulsions [57]. More recently, Shin et al. (2019) [58] reported that 4-vinylguaiacol (4-VG), (product of FA decarboxylation), exerted a strong antioxidant activity when added at 200 ppm in a 10% o/w emulsion for 50 days. Permin et al. (2019) [59], however, reported that FA showed no antioxidant activity in o/w emulsions rich in ω-3 fatty acids and stabilised by whey proteins.

2.6. p-Coumaric Acid (p-CA)

A large number of natural plants sources have been reported to be rich in p-CA such as fungi, peanuts, navy beans, tomatoes, carrots, basil, and garlic [60]. The substance p-CA is abundant in most fruits (especially pears and berries) and cereals [20,61], as well as in honey at a concentration range 1.7–4.7 mg/kg [62]. Kannan et al. (2013) [63], by using HPLC analysis, reported that the extracts of *Halodule pinifolia* and *Clytra rotundata* are rich in p-CA, a fact that may account for their high biological activity. The same authors observed that p-CA is present in high amounts in a few mushroom species. In addition, a few researchers noted that p-CA is present in extracts derived from Amaranth leaves and stem at a concentration range of 28–44 mg/kg [64,65]. Oh et al. (2015) [66] identified p-CA as the main phenolic constituent present in the aqueous extracts of hulled barley (*Hoerdeum vulgare* L.).

In addition, a recent body of research evidence has focused on antioxidant activities of p-CA [67] along with its identification in various natural sources [68]. Very recently, Park et al. (2019) [69] conducted HPLC and NMR analysis of aqueous and ethanolic extracts of roasted rice hulls and identified p-CA, VA, and FA as the dominant phenolic compounds. The authors reported that added roasted rice hull extracts, particular rich in p-CA, protected against the oxidative deterioration of o/w emulsions, at 60 °C.

2.7. Vanillic Acid (VA)

VA is a dihydroxybenzoic acid derivative commonly used as a flavoring agent. It is found in several fruits, olives, and cereal grains (e.g., whole wheat), as well as in wine, beer, and cider [70,71]. Kim et al. (2019) [72] performed an identification of the main phenolic constituents in potatoes samples (*Solanum tuberosum* L.) and quantified VA at a concentration between 0.02 and 0.04 g/kg. Espinosa et al. (2015) [73] analysed an extract of red propolis and reported VA as being the major phenolic constituent. VA was also found in fruit extract of the açaí palm plant (*Euterpe oleracea*) [74] and was identified by Zhao et al. [75] in the root of *Angelica sinensis* (an herb indigenous to China) at concentrations between 1.1 and 1.3 g/kg. Furthermore, Radmanesh et al. (2017) [76] reported that VA is present in various botanical sources including *Juglans regia* L., *Chenopodiastrum murale*, orchard grass, and *Melilotus messanensis*.

Keller et al. (2016) [77] observed a strong antioxidant character of VA during autoxidation of Tween 40-based o/w systems, at pH 3.5. Furthermore, Vishnu et al. (2017) [78] evaluated the antioxidant activity of VA grafted chitosan (Va-g-Ch) during the microencapsulation of polyunsaturated fatty acid-rich sardine oil in o/w emulsions. After four weeks of storage, a decrease of peroxide values demonstrated good oxidative stability and encapsulation efficiency of Va-g-Ch.

Table 1. Natural sources of the examined phenolic acids along with the most recent literature references.

Phenolic Acid Structure	Natural Source	Amount(g/kg, Dry Basis)	Literature
Caffeic acid	Coffee	0.90 [a]	[22]
	Blueberry	1.47 [a]	[22]
	Yerba mate	1.50	[18]
	Banana	0.23–0.31	[70]
	Mango	1.00–1.76	[70]
	Eucalyptus globulus	0.2–2.9	[21,79]
Gallic acid	Black tea	0.8	[29]
	banana	1.10–1.24	[70]
	mango	11.45–34.49	[70]
	berries	0.03–0.09 [a]	[20]
	chestnut	3.50–9.10 [a]	[20]
Rosmarinic acid	Rosmarinus officinalis	0.16–12.86	[38,39]
	Salvia officinialis	1.18 [a]–21.86	[39]
	oregano species	0.05–25.63	[38,39]
	thyme	0.08–6.81	[38,39]
	Sweet basil	10.86	[39]
	Pink savory (S. thymbra)	19.50	[40]
Carnosic acid	Rosemary leaves (R. officinalis)	40–100	[51]
	salvia species	0.1–21.8	[51]
	Sage (S. officinalis) leaves	15–25	[45,50]

Table 1. *Cont.*

Phenolic Acid Structure	Natural Source	Amount(g/kg, Dry Basis)	Literature
Ferulic acid	Cereal grains	Up to 2	[80]
	Cell walls of grains	13.51–33.00	[52,81]
	Flaxseed	4.10 (as glucoside)	[53]
	artichoke	2.75	[52,79]
	coffee	0.09–0.14	[53]
	eggplant, redbeet, spinach, peanut grapefruit, and orange	0.07–0.35	[53]
	Banana	0.49–0.53	[70]
	Mango	0.75	[70]
	Beans	0.8	[55]
	Acai (*Euterpe oleracea*) oil	0.10	[71]
p-Coumaric acid	Strawberries	1.11	[20]
	Berries	0.01–0.95	[20,61]
	Pear	0.01–0.45	[61]
	Banana	1.05	[70]
	Mango	0.90	[70]
	Peanuts	1.03 [a]	[20]
	Onion peel	0.58	[61]
	Honey	0.002–0.005	[62]
	Mushrooms	Traces–3.70	[63]
	Amaranthus cruentus	0.028–0.042	[65]
Vanillic acid	Banana	0.12–0.37	[70]
	Mango	0.47–3.76	[70]
	Acai (*Euterpe oleracea*)	0.002	[74]
	Angelica sinensis	1.1–1.3	[75]
	Potato tuber (*Solanum tuberosum*)	0.02-0.04	[72]

[a]: on fresh weight basis.

3. Health and Biochemical Properties of Phenolic Acids and Their Natural Extracts

3.1. General Health Aspects of Phenolic Acids

Section 2 of the manuscript provided an overview of several natural sources of the phenolic acids while also reported on a number of research findings most of which revealed their clear antioxidant character against the lipid oxidation of o/w emulsions. As also discussed in the introduction of the manuscript, the emulsions could offer useful in vitro model systems as a basis for further investigation of the phenolic activity in interfacial biological systems.

This section focuses on the main target of this review, which is to provide an overview of the latest literature concerning the health properties of the examined phenolic acids. The authors have reviewed a large number of studies investigating into individual phenolic acids and/or their mixtures extracted from natural plant sources. A body of research evidence focuses on the activity of various phenolic acids against cancer and the main mechanisms by which they may exert their effects such as: scavenging of free radicals, induction of enzymes, DNA damage repair, cell proliferation, and apoptosis [82].

Rosa et al. (2016) [83] supported that phenolic acids have been a prime source for the treatment of various forms of cancer, with focus on colon cancer in human colon adenocarcinoma cells. Vinayagam et al. (2016) [84] reviewed the properties of phenolic acids to improve glucose and lipid profiles linked pathologic conditions (diabetes, cardiovascular diseases etc.). A diet rich in phenolic acids has been also reported to protect against certain allergies and slow down the development of Alzheimer's disease [85].

Table 2 provides an overview of recent in vitro and in vivo clinical studies on the health/biochemical properties of the examined phenolic acids. More specific information per phenolic acid is presented in the following paragraphs.

3.2. Caffeic Acid (CA) and its Esters

CA presents a remarkable antioxidant potential and also demonstrates in vitro antimicrobial properties [86]. Further to its well-established antioxidant and anti-aging activities, CA has been reported to own strong antimicrobial properties and protect against dermal diseases [87]. De Oliveira et al. (2012) [88] designed a drug delivery system based on o/w emulsions with CA containing microparticles, developed in order to ensure a prolonged CA release in the target cells and thereby treat the folliculitis skin disease. Similarly, Paulo and Santos (2019) [89] examined how incorporation of caffeic–ethyl cellulose microparticles in skin care products can offer anti-aging protection. Furthermore, a body of recent research evidence has demonstrated that caffeic acid phenyl ester (CAPE) is a natural compound with anticancer activities. The chemical structure of CA (presence of free phenolic hydroxyls) is believed to strongly account for its antioxidant capacities that, in turn, link to certain anti-carcinogenic properties [90]. Dietary supplementation of rats, with CA and CAPE (5 mg/kg body wt subcutaneous or 20 mg/kg oral), was shown to inhibit tumor growth in HCC cells (HepG2) and reduce the tumor invasion at a liver metastatic site [91]. Another clinical study [92] has reported a clear chemoprotective effect of CAPE and its analogs (20 mg/kg body wt) against lipid peroxidation and subsequent cell proliferation of hepatic tumors (HCC) in rats.

Guan et al. (2019) [93] used sucrose fatty acid ester to nano-encapsulate CAPE in aqueous propylene glycol with a temperature-cycle method and reported that nano-encapsulation enhanced cytotoxicity of CAPE against colon cancer HCT-116, and breast cancer MCF-7 cells. Another recent medical study [94] reported clear inhibitory effects of CAPE derivatives against acetylcholinesterase, an enzyme linked with the development of Alzheimer's disease. CA and its derivatives, such as CAPE, have been reported to act against colon cancer through their cytotoxic to tumors but not to normal cells [95]. In addition, Zhang et al. 2017 [96] examined the action of CA (100 mg/kg) on structural changes caused by HCC in the rat microbiota. The authors concluded that this phenolic compound reduces certain biomarkers that indicate liver injury (among other alanine, transaminase, aspartate aminotransferase, alkaline phosphatase, total bile acid and total cholesterol).

3.3. Carnosic Acid (CarA)

Since its first extraction from various natural sources (e.g., *Salvia* and *Rosmarinus* species) and given its well reported functional and antioxidant properties, CarA has been used in a range of cosmetic and pharmaceutical applications [50,51].

Several researchers have focused on the liver protective effect of CarA. In an interesting placebo clinical trial [97] a male *ob/ob* mice (model for NAFLD (non-alcoholic fat liver disease)) followed a diet with CarA for 5 weeks and compared to placebo experienced weight loss and reduced visceral adiposity. The authors concluded that CarA could be considered for the development of new drugs against the NAFLD liver syndrome. Dickmann et al. (2012) [98] explored the hepatotoxicity potential of CarA (at varying concentrations of 4–10 μM) in primary human hepatocytes and microsomes. While CarA did not exhibit any significant time-dependent enzyme inhibition at 4 mM, it even increased enzyme activity at 10 μM, compared with Phenobarbital and Rifampicin drugs. According to the authors, the results indicate potential CarA interaction with drugs, thereby a need for its appropriate safety assessment before its further use as a weight loss supplement.

Bahri et al., 2016 [99] noted that CarA can have a protective effect against chronic neurodegenerative conditions, like Parkinson's disease, via a mechanism that links to the transcriptional activation of antioxidant Nrf2/ARE pathway.

Einbond et al. (2012) [100], after in vitro experiments in human breast cancer cells, have observed that treatment with CarA at 20 μg/mL resulted in the prevention of ER-negative breast cancer via an activation of expression of antioxidant and apoptosis genes. A more recent study by Solomonov et al. (2018) [101] demonstrated a significant anti-inflammatory effect of CarA combined with astaxanthin and a lycopene-rich tomato extract in a nutrient supplementation.

However, Raes et al. (2015) [45] did not report any effect of CarA, against lipid and protein oxidation in an in vitro simulated gastric digestion model.

3.4. Gallic Acid

Over the last few years, a body of research evidence had reported cardioprotective, neuroprotective, and anticancer properties of GA and gallates that are mostly attributed to their antioxidative properties against the reactive oxygen species (ROS) signaling networks [102]. Sourani et al. (2016) [103] reported that GA inhibits proliferation and induces apoptosis in lymphoblastic leukemia cell line. In a very recent study [104] the ability of GA to potentiate the anti-cancer effects of chemotherapeutic drugs (e.g., Paclitaxel, Carboplatin) was examined in human HeLa cells. The authors reported that a Paclitaxel/GA combination could represent a promising alternative with lower side effects for Paclitaxel/Carboplatin combinations in treatment of cervical cancer. Recent pharmacokinetic human and animal clinical studies were based on Chinese GA-based patented medicines but further investigation is needed on the GA kinetic profile after dietary supplementation before drawing any conclusion for its efficacy against pathological conditions [105]. Paolini et al. (2015) [106] performed a study to explore the potential of GA as a promising new anticancer drug. The authors treated T98G human glioblastoma cell lines for 24 h with increasing concentrations of GA (ranging from 1 to 100 μg/mL). According to the results, GA exerts a protective or an anti-proliferative effect on glioma T98G cells via dose-dependent epigenetic regulation mediated by miRNAs.

Yu et al. (2018) [107] contacted a clinical study on myocardial infarcted rats with an oral administration of GA monohydrate at a dose of 50 and 100 mg/kg body wt. The authors observed that myocardial infarction could modify the pharmacokinetic process of GA and thereby determine its potential activity. Similarly, Nwokocha et al. [108] concluded that GA can present negative chronotropic and inotropic effects in isoproterenol induced myocardial damage.

3.5. Ferulic Acid (FA)

Over the last few years, a number of clinical studies have demonstrated that FA can exert in vivo antioxidant effects by scavenging free radicals and enhancing the cell stress response through the upregulation of cytoprotective systems [80,109]. Based on its antioxidant and anti-inflammation functions, FA is widely considered as a phenolic compound with well documented protective actions against many pathologic conditions (e.g., types of cancer, cardiovascular diseases, diabetes mellitus and skin problems) [110]. Sgarbossa et al. (2015) [79] reviewed the health benefits of FA and noted its protective role against neurotoxicity based on a number of in vitro and in vivo animal clinical studies. Sung et al. (2014) [111] treated Dawley rats (male, 210–230 g) with FA (100 mg/kg body wt) and reported a clear neuroprotective role. The above-indicated findings recommend the use of FA for drugs development against neurodegenerative diseases, although a few questions are still open before its clinical development and application in patients.

Chowdhury et al. (2016) [112] performed a clinical study that involved oral administration of diabetic rats with FA (at a dose of 50 mg/kg body wt, orally for eight weeks). The authors concluded a protective role of FA against streptozotocin-induced cellular stress in the cardiac tissues. Baeza et al. (2017) [113] reported a strong inhibitory effect of dihydroferulic acid against in vitro platelet activation.

Ambothi et al. (2014) [114] concluded that FA (in the concentration range 10–40 µg/mL) can prevent the ultraviolet-B radiation (290–320 nm) induced oxidative DNA damage in human dermal fibroblasts. The same researchers conducted another clinical study [115] reporting that FA protected against carcinogenesis and tumor formation induced via chronic UVB exposure (180 mJ/cm^2 for 30 weeks) in the skin of Swiss albino mice. Russo et al. (2017) [116] conducted a population-based case-control study in South Italy to examine any association between dietary phenolic acid consumption and prostate cancer. From a sample of 2044 individuals, 118 histopathological-verified prostate cancer cases were collected, and multivariate logistic regression showed that both CA and FA were associated with reduced risk of this cancer type.

3.6. p-Coumaric Acid (p-CA)

p-CA has been reported to decrease the peroxidation of low-density lipoproteins (LDL) and exert anti-mutagenesis, anti-genotoxicity, and anti-microbial activities [117]. Very recently, Ferreira et al. (2019) [118] gave a literature overview of certain biochemical properties of CA (including radical scavenging and tumor suppression activities) that link to its claimed pharmacological effects. Boo (2019) [119] has highlighted the anti-melanogenic effects of p-CA by focusing on its inhibitory action against melanin synthesis as observed in human epidermal melanocytes. Neog and Rasool (2018) [120] supported that dietary p-CA could intervene in the osteoclast formation and thereby alleviate the effect of rheumatoid arthritis, a finding also supported by Trisha (2016) [60].

Pei et al. (2015) [61] focused on p-CA and its conjugates reviewing their dietary sources, and biological activities. The authors concluded that future studies should focus on pharmacokinetic properties of p-CA in order to further promote its use in the food and cosmetic applications. Janicke et al. (2011) [121] treated Caco-2 cells with 150 µM p-CA for 24 h and noticed a protective effect against the development of colon cancer by retarding the cell cycle progression. In addition, Sharma et al. (2017) [122] conducted a study to evaluate the chemo-preventive potential of p-CA in rats challenged with the colon-specific procarcinogen DMH. According to the results, p-CA presented a concentration-dependent anti-carcinogenic effect since it acted more efficiently at a dose of 100 mg/kg body wt, compared to 50 mg/kg body wt. Amalan et al. (2016) [123] reported that p-CA inhibited the development of oxidative stress by increasing the endogenous antioxidant capacity (level of glutathione-GSH) in the livers of diabetic rats. In addition, Vauzour et al. (2010) [124] compared the neuroprotection capacities of various phenolic compounds in primary cultures of mice cortical neuron. The authors concluded a stronger protective effect of p-CA at 1 mM concentration than those of CA and GA. Very recently, Sunitha et al. (2018) [125] reported that p-CA mediated the protection of H9c2 cells from Doxorubicin-induced cardiotoxicity.

3.7. Rosmarinic Acid (RA)

Yang et al. (2013) [126] reported the health protective effects of RA on high mobility group box1 (HMGB1) protein-induced inflammation that mediates responses to infection and injury cases. Similarly, Tsung et al. (2013) [127] observed that RA can suppresses *Propionibacterium acnes*–induced inflammatory responses. Concerning the anti-inflammatory mechanism, Ku et al. (2013) [128] observed that RA down-regulates endothelial protein C receptor shedding, in vitro and in vivo. Braidy et al. (2016) [129] investigated whether RA (0.01–0.1 mg/mL) can protect against CTX-mediated toxicity in primary human neurons. According to the results pre-treatment with RA at 0.01 mg/mL (but not higher) exerted a neuroprotective effect, generating significant decrease in CTX-mediated extracellular LDH activity, NAD decline, and DNA damage, compared to CTX treated cells alone.

Nunes et al. (2017) [130] noted that RA displays several health beneficial effects (including antimicrobial and anti-carcionogenic properties) the magnitude of which depends greatly on both its intake and bioavailability. Hossan et al. (2014) [131] have more specifically focused on the anticarcinogenic properties of RA proposing various mechanisms of anticancer activity including antioxidant actions along with proliferation and apoptosis of cancer cells.

Stansbury (2014) [37] summarised the clinical trials that have demonstrated RA activities against allergic immunoglobulin and inflammatory responses of polymorphonuclear leukocytes, thereby being effective in the treatment of allergic disorders. Alagawany et al. (2019) [132] have also reviewed the mode of action, and health benefits of RA.

Domitrović et al. (2013) [133] reported that RA can protect against acute liver damage in intoxicated mice by exerting certain antioxidant, anti-inflammatory, and anti-apoptotic activities. More recently, De Oliveira et al. (2019) [134] examined the protective effects of RA against ethanol-induced DNA damage in mice and reported a clear antigenotoxic capacity in a concentration of 100 mg/kg body wt by using the comet assay. Luno et al. (2014) [135] concluded that RA at 105 μM concentration improves function and in vitro fertilising ability of boar sperm, by inhibiting oxidative stress during cryopreservation. Furthermore, Venkatachalam et al. (2016) [136] investigated the mode and molecular mechanisms that govern the chemoprotective action of RA against colon cancer in rats. The authors reported that supplementation with RA (5–20 mg/kg body wt) protected treated rats from the deleterious effects caused by the colon carcinogenic 1,2-dimethylhydrazine.

3.8. Vanillic Acid (VA)

VA has been reported to confer certain health beneficial effects, via antioxidative, anti-mutagenic, anti-cancer, anti-inflammatory, and neuroprotective activities [137,138]. In a recent study [76], male rats (separated in groups of 10) were supplemented with varying concentrations of VA (0–10 mg/kg body wt) for a period of 10 days. The results have shown a clear effect of VA against the risk of myocardial dysfunction. Similarly, Dianat et al. (2014) [139] demonstrated the effectiveness of VA against lipid peroxidation, indicated by malondialdehyde (MDA) reduction, and endogenous antioxidant enzymes improvement, in isolated rat hearts exposed to ischemia-reperfusion. Kim et al. (2010) [140] following a clinical trial in rats reported beneficial effects of VA in the treatment of ulcerative colitis. Erdem et al. (2012) [141] examined the potential effect of VA against mitomycin C-induced genomic damage in human lymphocytes in vitro. Interestingly, VA (at 1 μg/mL) significantly reduced DNA damage cells but at a higher concentration (2 μg/mL) exerted a genotoxic effect on DNA. On the contrary, Krga et al. [142] (2018) reported that VA at 2 μM did not significantly decrease biomarkers of platelet activation development of cardiovascular diseases.

Table 2. Overview of recent in vitro and in vivo clinical studies on the beautiful health/biochemical properties of the examined phenolic acids.

Phenolic Acid.	Experimental Conditions	Conclusion of Study/Health Effect	Reference
Caffeic acid (CA) & caffeic acid phenyl ester (CAPE)	Treatment of rats with CA (20 mg/kg body wt).	CA caused suppression of tumor growth in HCC cells (HepG2)/reduction of tumor invasion at liver metastasis.	[91]
	CAPE and its analogs (20 mg/kg body wt) in rats.	CA chemoprotective effect on cell proliferation, p56 activation of hepatic tumors (HCC).	[92]
	CA (100 mg/kg) in the rat microbiota	CA reduce certain biomarkers that indicate liver injury	[96]
Carnosic acid (CarA)	Treatment of human breast cancer cells with 20 µg/mL CarA	CarA activated the expression of antioxidant/apoptosis genes resulting in protection against breast cancer.	[100]
	P450 enzyme inhibition was examined in human hepatocytes and microsomes at presence of 4–10 µM of CarA	Increased enzyme activity at 10 mM of CarA, compared to drugs/need for CarA safety assessment before its use against hepatotoxicity	[98]
Gallic acid (GA)	Treatment of T98G human cells for 24 h with GA (in the range 1-100 µg/mL).	GA exerts a protective anti-proliferative effect on glioma T98G cells via dose-dependent epigenetic regulation mediated by miRNAs.	[106]
	Oral administration of GA monohydrate (50 and 100 mg/kg body wt) in normal myocardial infracted rats.	Cardioprotective effect of GA.	[107]
Ferulic acid (FA)	Treatment of Dawley rats with FA (100 mg/kg body wt).	FA exerted a neuroprotective role by attenuating decreases of peroxiredoxin-2 and thioredoxin levels in neuronal cell injury.	[111]
	Treatment in the skin of salbino mice exposed to UVB (180 mJ/cm^2) for 30 weeks.	FA protected against carcinogenesis and tumor formation.	[115]
p-coumaric (p-CA)	Treatment of Caco-2 cells with 150 µM p-CA for 24 h,	p-CA protective effect against the development of colon cancer retarding the cell cycle progression	[121]
	Treatment of rats (50-200 mg/kg body wt) challenged with colon specific procarcinogen DMH.	p-CA exhibits a significant chemo-preventive potential at 100 mg/kg	[123]
	Treatment of cultures of mice cortical neuron with p-CA (1 mM) against cysteinyldopamine-induced neurotoxicity.	p-CA provided the best neuroprotection, compared to other phenolics (CA and GA).	[124]

Table 2. *Cont.*

Phenolic Acid.	Experimental Conditions	Conclusion of Study/Health Effect	Reference
Rosmarinic Acid (RA)	Study of RA effect (0.01 mg/mL) on cell viability and normal cellular function in human neuronal cells	RA at 0.01 mg/mL (but not higher) exerted a neuroprotective effect generating significant decrease in CTX-mediated extracellular LDH activity compared to control.	[129]
	Treatment of mice (100 mg/kg body wt) and study of the effect on ethanol-induced DNA damage.	Anti-genotoxic capacity of RA against DNA damage (via comet assay)	[134]
	Supplementation of rats with RA (5–20 mg/kg body wt).	RA protected treated rats from the deleterious effects caused by colon carcinogen, 1,2-dimethylhydrazine.	[136]
Vanillic acid (VA)	Supplementation of male rats (0–10 mg/kg body wt) for 10 days.	VA was effective against the risk of myocardial dysfunction.	[76]
	In vitro examination of the effect on mitomycin C-induced genomic damage in human lymphocytes.	VA (at 1 µg/mL) significantly reduced DNA damage cells but at a higher (2 µg/mL) itself exerted genotoxic effects on DNA.	[137]
	Supplementation of 5 groups of mice with VA (5–100 mg/kg body wt) for 28 days	VA at 50 and 100 mg/kg dose significantly ($p < 0.001$) improved the habituation memory of mice, and increased the antioxidant capacity.	[143]

In a clinical trial by Chellammal et al. (2015) [143], five groups of mice were treated as control or active groups supplemented with VA in the concentration range 5–100 mg/kg for 28 days. The results showed that VA at 50 and 100 mg/kg dose significantly ($p < 0.001$) improved the habituation memory, decreased the AChE, corticosterone, and increased the antioxidant capacity of the mice. Furthemore, Yemis et al. (2011) [144] reported a pH-dependent antimicrobial effect of VA that was found to inhibit the growth and heat resistance of Cronobacter bacterial species, a conclusion that could lead to the use of VA for new food storage applications.

3.9. Natural Extracts as Mixtures of Phenolic Acids

Nature has generously offered a wide range of herbs (e.g., thyme, oregano, rosemary, sage, mint) that are rich in many phenolic compounds with strong antioxidant biochemical and anti-inflammatory properties [145,146] including protection of DNA from oxidative damage [147]. More specifically, bael (Aegle marmelos) flower (rich in p-CA, CA, and VA) and tulsi (Ocimum tenuiflorum) seeds (rich in GA and p-CA) have been reported to present a strong antioxidant character again DNA damage [148,149].

Findings from recent nutritional intervention studies with natural extracts rich in phenolic acids suggest that they can exert a clear cardio-protective effect through modulations of platelet function [150]. Padmanabhan and Geetha (2015) [151] reported a clear hypo-lipidemic and anti-obesity effect of hydro-alcoholic fruit extract of avocado (particularly rich in GA and VA) in rats fed with high fat diet (co-administered with 100 mg/kg body wt of HFEA for 14 weeks).

Extensive research has been conducted in the last decade about rosemary extracts that are particularly rich in RA and CarA. Chkhikvishvili et al. (2013) [152] demonstrated that a rosemary extract (RE) can protect Jurkat cells from oxidative stress induced by hydrogen peroxide. Very recently, Pérez-Sánchez et al. (2019) [153] investigated the antitumor activity of RE obtained by using supercritical fluid extraction, through its capacity to inhibit various signatures of cancer progression and metastasis. Ulbricht et al. (2010) [154] has published an evidence-based systematic review on RE by examining various aspects of their health properties including also information on their adverse effects and toxicology. In a recent study, Sánchez Salcedo et al. (2015) [155] demonstrated that RE can exert an in vivo anti-tumor action through a reactive oxygen species-initiated cell death. Andrade et al. (2018) [156] reported a clear protective role of RE in preventing colds, rheumatism, and pain of muscles and joints.

De Oliveira et al. (2019) [134] reviewed the in vivo and in vitro studies of *R. officinalis* highlighting the therapeutic and prophylactic effects of RE on some physiological disorders caused by various biochemical agents. Moore et al. (2016) [157] reviewed the phytochemical biological activities and anti-carginogenic properties of *R. officinalis*.

Moreover, p-CA rich methanolic extracts of *Amaranthus spinosus* and of *Amaranthus caudatus L.* were shown to possess significant central and peripheral anti-nociceptive potential and anti-inflammatory activity, in mouse model [36]. Jeong et al. (2017) [158] observed clear therapeutic effects of polyphenolic mixtures (containing among others GA, p-CA and ellagic acid) against cell lung cancer. Hydroxycinnamic acid derivatives of mulberry fruits were reported to increase the production of reactive oxygen species production by acting as pro-oxidants and hence killing the cancer cells [159]. Hilbig et al. (2017) [160] reported that an aqueous extract from pecan nut (particularly rich in GA, CA and VA) showed clear inhibitory effects against breast cancer cell line MCF-7, as well as against tumor growth in Balb-C mice. Simin et al. (2019) [161] provided an overview of the beneficial biological activities of less known wild onions (*A. sect. Codonoprasum*), which are particularly rich in the common phenolic acids. The same group [162] has concluded that a methanolic extract of small yellow onion (*Allium flavum*), particularly rich in FA, *p*-CA, CA, and VA, can exert selective inhibitory action towards cervix epithelioid carcinoma and colon adenocarcinoma cells.

4. Conclusions and Future Challenges

Based on the analysis of this manuscript on in vitro and in vivo biochemical activities of phenolic acids, the authors have drawn the following conclusions along with a few recommendations for future investigation in this field:

4.1. Activities of Phenolic Acids in O/W Emulsions

Over the last few years, an increasing number of researchers have reported well documented antioxidant activities of naturally occurring phenolic acids in o/w model systems. The findings of the recent studies on phenolic acids antioxidant activity in emulsion formulations could offer a basis for innovative insights in a wide range of food and cosmetic relevant products.

Controlling the interfacial concentrations of antioxidants in o/w emulsions could be regarded as a reasonable approach to monitor more systematically their activity against lipid oxidation Further work in this field could focus on estimating distribution constants of the phenolic acids in emulsions as a factor that would further elucidate and monitor their antioxidant efficiencies in interfacial systems.

Future challenges may include the development of nano-based emulsion systems to enable the delivery of functional bio-constituents (e.g., phenolic acids) and thereby promote their applications in innovative dietary supplements or even drug formulations.

4.2. Health Biochemical Properties of Phenolic Acids

This analysis presented a summary of the most recent clinical (mainly animal) studies on phenolic acids. The findings overall offer sufficient evidence to support that each of the examined phenolic acids, through their dietary supplementation, could exert health protective effects against a wide range of pathogenic conditions including cancer, bacterial infections, cardiovascular, inflammatory, and neurodegenerative diseases.

Mixtures of phenolic acids (most commonly present in a wide range of botanical extracts) have been reported by latest research evidence to possess a number of beneficial health properties. The strong antioxidant and biochemical potential of these natural plant preparations may more specifically link to the synergistic effect of their individual phenolic compounds.

Although a few phenolic acids are well-known as efficient bioactive dietary ingredients, their pharmacokinetics and metabolic properties are not fully elucidated yet. This is a factor that limits their current use and therapeutic potential and requires further clinical investigations to support and optimise their future use in nutritional and pharmaceutical applications.

Author Contributions: Conceptualization, S.K., C.P. and V.O.; resources, S.K. and V.O.; writing—original draft preparation, S.K.; writing—review and editing, C.P. and V.O.; supervision, V.O. All authors have read and agreed to the published version of the manuscript. Authors have contributed substantially to the work reported.

Funding: This research received no external funding.

Conflicts of Interest: The authors declare no conflict of interest.

References

1. Kiokias, S.; Varzakas, T. Activity of flavonoids and beta-carotene during the auto-oxidative deterioration of model food oil-in water emulsions. *Food Chem.* **2014**, *150*, 280–286. [CrossRef] [PubMed]
2. Liudvinaviciute, D.; Rutkaite, R.; Bendoraitiene, J.; Klimaviciute, R. Thermogravimetric analysis of caffeic and rosmarinic acid containing chitosan complexes. *Carbohyd. Polym.* **2019**, *222*, 115003. [CrossRef]
3. Lisete-Torres, P.; Losada-Barreiro, S.; Albuquerque, H.; Sánchez-Paz, V.; Paiva-Martins, F.; Bravo-Díaz, C. Distribution of hydroxytyrosol and hydroxytyrosol acetate in olive oil emulsions and their antioxidant efficiency. *J. Agric. Food Chem.* **2012**, *60*, 7318–7325. [CrossRef] [PubMed]
4. Terpinc, P.; Polak, T.; Šegatin, N.; Hanzlowsky, A. Antioxidant properties of 4-vinyl derivatives of hydroxycinnamic acid. *Food Chem.* **2011**, *128*, 62–69. [CrossRef]

5. Phonsatta, N.; Deetae, P.; Luangpituksa, P.; Grajeda-Iglesias, C.; Figueroa-Espinoza, M.C.; Le Comte, J.; Villeneuve, P.; Decker, E.A.; Visessanguan, W.; Panya, A. Comparison of antioxidant evaluation assays for investigating antioxidative activity of gallic acid and its alkyl esters in different food matrices. *J. Agric. Food Chem.* **2017**, *65*, 7509–7518. [CrossRef] [PubMed]

6. Waraho, T.; McClements, D.-J.; Decker, E.-A. Mechanisms of lipid oxidation in food dispersions. *Trends Food Sci. Technol.* **2011**, *22*, 3–13. [CrossRef]

7. Dimakou, C.; Oreopoulou, V. Antioxidant activity of carotenoids against the oxidative destabilization of sunflower oil-in-water emulsions. *LWT-Food Sci. Technol.* **2012**, *46*, 393–400. [CrossRef]

8. Charoen, R.; Jangchud, A.; Jangchud, K.; Harnsilawat, T.; Naivikul, O.; McClements, D.-J. Influence of interfacial composition on oxidative stability of oil-in-water emulsions stabilized by biopolymer emulsifiers. *Food Chem.* **2012**, *131*, 1340–1346. [CrossRef]

9. Poyato, C.; Navarro-Blasco, I.; Calvo, M.I.; Cavero, R.Y.; Astiasarán, I.; Ansorena, D. Oxidative stability of O/W and W/O/W emulsions: Effect of lipid composition and antioxidant polarity. *Food Res. Intern.* **2013**, *51*, 132–140. [CrossRef]

10. Zhong, Y.; Shahidi, F. Antioxidant behavior in bulk oil: Limitations of polar paradox theory. *J. Agric. Food Chem.* **2012**, *60*, 4–6. [CrossRef]

11. Kiokias, S.; Varzakas, T. Innovative applications of food related emulsions. *Crit. Rev. Food Sci. Nutr.* **2017**, *57*, 3165–3172. [CrossRef] [PubMed]

12. Kiokias, S.; Gordon, M.; Oreopoulou, V. Compositional and processing factors that monitor oxidative deterioration of food relevant protein stabilised emulsions. *Crit. Rev. Food Sci. Nutr.* **2017**, *57*, 549–558. [CrossRef] [PubMed]

13. Axmann, M.; Strobl, W.-M.; Plochberger, B.; Stangl, H. Cholesterol transfer at the plasma membrane. *Atherosclerosis* **2019**, *290*, 111–117. [CrossRef]

14. Kiokias, S.; Proestos, C.; Oreopoulou, V. Effect of natural Food Antioxidants against LDL and DNA Oxidative damages. *Antioxidants* **2018**, *7*, 133. [CrossRef]

15. Kiokias, S. Antioxidant effects of vitamins C, E and provitamin A compounds as monitored by use of biochemical oxidative indicators linked to atherosclerosis and carcinogenesis. *Int. J. Nutr. Res.* **2019**, *1*, 1–13.

16. Dalbem, L.; Costa Monteiro, C.-M.; Anderson, J.-T. Anticancer properties of hydroxycinnamic acids-A Review. *Canc. Clinic. Oncol.* **2012**, *1*, 109–121.

17. Bojić, M.; HaasŠ, V.-S.; arić, D.; Maleš, Z. Determination of flavonoids, phenolic acids, and xanthines in Mate tea (Ilex paraguariensis St.-Hil.). *J. Anal. Methods Chem.* **2013**, *6*, 1–6.

18. Berté, K.-A.; Beux, M.-R.; Spada, P.-K.; Salvador, M.; Hoffmann-Ribani, R. Chemical composition and antioxidant activity of yerba-mate (*Ilex paraguariensis* A. St.-Hil., Aquifoliaceae) extract as obtained by spray drying. *J. Agric. Food Chem.* **2011**, *59*, 5523–5527. [CrossRef]

19. Zheng, W.; Wang, S.-Y. Antioxidant activity and phenolic compounds in selected herbs. *J. Agric. Food Chem.* **2001**, *49*, 5165–5170. [CrossRef]

20. Da Silva, A.B.; Koistinen, V.M.; Mena, P.; Bronze, M.R.; Hanhineva, K.; Sahlstrøm, S.; Kitrytė, V.; Moco, S.; Aura, A.-M. Factors affecting intake, metabolism and health benefits of phenolic acids: Do we understand individual variability? *Eur. J. Nutr.* **2019**, 1–19. [CrossRef]

21. Dezsi, S.; Bădărău, A.-S.; Bischin, C.; Vodna, D.-C.; Dumitrescu, R.-S.; Gheldiu, A.-M.; Mocan, A.; Laurian, V. Antimicrobial and antioxidant activities and phenolic profile of *Eucalyptus globulus*. *Molecules* **2015**, *20*, 4720–4734. [CrossRef] [PubMed]

22. Deotale, S.M.; Dutta, S.; Moses, J.A.; Anandharamakrishnan, C. Coffee oil as a natural surfactant. *Food Chem.* **2019**, *295*, 180–188. [CrossRef] [PubMed]

23. Boke, S.; Goren, A.C.; Kirmizigul, S. Simultaneous determination of several flavonoids and phenolic compounds in nineteen different *Cephalaria* species by HPLC-MS/MS. *J. Pharm. Biomed. Anal.* **2019**, *173*, 120–125. [CrossRef] [PubMed]

24. Conde, E.; Gordon, M.-H.; Moure, A.; Dominguez, H. Effects of caffeic acid and bovine serum albumin in reducing the rate of development of rancidity in oil-in-water and water-in-oil emulsions. *Food Chem.* **2011**, *129*, 1652–1659. [CrossRef]

25. Sørensen, A.D.; Villeneuve, P.; Jacobsen, C. Alkyl caffeates as antioxidants in O/W emulsions: Impact of emulsifier type and endogenous tocopherols. *Eur. J. Lipid Sci. Technol.* **2017**, *119*, 6. [CrossRef]

26. Pandurangan, A.-K.; Mohebali, N.; Norhaizan, M.-E.; Looi, C.-Y. Gallic acid attenuates dextran sulfate sodium-induced experimental colitis in BALB/c mice". *Drug Des. Dev. Ther.* **2015**, *9*, 3923–3934. [CrossRef]

27. Liu, Y.; A Carver, J.; Calabrese, A.N.; Pukala, T.L. Gallic acid interacts with α-synuclein to prevent the structural collapse necessary for its aggregation. *Biochim. Biophys. Acta (BBA) Proteins Proteom.* **2014**, *1844*, 1481–1485. [CrossRef]

28. Zucca, P.; Antonella, R.; Tuberoso, C.; Piras, A.; Rinaldi, A.; Sanjust, E.; Dessì, M.; Rescigno, A. Evaluation of Antioxidant Potential of "Maltese Mushroom" (Cynomorium coccineum) by means of multiple chemical and biological assays". *Nutrients* **2013**, *5*, 149–161. [CrossRef]

29. Souza, M.-C.; Santos, M.-P.; Sumere, B.-P.; Silva, L.-C.; Cunha, D.-T.; Martínez, J.; Barbero, F.-G.; Rostagno, M.-A. Isolation of gallic acid, caffeine and flavonols from black tea by on-line coupling of pressurized liquid extraction with an adsorbent for the production of functional bakery products. *LWT Food Sci. Technol.* **2020**, *117*, 108661. [CrossRef]

30. Bou, R.; Boo, C.; Kwek, A.; Hidalgo, D.; Decke, E.A. Effect of different antioxidants on lycopene degradation in oil-in-water emulsions. *Eur. J. Lipid Sci. Technol.* **2011**, *113*, 724–729. [CrossRef]

31. Alavi Rafiee, S.; Farhoosh, R.; Sharif, A. Antioxidant activity of gallic acid as affected by an extra carboxyl group than pyrogallol in various oxidative environments. *Eur. J. Lipid Sci. Technol.* **2018**, *120*, 1800319. [CrossRef]

32. Di Mattia, C.; Sacchetti, G.; Pittia, P.; Mastrocola, D. Effect of phenolic antioxidants on the dispersion state and chemical stability of olive oil O/W emulsions. *Food Res. Int.* **2009**, *42*, 1163–1170. [CrossRef]

33. Zhu, M.; Wu, C.; Chen, Y.; Xie, M. Antioxidant effects of different polar gallic acid and its alkyl esters in oil-in-water emulsions. *J. Chin. Inst. Food Sci. Technol.* **2019**, *19*, 13–22.

34. Wang, Y.; Wu, C.; Zhou, X.; Zhang, M.; Chen, Y.; Nie, S.; Xie, M. Combined application of gallate ester and α-tocopherol in oil-in-water emulsion: Their distribution and antioxidant efficiency. *J. Dispers. Sci. Technol.* **2019**, 1–9. [CrossRef]

35. Evageliou, V.; Panagopoulou, E.; Mandala, I. Encapsulation of EGCG and esterified EGCG derivatives in double emulsions containing whey protein isolate, bacterial cellulose and salt. *Food Chem.* **2019**, *281*, 171–177. [CrossRef]

36. Oreopoulou, A.; Papavassilopoulou, E.; Bardouki, H.; Vamvakias, M.; Bimpilas, A.; Oreopoulou, V. Antioxidant recovery from hydrodistillation residues of selected Lamiaceae species by alkaline extraction. *J. Appl. Res. Med. Aromat. Plants* **2018**, *8*, 83–89. [CrossRef]

37. Stansbury, J. Rosmarinic acid as a novel agent in the treatment of allergies and asthma. *J. Restor. Med.* **2014**, *3*, 121. [CrossRef]

38. Vallverdú-Queralt, A.-L.; Rejueiro, J.; Martínez-Huélamo, M.; Alvarenga, J.-F.; Leal, L.-N.; Lamuela-Raventos, R.-M. A comprehensive study on the phenolic profile of widely used culinary herbs and spices: Rosemary, thyme, oregano, cinnamon, cumin and bay. *Food Chem.* **2014**, *154*, 299–307. [CrossRef]

39. Yashin, A.; Yashin, Y.; Xia, X.; Nemzer, B. Antioxidant activity of spices and their impact on human health: A review. *Antioxidants* **2017**, *6*, 70. [CrossRef]

40. Tsimogiannis, D.; Choulitoudi, E.; Bimpilas, A.; Mitropoulou, G.; Kourkoutas, Y.; Oreopoulou, V. Exploitation of the biological potential of Satureja thymbra essential oil and distillation by-products. *J. Appl. Res. Med. Aromat Plants* **2016**, *4*, 12–20. [CrossRef]

41. Alamed, J.; Chaiyasit, W.; McClements, D.-J.; Decker, E.-A. Relationships between free radical scavenging and antioxidant activity in foods. *J. Agric. Food Chem.* **2009**, *57*, 2969–2976. [CrossRef] [PubMed]

42. Panya, A.; Kittipongpittaya, K.; Laguerre, M.; Bayrasy, C.; Lecomte, J.; Villeneuve, P.; Decker, E.-A. Interactions between α-tocopherol and rosmarinic acid and its alkyl esters in emulsions: Synergistic, additive, or antagonistic effect? *J. Agric. Food Chem.* **2012**, *60*, 10320–10330. [CrossRef] [PubMed]

43. Bakota, E.-L.; Winkler-Moser, J.-K.; Berhow, M.-A.; Eller, F.-J.; Vaughn, S.-F. Antioxidant activity and sensory evaluation of a rosmarinic acid-enriched extract of *Salvia officinalis*. *J. Food Sci.* **2015**, *80*, 711–717. [CrossRef] [PubMed]

44. Loussouarn, M.; Krieger-Liszkay, A.; Svilar, L.; Bily, A.; Birtić, S.; Havaux, M. Carnosic acid and carnosol, two major antioxidants of rosemary, act through different mechanisms. *Plant Physiol.* **2017**, *175*, 381–1394. [CrossRef]

45. Raes, K.; Doolaege, E.H.; Deman, S.; Vossen, E.; De Smet, S. Effect of carnosic acid, quercetin and α-tocopherol on lipid and protein oxidation in anin vitrosimulated gastric digestion model. *Int. J. Food Sci. Nutr.* **2015**, *66*, 216–221. [CrossRef]

46. Li, Z.; Feng, Y.; Lei, Y.; Yuan, Z. Comparison of the antioxidant effects of carnosic acid and synthetic antioxidants on tara seed oil. *Chem. Cent. J.* **2018**, *12*, 37. [CrossRef]

47. Frankel, E.N.; Huang, S.-W.; Aeschbach, R.; Prior, E. Antioxidant Activity of a Rosemary Extract and Its Constituents, Carnosic Acid, Carnosol, and Rosmarinic Acid, in Bulk Oil and Oil-in-Water Emulsion. *J. Agric. Food Chem.* **1996**, *44*, 131–135. [CrossRef]

48. Gallego, G.; Gordon, M.H.; Segovia, F.J.; Skowyra, M.; Almajano, M.P. Antioxidant Properties of Three Aromatic Herbs (Rosemary, Thyme and Lavender) in Oil-in-Water Emulsions. *J. Am. Oil Chem. Soc.* **2013**, *90*, 1559–1568. [CrossRef]

49. Gallego, M.-G.; Hakkarainen, M.; Almajano, M.-P. Stability of O/W emulsions packed with PLA film with incorporated rosemary and thyme. *Eur. Food Res. Technol.* **2017**, *243*, 1249–1259. [CrossRef]

50. Lei, C.; Tang, X.; Chen, M.; Chen, H.; Yu, S. Alpha-tocopherol-based microemulsion improving the stability of carnosic acid and its electrochemical analysis of antioxidant activity. *Coll. Surf. A Phys. Eng. Aspects* **2019**, *580*, 123708. [CrossRef]

51. Birtić, S.; Dussort, P.; Pierre, F.-X.; Bily, A.-C.; Roller, M. Carnosic acid. *Phytochemistry* **2015**, *115*, 9–19. [CrossRef] [PubMed]

52. Kumar, N.; Vikas, P. Potential applications of ferulic acid from natural sources. *BioTechnol. Rep.* **2014**, *4*, 86–93. [CrossRef] [PubMed]

53. Bagchi, D.; Moriyama, H.; Swaroop, A. *Green Coffee Bean Extract in Human Health*; CRC Press: Boca Raton, FL, USA, 2016; p. 92.

54. Mojica, L.; Meyer, A.; Berhow, M.; González, E. Bean cultivars (*Phaseolus vulgaris* L.) have similar high antioxidant capacity, in vitro inhibition of α-amylase and α-glucosidase while diverse phenolic composition and concentration. *Food Res. Intern.* **2015**, *69*, 38–48. [CrossRef]

55. Luthria, D.-L.; Pastor-Corrales, A.-M. Phenolic acids content of fifteen dry edible bean (*Phaseolus vulgaris* L.) varieties. *J. Food Compos. Anal.* **2006**, *19*, 205–211. [CrossRef]

56. Oehlke, K.; Heins, A.; Stöckmann, H.; Schwarz, K. Impact of emulsifier microenvironments on acid–base equilibrium and activity of antioxidants. *Food Chem.* **2010**, *118*, 48–55. [CrossRef]

57. Maqsood, S.; Benjakul, S. Comparative studies of four different phenolic compounds on in vitro antioxidative activity and the preventive effect on lipid oxidation of fish oil emulsion and fish mince. *Food Chem.* **2010**, *119*, 123–132. [CrossRef]

58. Shin, J.-A.; Jeong, S.-H.; Jia, C.-H.; Hong, S.-T.; Lee, K.-T. Comparison of antioxidant capacity of 4-vinylguaiacol with catechin and ferulic acid in oil-in-water emulsion. *Food Sci. Biotech.* **2019**, *28*, 35–41. [CrossRef]

59. Permin, A.; Bosc, V.; Soto, P.; Le Roux, E.; Maillard, M.-N. Lipid oxidation in oil-in-water emulsions rich in omega-3: Effect of aqueous phase viscosity, emulsifiers, and antioxidants. *Eur. J. Lipid Sci. Technol.* **2019**, *121*, 9.

60. Trisha, S. Role of hesperdin, luteolin and coumaric acid in arthritis management: A Review. *Inter. J. Phys. Nutr. Phys. Educ.* **2018**, *3*, 1183–1186.

61. Kehan, P.; Juanying, O.; Shiyi, O. p-Coumaric acid and its conjugates: Dietary sources, pharmacokinetic properties and biological activities. *J. Sci. Food Agric.* **2016**, *96*, 2952–2962.

62. Nesovic, M.; Tosti, T.; Trifkovi, J.; Baosic, R.; Blagojevic, S.; Ignjatovic, L.; Tesic, Z. Physicochemical analysis and phenolic profile of polyfloral and honeydew honey from Montenegro. *RSC Adv.* **2020**, *10*, 2462. [CrossRef]

63. Kannan, R.-R.; Arumugam, R.; Thangaradjou, R.; Thirunavukarasu, T.; Perumal, A. Phytochemical constituents, antioxidant properties and p-coumaric acid analysis in some seagrasses. *Food Res. Intern.* **2013**, *54*, 1229–1236. [CrossRef]

64. Kavita, P.; Puneet, G. Rediscovering the therapeutic potential of *Amaranthus species*: A review. *Egyp. J. Basic Appl. Sci.* **2017**, *3*, 196–205.

65. Pasko, P.; Sajewicz, M.; Corinstein, S.; Zachwieja, Z. Analysis of selected phenolic acids and flavonoids in *Amaranthus cruentus* and *Chenopodium quinoa* seeds and sprouts by HPLC. *Acta Chromatogr.* **2008**, *20*, 661–672. [CrossRef]

66. Oh, S.; Kim, M.-J.; Park, K.-W.; Lee, J.-H. Antioxidant properties of aqueous extract of roasted hulled barley in bulk oil or oil-in-water emulsion matrix. *J. Food Sci.* **2015**, *80*, 11. [CrossRef] [PubMed]

67. Kiliç, I.; Yeşiloğlu, Y. Spectroscopic studies on the antioxidant activity of p-coumaric acid. *Spectrochim Acta A Mol. Biomol. Spectrosc.* **2013**, *115*, 719–724. [CrossRef] [PubMed]

68. Zhigang, H.; Shengguan, C.; Xuelei, Z.; Qiufeng, Q.; Huang, Y.; Dai, F.; Guoping, Z. Development of predictive models for total phenolics and free *p*-coumaric acid contents in barley grain by near-infrared spectroscopy. *Food Chem.* **2017**, *227*, 342–348.

69. Park, J.; Gim, S.-Y.; Jeon, J.-Y.; Kim, M.-J.; Choi, H.-K.; Lee, J. Chemical profiles and antioxidant properties of roasted rice hull extracts in bulk oil and oil-in-water emulsion. *Food Chem.* **2019**, *272*, 242–250. [CrossRef]

70. Siriamornpun, S.; Kaewseejan, N. Quality, bioactive compounds and antioxidant capacity of selected climacteric fruits with relation to their maturity *Sci. Hortic.* **2017**, *221*, 33–42. [CrossRef]

71. European Medicinal Agency. Assessment Report on *Angelica sinensis* (Oliv.) Diels, radix. Committee on Herbal Medicinal Products (HMPC). 2013. EMA/HMPC/614586/2012. Available online: https://www.ema.europa.eu/en/documents/herbal-report/final-assessment-report-angelica-sinensis-oliv-diels-radix-first-version_en.pdf (accessed on 23 March 2019).

72. Kim, J.; Soh, S.-Y.; Bae, H.; Nam, S.-Y. Antioxidant and phenolic contents in potatoes (*Solanum tuberosum* L.) and micropropagated potatoes. *Appl. Biol. Chem.* **2019**, *62*, 1. [CrossRef]

73. Espinosa, R.R.; Inchingolo, R.; Alencar, S.-M.; Rodriguez-Estrada, M.-T.; Castro, I.A. Antioxidant activity of phenolic compounds added to a functional emulsion containing omega-3 fatty acids and plant sterol esters. *Food Chem.* **2018**, *182*, 95–104. [CrossRef] [PubMed]

74. Pacheco-Palencia, L.A.; Mertens-Talcott, S.U.; Talcott, S.T. Chemical Composition, Antioxidant Properties, and Thermal Stability of a Phytochemical Enriched Oil from Açai (Euterpe oleraceaMart.). *J. Agric. Food Chem.* **2008**, *56*, 4631–4636. [CrossRef] [PubMed]

75. Zhao, C.; Yuan Jia, Y.; Lu, F. Angelica stem: A potential low-cost source of bioactive phthalides and phytosterols. *Molecules* **2018**, *23*, 3065. [CrossRef] [PubMed]

76. Radmanesh, E.; Dianat, M.; Badavi, M.; Goudarzi, G.; Mard, S.A. The cardio protective effect of vanillic acid on hemodynamic parameters, malondialdehyde, and infract size in ischemia-reperfusion isolated rat heart exposed to PM. *Iran J. Basic Med. Sci.* **2017**, *20*, 760–768.

77. Keller, S.; Locquet, N.; Cuvelier, M.-E. Partitioning of vanillic acid in oil-in-water emulsions: Impact of the Tween®40 emulsifier. *Food Res. Int.* **2016**, *88*, 61–69. [CrossRef]

78. Vishnu, K.-V.; Chatterjee, N.-S.; Ajeeshkumar, K.-K.; Lekshmi, R.-K.; Tejpal, C.-S.; Mathew, S.; Ravishankar, C.N. Microencapsulation of sardine oil: Application of vanillic acid grafted chitosan as a bio-functional wall material. *Carbohydr. Polym.* **2017**, *174*, 540–548. [CrossRef]

79. Mancuso, C.; Santangelo, R. Ferulic acid: Pharmacological and toxicological aspects. *Food Chem. Toxicol.* **2014**, *65*, 185–195. [CrossRef]

80. Sgarbossa, A.; Giacomazza, D.; Martadi, C. Ferulic acid: A hope for Alzheimer's disease therapy from plants. *Nutrients* **2015**, *7*, 5764–5782. [CrossRef]

81. Ferreira, C.-S.; Pereyra, A.; Patriarca, A.; Mazzobre, M.-F.; Polak, T.; Abram, V.; Buera, M.-P.; Ulrih, N.-P. Phenolic compounds in extracts from *Eucalyptus globulus* leaves and *Calendula officinalis* flowers. *J. Nat. Prod. Res.* **2016**, *2*, 53–55.

82. Manuja, R.; Sachdeva, S.; Jain, A.; Chaudhary, J. A comprehensive review on biological activities of *p*-hydroxy benzoic acid and its derivatives. *Int. J. Pharm. Sci. Rev. Res.* **2013**, *22*, 109–115.

83. Rosa, L.-S.; Silva, N.-J.; Soares, N.-C.; Monteiro, M.-C.; Teodoro, A.-J. Anticancer properties of phenolic acids in colon cancer–A review. *J. Nutr. Food Sci.* **2016**, *6*, 2.

84. Vinayagam, R.; Jayachandran, M.; Xu, B. Antidiabetic effects of simple phenolic acids: A comprehensive review. *Phytother. Res.* **2016**, *30*, 184–199. [CrossRef]

85. Shahidi, F.; Ju Dong, Y. Bioactivities of phenolics by focusing on suppression of chronic diseases: A Review. *Int. J. Mol. Sci.* **2018**, *19*, 1573. [CrossRef] [PubMed]

86. Chisvert, A.; Salvador, A. Cosmetic ingredients: From the cosmetic to the human body and the environment. *Anal. Methods* **2012**, *5*, 309–310. [CrossRef]

87. Magnani, C.; Isaac, V.-L.; Correa, M.-A.; Salgado, H.-R. Caffeic acid: A review of its potential use in medications and cosmetics. *Anal. Methods* **2014**, *6*, 3203–3210. [CrossRef]

88. De Oliveira, N.C.; Sarmento, M.S.; Nunes, E.; Porto, C.M.; Rosa, D.P.; Bona, S.; Rodrigues, G.; Marroni, N.P.; Pereira, P.; Picada, J.N.; et al. Rosmarinic acid as a protective agent against genotoxicity of ethanol in mice. *Food Chem. Toxicol.* **2012**, *50*, 1208–1214. [CrossRef]

89. Paulo, F.; Santos, L. Microencapsulation of caffeic acid and its release using a w/o/w double emulsion method: Assessment of formulation parameters. *Dry. Technol.* **2019**, *37*, 950–961. [CrossRef]

90. Sidoryk, K.; Jaromin, A.; Filipczak, N.; Cmoch, P.; Cybulski, M. Synthesis and antioxidant activity of caffeic acid derivatives. *Molecules* **2018**, *23*, 2199. [CrossRef]

91. Espíndola, K.M.M.; Ferreira, R.G.; Narvaez, L.E.M.; Rosario, A.C.R.S.; Da Silva, A.H.M.; Silva, A.G.B.; Vieira, A.P.O.; Monteiro, M.C. Chemical and pharmacological aspects of caffeic acid and its activity in hepatocarcinoma. *Front. Oncol.* **2019**, *9*, 541. [CrossRef]

92. Macías-Pérez, J.R.; Ramírez-Bello, J.; Vásquez-Garzón, V.R.; Salcido-Neyoy, M.E.; Martínez-Soriano, P.A.; Ruiz-Sánchez, M.B.; Angeles, E.; Villa-Treviño, S. The effect of caffeic acid phenethyl ester analogues in a modified resistant hepatocyte model. *Anticancer Drugs* **2013**, *24*, 394–405. [CrossRef]

93. Guan, Y.; Chen, H.; Zhong, Q. Nanoencapsulation of caffeic acid phenethyl ester in sucrose fatty acid esters to improve activities against cancer cells. *J. Food Eng.* **2019**, *246*, 125–133. [CrossRef]

94. Gießel, J.-M.; Loesche, A.; Csuk, R. Caffeic acid phenethyl ester (CAPE)-derivatives act as selective inhibitors of acetylcholinesterase. *Eur. J. Med. Chem.* **2019**, *177*, 259–268. [CrossRef]

95. Koru, F.; Avcu, M.; Tanyuksel, A.; Ural, U.R.; Araz, E.; Sener, K. Cytotoxic effects of caffeic acid phenethyl ester (CAPE) on the human multiple myeloma cell line. *Turk. J. Med. Sci.* **2009**, *39*, 863–870.

96. Zhang, Z.; Wang, D.; Qiao, S.; Wu, X.; Cao, S.; Wang, L.; Su, X.; Li, L. Metabolic and microbial signatures in rat hepatocellular carcinoma treated with caffeic acid and chlorogenic acid. *Sci. Rep.* **2017**, *7*, 4508. [CrossRef] [PubMed]

97. Greenhill, C. Carnosic acid could be a new treatment option for patients with NAFLD or the metabolic syndrome. *Nat. Rev. Gastroenterol. Hepatol.* **2011**, *8*, 122. [CrossRef]

98. Dickmann, L.-J.; VandenBrin, B.-M.; Lin, Y.-S. In vitro hepatotoxicity and cytochrome P450 induction and inhibition characteristics of carnosic acid, a dietary supplement with antiadipogenic properties. *Drug Metab. Dispos.* **2012**, *40*, 1263–1267. [CrossRef]

99. Bahri, S.; Jameleddine, S.; Shlyonsky, V. Relevance of carnosic acid to the treatment of several health disorders: Molecular targets and mechanisms. *Biomed. Pharmacother.* **2016**, *84*, 569–582. [CrossRef]

100. Einbond, L.-S.; Wu, H.-A.; Kashiwazaki, R.; He, K.; Roller, M.; Su, T.; Wang, X.; Goldsberry, S. Carnosic acid inhibits the growth of ER-negative human breast cancer cells and synergizes with curcumin. *Fitoterapia* **2012**, *83*, 1160–1168. [CrossRef]

101. Solomonov, Y.; Hadad, N.; Levy, R. The combined anti-inflammatory effect of astaxanthin, lyc-O-mato and carnosic acid in vitro and in vivo in a mouse model of peritonitis. *J. Nutr. Food Sci.* **2018**, *8*, 653. [CrossRef]

102. Kosuru, R.Y.; Roy, A.; Das, S.K.; Bera, S. Gallic acid and gallates in human health and disease: Do mitochondria hold the key to success? *Mol. Nutr. Food Res.* **2017**, *10*, 62. [CrossRef]

103. Sourani, Z.-M.; Pourgheysari, B.-P.; Beshkar, P.-M.; Shirzad, H.-P.; Shirzad, M.-M. Gallic acid inhibits proliferation and induces apoptosis in lymphoblastic leukemia cell line (C121). *Iran J. Med. Sci.* **2016**, *41*, 525–530.

104. Aborehab, N.-M.; Nada, O. Effect of gallic acid in potentiating chemotherapeutic effect of Paclitaxel in HeLa cervical cancer cells. *Canc. Cell Intern.* **2019**, *19*, 1–13. [CrossRef] [PubMed]

105. Xu, C.; Yu, Y.; Ling, L.; Wang, Y.; Zhang, J.; Li, Y.; Duan, G. A C8-modified Graphene@mSiO2 composites based method for quantification of gallic acid in rat plasma after oral administration of Changtai granule and its application to pharmacokinetics. *Biol. Pharm. Bull.* **2017**, *40*, 1021–1028. [CrossRef] [PubMed]

106. Paolini, A.; Curti, V.; Pasi, F.; Mazzini, G.; Nano, R.; Capelli, E. Gallic acid exerts a protective or an anti-proliferative effect on glioma T98G cells via dose-dependent epigenetic regulation mediated by mi RNAs. *J. Oncol.* **2015**, *46*, 1491–1497. [CrossRef] [PubMed]

107. Yu, Z.; Song, F.; Jin, Y.-C.; Zhang, W.-M.; Zhang, Y.; Liu, E.-J.; Zhou, D.; Bi, L.-L.; Yang, Q.; Li, H.; et al. Comparative pharmacokinetics of gallic acid after oral administration of gallic acid monohydrate in normal and isoproterenol-induced myocardial infarcted rats. *Front. Pharmacol.* **2018**, *6*, 328. [CrossRef] [PubMed]

108. Nwokocha, C.; Palacios, J.; Simirgiotis, M.J.; Thomas, J.; Nwokocha, M.; Young, L.; Thompson, R.K.; Cifuentes, F.; Paredes, A.; Delgoda, R. Aqueous extract from leaf of *Artocarpus altilis* provides cardio-protection from isoproterenol induced myocardial damage in rats: Negative chronotropic and inotropic effects. *J. Ethnoph.* **2017**, *203*, 163–170. [CrossRef]

109. Zamora-Ros, R.; Rothwell, J.-A.; Scalbert, A.; Knaze, V.; Romieu, I.; Slimani, N. Dietary intakes and food sources of phenolic acids in the European Prospective Investigation into Cancer and Nutrition (EPIC) study. *Br. J. Nutr.* **2013**, *110*, 1500–1511. [CrossRef]

110. Uraji, M.; Kimura, M.; Inoue, Y.; Kawakami, K.; Kumagay, Y.; Harazono, K.; Hatanaka, T. Enzymatic production of ferulic acid from defatted rice bran by using a combination of bacterial enzymes. *Appl. Biochem. Biotechnol.* **2013**, *171*, 1085–1093. [CrossRef]

111. Sung, J.-H.; Gim, S.-A.; Koh, P.-O. Ferulic acid attenuates the cerebral ischemic injury-induced decrease in peroxiredoxin-2 and thioredoxin expression. *Neurosci. Lett.* **2014**, *30*, 88–92. [CrossRef]

112. Chowdhury, S.; Ghosh, S.; Rashid, K.; Sil, P.-C. Deciphering the role of ferulic acid against streptozotocin-induced cellular stress in the cardiac tissue of diabetic rats. *Food Chem. Toxicol.* **2016**, *97*, 187–198. [CrossRef]

113. Baeza, G.; Bachmair, E.-M.; Wood, S.; Mateos, R.; Bravo, L.; de Roos, B. The colonic metabolites dihydrocaffeic acid and dihydroferulic acid are more effective inhibitors of in vitro platelet activation than their phenolic precursors. *Food Funct.* **2017**, *8*, 1333–1342. [CrossRef] [PubMed]

114. Ambothi, N.; Rajendra, P.; Agilan, B. Ferulic acid prevents ultraviolet-B radiation induced oxidative DNA damage in human dermal fibroblasts. *Int. J. Nutr. Pharmacol. Neurol. Dis.* **2014**, *4*, 203–213.

115. Ambothi, K.; Prasad, N.-R.; Balupillai, A. Ferulic acid inhibits UVB-radiation induced photocarcinogenesis through modulating inflammatory and apoptotic signaling in Swiss albino mice. *Food Chem. Toxicol.* **2015**, *82*, 72–78. [CrossRef] [PubMed]

116. Russo, G.I.; Campisi, D.; Di Mauro, M.; Regis, F.; Reale, G.; Marranzano, M.; Ragusa, R.; Solinas, T.; Madonia, M.; Cimino, S.; et al. Dietary consumption of phenolic acids and prostate cancer: A case-control study in Sicily. *Molecules* **2017**, *22*, 2159. [CrossRef]

117. Akdemir, F.E.; Albayrak, M.; Çalik, M.; Bayir, Y.; Gulcin, I. The protective effects of *p*-coumaric acid on acute liver and kidney damages induced by cisplatin. *Biomedicines* **2017**, *5*, 18. [CrossRef] [PubMed]

118. Ferreira, P.-S.; Victorelli, F.-D.; Fonseca-Santos, B.; Chorilli, M. A Review of analytical methods for p-coumaric acid in plant-based products, beverages, and biological matrices. *Crit. Rev. Anal. Chem.* **2019**, *49*, 21–31. [CrossRef]

119. Boo, Y.-C. p-Coumaric acid as an active ingredient in cosmetics: A review focusing on its antimelanogenic effects. *Antioxidants* **2019**, *8*, 275. [CrossRef]

120. Neog, M.-K.; Rasool, M. Targeted delivery of p-coumaric acid encapsulated mannosylated liposomes to the synovial macrophages inhibits osteoclast formation and bone resorption in the rheumatoid arthritis animal model. *Eur. J. Pharm. Biopharm.* **2018**, *133*, 162–175. [CrossRef]

121. Janicke, B.; Hegardt, C.; Krogh, M.; Onning, G.; Akesson, B.; Cirenajwis, H.M.; Oredsson, S.M. The antiproliferative effect of dietary fiber phenolic compounds ferulic acid and p-coumaric acid on the cell cycle of Caco-2 cells. *Nutr. Cancer.* **2011**, *63*, 611–622. [CrossRef]

122. Sharma, S.-H.; Chellappan, D.-R.; Chinnaswamy, P.; Nagarajan, S. Protective effect of p-coumaric acid against 1,2 dimethylhydrazine induced colonic preneoplastic lesions in experimental rats. *Biomed. Pharmacother.* **2017**, *94*, 577–588. [CrossRef]

123. Amalan, V.; Natesan, V.; Dhananjayan, I.; Arumugam, R. Antidiabetic and antihyperlipidemic activity of p-coumaric acid in diabetic rats, role of pancreatic GLUT 2: In vivo approach. *Biomed. Pharmacother.* **2016**, *84*, 230–236. [CrossRef] [PubMed]

124. Vauzour, D.; Corona, G.; Spencer, J.P. Caffeic acid, tyrosol and p-coumaric acid are potent inhibitors of 5-S-cysteinyl-dopamine induced neurotoxicity. *Arch. Biochem. Biophys.* **2010**, *501*, 106–111. [CrossRef]

125. Sunitha, M.-C.; Dhanyakrishnan, R.; PrakashKumar, B.; Nevin, K.-G. p-Coumaric acid mediated protection of H9c2 cells from Doxorubicin-induced cardiotoxicity: Involvement of augmented Nrf2 and autophagy. *Biomed. Pharmacother.* **2018**, *102*, 823–832. [CrossRef] [PubMed]

126. Yang, E.-J.; Ku, S.-K.; Lee, W.; Lee, S.; Lee, T.; Song, K.-S.; Bae, J.-S. Barrier protective effects of rosmarinic acid on HMGB1 induced inflammatory responses in vitro and in vivo. *J. Cell. Physiol.* **2013**, *228*, 975–982. [CrossRef] [PubMed]

127. Tsung-Hsien, T.; Chuang, L.-T.; Lien, T.-J.; Liing, Y.-R.; Chen, W.-Y.; Tsai, P.-J. *Rosmarinus officinalis* extract suppresses *Propionibacterium acnes*–induced inflammatory responses. *J. Med. Food* **2013**, *16*, 324–333.

128. Ku, S.-K.; Yang, E.-J.; Song, K.-S.; Bae, J.-S. Rosmarinic acid down-regulates endothelial protein C receptor shedding in vitro and in vivo. *Food Chem. Toxicol.* **2013**, *59*, 311–315. [CrossRef] [PubMed]

129. Braidy, N.; Matin, A.; Rossi, F.; Chinain, M.; Laurent, D.; Guillemin, G.J. Neuroprotective effects of rosmarinic acid on ciguatoxin in primary human neurons. *Neurotox. Res.* **2014**, *25*, 226–234. [CrossRef]

130. Nunes, S.; Madureira, A.-R.; Campos, D.; Sarmento, B.; Gomes, A.-R.; Pintado, M.; Reis, F. Therapeutic and nutraceutical potential of rosmarinic acid. Cytoprotective properties and pharmacokinetic profile. *Crit. Rev. Food Sci. Nutr.* **2017**, *57*, 1799–1806. [CrossRef]

131. Hossan, M.S.; Rahman, S.; Bashar, A.B.M.A.; Jahan, R.; Al-Nahain, A.; Rahmatullah, M. Rosmarinic acid: A review of its anticancer action. *World J. Pharmac. Sci.* **2014**, *3*, 57–70.

132. Alagawany, M.; El-Hack, M.E.A.; Farag, M.R.; Gopi, M.; Karthik, K.; Malik, Y.S.; Dhama, K. Rosmarinic acid: Modes of action, medicinal values and health benefits. *Anim. Health Res. Rev.* **2019**, *1*, 1–11. [CrossRef]

133. Domitrović, R.; Skoda, M.; Marchesi, V.; Cvijanović, O.; Pernjak Pugel, E.; Stefan, M.-B. Rosmarinic acid ameliorates acute liver damage and fibrogenesis in carbon tetrachloride-intoxicated mice. *Food Chem. Toxicol.* **2013**, *51*, 370–378. [CrossRef]

134. De Oliveira, J.-R.; Afonso CamargoPorto, S.-E.; De Oliveira, L.D. *Rosmarinus officinalis* L. (rosemary) as therapeutic and prophylactic agent. *J. Biomed. Sci.* **2019**, *26*, 5. [CrossRef]

135. Luño, V.; Gil, L.; Olaciregui, M.; González, N.; Jerez, R.-A.; de Blas, I. Rosmarinic acid improves function and in vitro fertilising ability of boar sperm after cryopreservation. *Cryobiology* **2014**, *69*, 157–162. [CrossRef]

136. Venkatachalam, K.; Gunasekaran, S.; Namasivayam, N. Biochemical and molecular mechanisms underlying the chemopreventive efficacy of rosmarinic acid in a rat colon cancer. *Eur. J. Pharmacol.* **2016**, *15*, 37–50. [CrossRef]

137. Mathew, S.; Halamaa, A.; Kadera, S.-A.; Choea, M.; Mohney, R.-P.; Malekc, J.-A.; Suhrea, K. Metabolic changes of the blood metabolome after a date fruit challenge. *J. Funct. Foods* **2018**, *49*, 267–276. [CrossRef]

138. Wang, L.; Sweet, D.H. Potential for food-drug interactions by dietary phenolic acids on human organic anion transporters 1 (SLC22A6), 3 (SLC22A8), and 4 (SLC22A11). *Biochem. Pharmacol.* **2012**, *15*, 1088–1095. [CrossRef]

139. Dianat, M.; Hamzavi, G.-H.; Badavi, M.; Samarbafzadeh, A. Effects of Losartan and vanillic acid co-administration on ischemia-reperfusion-induced oxidative stress in isolated rat heart. *Iran Red. Crescent Med. J.* **2014**, *16*, 1–7. [CrossRef]

140. Kim, S.-J.; Kim, M.-K.; Um, J.; Hong, S.-H. The beneficial effect of vanillic acid on ulcerative colitis. *Molecules* **2010**, *15*, 7208–7217. [CrossRef]

141. Erdem, M.-G.; Cinkilic, N.; Vatan, O.; Yilmaz, D.; Bagdas, D.; Bilaloglu, R. Genotoxic and anti-genotoxic effects of vanillic acid against mitomycin C-induced genomic damage in human lymphocytes in vitro. *Asian Pac. J. Cancer Pre.* **2012**, *13*, 4993–4998. [CrossRef]

142. Krga, I.; Vidovica, N.; Milenkovicb, D.; Ristica, A.; Stojanovica, F.; Morand, C.; Glibeti, M. Effects of anthocyanins and their gut metabolites on adenosine diphosphateinduced platelet activation and their aggregation with monocytes and neutrophils. *Arch. Biochem. Biophys.* **2018**, *645*, 34–41. [CrossRef]

143. Chellammal, J.; Singh, H.; Kakalij, R.-M.; Kshirsagar, R.-P.; Kumar, B.-H.; Komakula, S.-B.; Diwan, P.M. Cognitive effects of vanillic acid against streptozotocin-induced neurodegeneration in mice. *Pharmac. Biol.* **2015**, *53*, 630–636.

144. Yemis, C.-P.; Pagotto, F.; Bach, S.; Delaquis, P. Effect of vanillin, ethyl vanillin, and vanillic acid on the growth and heat resistance of Cronobacter species. *J. Food Prot.* **2011**, *74*, 2062–2069. [CrossRef]

145. Jungbauer, A.; Medjakovic, S. Anti-inflammatory properties of culinary herbs and spices that ameliorate the effects of metabolic syndrome. *Maturitas* **2012**, *71*, 227–239. [CrossRef]

146. Khalifa, I.; Zhu, W.; Li, K.K.; Li, C.M. Polyphenols of mulberry fruits as multifaceted compounds: Compositions, metabolism, health benefits, and stability—A structural review. *J. Funct. Foods* **2018**, *40*, 28–43. [CrossRef]

147. Fernandes, A.-S.; Mazzei, J.-L.; Evangelista, H.; Marques, M.-R.; Ferraz, E.-R.; Felzenszwalb, I. Protection against UV-induced oxidative stress and DNA damage by Amazon moss extracts. *J. Photochem. Photobiol. B* **2018**, *183*, 331–341. [CrossRef]

148. Chandrasekara, A.; Daugelaite, J.; Shahidi, F. DNA scission and LDL cholesterol oxidation inhibition and antioxidant activities of Bael (*Aegle marmelos*) flower extracts. *J. Tradit. Complement. Med.* **2018**, *8*, 428–435. [CrossRef]

149. Xiao, Z.; Fang, L.; Niu, Y.; Yu, H. Effect of cultivar and variety on phenolic compounds and antioxidant activity of cherry wine. *Food Chem.* **2015**, *186*, 69–73. [CrossRef]

150. Thompson, K.; Pederick, W.; Singh, I.; Santhakumar, A.B. Anthocyanin supplementation in alleviating thrombogenesis in overweight and obese population: A randomized, double-blind, placebo-controlled study. *J. Funct. Foods* **2017**, *32*, 131–138. [CrossRef]

151. Padmanabhan, M.; Geetha, A. The modulating effect of *Persea americana* fruit extract on the level of expression of fatty acid synthase complex, lipoprotein lipase, fibroblast growth factor-21 and leptin—A biochemical study in rats subjected to experimental hyperlipidemia and obesity. *Phytomedicine* **2015**, *22*, 939–945.

152. Chkhikvishvili, I.; Sanikidze, T.; Gogia, N.; Mchedlishvili, T.; Enukidze, M.; Machavariani, M.; Vinokur, Y.; Rodov, V. Rosmarinic acid-rich extracts of summer savory (*Satureja hortensis* L.) protect Jurkat T cells against oxidative stress. *Oxid. Med. Cell. Long.* **2013**, *456253*, 1–9.

153. Pérez-Sánchez, A.; Barrajón-Catalán, E.; Ruiz-Torres, V.; Agulló-Chazarra1, L.; Herranz-López, M.; Valdés, A.; Cifuentes, A.; Micol, V. Rosemary (*Rosmarinus officinalis*) extract causes ROS-induced necrotic cell death and inhibits tumor growth in vivo. *Sci. Rep.* **2019**, *9*, 808. [CrossRef] [PubMed]

154. Ulbricht, C.; Tracee, R.-P.; Brigham, A.; Ceurvels, J.; Clubb, J.; Curtiss, W. An evidence-based systematic review of rosemary (*Rosmarinus officinalis*) by the Natural Standard Research Collaboration. *J. Diet. Suppl.* **2010**, *7*, 351–413. [CrossRef] [PubMed]

155. Sánchez-Salcedo, E.M.; Mena, P.; García-Viguera, C.; Martínez, J.J.; Hernández, F. Phytochemical evaluation of white (*Morus alba* L.) and black (*Morus nigra* L.) mulberry fruits, a starting point for the assessment of their beneficial properties. *J. Funct. Foods* **2015**, *12*, 399–408. [CrossRef]

156. Andrade, J.-M.; Faustino, C.; Garcia, C.; Ladeiras, D.; Reis, C.-P.; Rijo, P. *Rosmarinus officinalis* L. An update review of its phytochemistry and biological activity. *Future Sci. OA* **2018**, *4*, FSO283. [CrossRef]

157. Moore, J.; Yousef, M.; Tsiani, E. Anticancer effects of rosemary (*Rosmarinus officinalis* L.) extract and rosemary extract polyphenols. *Nutrients* **2016**, *8*, 731. [CrossRef] [PubMed]

158. Jeong, N.; Phan, H.; Jong-Whan, C. Anti-cancer effects of polyphenolic compounds in epidermal growth factor receptor tyrosine kinase inhibitor-resistant non-small cell lung cancer. *Pharmacogn. Mag.* **2017**, *13*, 595–599.

159. Trivellini, A.; Lucchesini, M.; Maggini, R.; Mosadegh, H.; Villamarin, T.S.S.; Vernieri, P.; Mensuali-Sodi, A.; Pardossi, A.; Mensuali-Sodi, A. Lamiaceae phenols as multifaceted compounds: Bioactivity, industrial prospects and role of "positive-stress". *Ind. Crops Prod.* **2016**, *83*, 241–254. [CrossRef]

160. Hilbig, J.; Policarpi, P.-B.; Grinevicius, V.-M.; Mota, N.-S.; Toaldo, I.-M.; Luiz, M.-T.; Pedrosa, R.-C.; Block, J.-M. Aqueous extract from pecan nut [*Carya illinoinensis* (Wangenh) C. Koch] shell show activity against breast cancer cell line MCF-7 and Ehrlich ascites tumor in Balb-C mice. *J. Ethnoph.* **2017**, *211*, 256–266. [CrossRef]

161. Simin, N.; Orčić, D.; Četojević-Simin, D.D.; Mimica-Dukic, N.; Anačkov, G.; Beara, I.; Zaletel, I.; Bozin, B. Phenolic profile, antioxidant, anti-inflammatory and cytotoxic activities of small yellow onion (*Allium flavum* L. subsp. *flavum*, Alliaceae). *LWT Food Sci. Technol.* **2013**, *54*, 139–146. [CrossRef]

162. Simin, N.; Dragana, M.-C.; Pavic, A.; Orcic, D.; Nemes, I.; Cetojevic-Simin, D. An overview of the biological activities of less known wild onions (genus *Allium* sect. Codonoprasum). *Biol. Serbic.* **2019**, *41*, 57–62.

 *foods*

Review

Berberis Plants—Drifting from Farm to Food Applications, Phytotherapy, and Phytopharmacology

Bahare Salehi [1], Zeliha Selamoglu [2], Bilge Sener [3], Mehtap Kilic [3], Arun Kumar Jugran [4],
Nunziatina de Tommasi [5], Chiara Sinisgalli [6], Luigi Milella [6], Jovana Rajkovic [7],
Maria Flaviana B. Morais-Braga [8], Camila F. Bezerra [8], Janaína E. Rocha [9],
Henrique D.M. Coutinho [9], Adedayo Oluwaseun Ademiluyi [10], Zabta Khan Shinwari [11,12],
Sohail Ahmad Jan [12], Ebru Erol [13], Zulfiqar Ali [14], Elise Adrian Ostrander [15],
Javad Sharifi-Rad [16,*], María de la Luz Cádiz-Gurrea [17,18], Yasaman Taheri [19,20],
Miquel Martorell [21,22], Antonio Segura-Carretero [17,18] and William C. Cho [23,*]

1   Student Research Committee, School of Medicine, Bam University of Medical Sciences, Bam 44340847, Iran;
    bahar.salehi007@gmail.com
2   Department of Medical Biology, Faculty of Medicine, Nigde Ömer Halisdemir University, Campus,
    Nigde 51240, Turkey; zselamoglu@ohu.edu.tr
3   Department of Pharmacognosy, Faculty of Pharmacy, Gazi University, Ankara 06330, Turkey;
    bilgesener11@gmail.com (B.S.); klcmehtap89@gmail.com (M.K.)
4   G.B. Pant National Institute of Himalayan Environment and Sustainable Development, Garhwal Regional
    Centre, Srinagar 246174, Uttarakhand, India; arunjugran@gmail.com
5   Department of Pharmacy, University of Salerno, Via Giovanni Paolo II 132, 84084 Fisciano, Italy;
    detommasi@unisa.it
6   Department of Science, University of Basilicata, Viale dell'Ateneo Lucano 10, 85100 Potenza, Italy;
    chiara.sinisgalli@gmail.com (C.S.); luigi.milella@unibas.it (L.M.)
7   Institute of Pharmacology, Clinical Pharmacology and Toxicology, Medical Faculty, University of Belgrade,
    11129 Belgrade, Serbia; jolarajkovic@yahoo.com
8   Laboratory of Applied Micology of Cariri—LMAC, Regional University of Cariri—URCA,
    Crato 63105-000, CE, Brazil; flavianamoraisb@yahoo.com.br (M.F.B.M.-B.); camilawasidi@gmail.com (C.F.B.)
9   Laboratory of Microbiology and Molecular Biology—LMBM, Regional University of Cariri—URCA,
    Crato 63105-000, CE, Brazil; janainaesmeraldo@gmail.com (J.E.R.); hdmcoutinho@gmail.com (H.D.M.C.)
10 Functional Foods, Nutraceuticals and Phytomedicine Unit, Department of Biochemistry, Federal University
    of Technology, Akure 340252, Nigeria; AOADEMILUYI@futa.edu.ng
11 Department of Biotechnology, Quaid-i-Azam University, Islamabad 45320, Pakistan;
    shinwari2008@gmail.com
12 Department of Biotechnology, Hazara University Mansehra, Khyber Pakhtunkhwa 21120, Pakistan;
    sjan.parc@gmail.com
13 Department of Chemistry, Faculty of Science, Mugla Sitki Kocman University, Mugla 48121, Turkey;
    e.ebrusimya@gmail.com
14 National Center for Natural Products Research, School of Pharmacy, University of Mississippi,
    Oxford, MS 38677, USA; zulfiqar@olemiss.edu
15 Medical Illustration, Kendall College of Art and Design, Ferris State University,
    Grand Rapids, MI 49501, USA; eliseadrianostrander@gmail.com
16 Department of Pharmacology, Faculty of Medicine, Jiroft University of Medical Sciences,
    Jiroft 7861756447, Iran
17 Department of Analytical Chemistry, Faculty of Sciences, University of Granada, Avda. Fuentenueva s/n,
    18071 Granada, Spain; mluzcadiz@ugr.es (M.d.l.C.-G.); ansegura@ugr.es (A.S.-C.)
18 Research and Development Functional Food Centre (CIDAF), Bioregión Building, Health Science
    Technological Park, Avenida del Conocimiento s/n, 188016 Granada, Spain
19 Phytochemistry Research Center, Shahid Beheshti University of Medical Sciences, Tehran 1991953381, Iran;
    taaheri.yasaman@gmail.com
20 Department of Pharmacology and Toxicology, School of Pharmacy, Shahid Beheshti University of Medical
    Sciences, Tehran 11369, Iran
21 Department of Nutrition and Dietetics, Faculty of Pharmacy, University of Concepcion,
    Concepcion 4070386, Chile; martorellpons@gmail.com

22   Unidad de Desarrollo Tecnológico, Universidad de Concepción UDT, Concepcion 4070386, Chile
23   Department of Clinical Oncology, Queen Elizabeth Hospital, 30 Gascoigne Road, Hong Kong, China
*   Correspondence: javad.sharifirad@gmail.com (J.S.-R.); chocs@ha.org.hk (W.C.C.);
     Tel.: +98-21-8820-0104 (J.S.-R.); +852-3506-6284 (W.C.C.)

Received: 1 October 2019; Accepted: 14 October 2019; Published: 22 October 2019

Abstract: The genus *Berberis* includes about 500 different species and commonly grown in Europe, the United States, South Asia, and some northern areas of Iran and Pakistan. Leaves and fruits can be prepared as food flavorings, juices, and teas. Phytochemical analysis of these species has reported alkaloids, tannins, phenolic compounds and oleanolic acid, among others. Moreover, *p*-cymene, limonene and ocimene as major compounds in essential oils were found by gas chromatography. *Berberis* is an important group of the plants having enormous potential in the food and pharmaceutical industry, since they possess several properties, including antioxidant, antimicrobial, anticancer activities. Here we would like to review the biological properties of the phytoconstituents of this genus. We emphasize the cultivation control in order to obtain the main bioactive compounds, the antioxidant and antimicrobial properties in order to apply them for food preservation and for treating several diseases, such as cancer, diabetes or Alzheimer. However, further study is needed to confirm the biological efficacy as well as, the toxicity.

Keywords: *Berberis*; food preservative; alkaloid; antioxidant; human health

---

1. Introduction

Berberis spp. are shrubs in the family *Berberidaceae*, native to central and southern Europe, western Asia, as well as northwest Africa [1]. About 500 species of these plants are found in most areas of central and southern Europe, the north-eastern region of United States, and Asia (including the northern area of Pakistan [2] and Iran [3]). The genus *Berberis* consists of spiny deciduous evergreen shrubs which are characterized by yellow wood and flowers [2], dimorphic long and short shoots (1–2 mm). Some *Berberis* fruits are small oblong berries 7–10 mm long and 3–5 mm broad and turn blue or red upon ripening during the late summer or autumn [1].

Berberis species are mainly consumed fresh, dried and used in juice production [4]. The fruits are very popular, known as *zereshk* in Iran where they are commonly used for cooking and in jam production, thus, encouraging the production of fresh edible seedless barberries fruits reaching about 22,000 tons per annum [5]. The fruits are also processed into beverages, drinks, syrups, candy and other confectionary products which are popular Iran. Furthermore, the leaves and fruits have also found applications in the production of food flavorings and teas. *Berberis* are popular due to their nutritional importance; however, they have found most usefulness in folk and traditional medicine where various parts, including roots, bark, leaves and fruits serve as major ingredients of herbal remedies in Ayurvedic, Iranian and Chinese medicine dating back at least 3000 years [6]. Currently, this species flower is popularly used amongst Tibetan speaking population in areas, such as Litang, China [7].

The effect of cold-pressed filtered oil of *Berberis* spp. seeds in delaying soybean oil oxidation in comparison to commercial antioxidants were carried out, and the study reported that *Berberis* oil contributed to oxidative stability of soybean oil comparably to commercial antioxidants [8]. Antioxidant and antibacterial activity of water extract of barberry has suggested their possible application as preservatives in food industries [9].

Isoquinoline alkaloids are the major bioactive constituents in *Berberis* [10]. Protoberberines and bisbenzyl-isoquinoline alkaloids, such as berbamine, tetrandrine and chondocurine, which have been known for their anti-inflammatory and immunosuppressive properties, have been detected by phytochemical analysis of the root and stem back extracts of *B. vulgaris*. Berberine (an isoquinoline alkaloid) and berbamine are the most abundant phytochemicals of *Berberis* species [2]. The fruits contain a high amount of alkaloids, tannins, phenolic compounds and oleanolic acid [3,11], gum, pectin, oleoresins, organic acids, anthocyanins and carotenoids. In addition, palmitine [10], stigmasterol and its glycoside [12] have all been detected in various species of the *Berberis* plant.

Some *Berberis* fruits have been employed in the treatment of guts [13] kidney stones [14] and liver [15] and gall bladder [10] conditions. The root bark and stem of the *Berberis* have found usage as a diuretic, febrifuge, cathartic and antiseptic. Furthermore, preparations of the stem and root bark have been used to treat mouth and stomach ulcers [16]. Several parts of the plant have been reported to possess astringent and antiseptic properties, while the stem bark and flowers were found to be anti-rheumatic [17]. The alkaloid rich root bark of the plant has also been used as purgative and treatment for both diarrhea and rheumatism [18]. The berberine-rich rhizomes of *Berberis* species possess marked antibacterial and antitumor properties, with reported efficacies in treatment of various eye conditions [10,19]. Furthermore, the anti-inflammatory activity of berberine has been extensively studied amongst other pharmacological actions [10,20].

Berberine sulphate which is an alkaloid extracted from the roots and bark of various *Berberis* spp. Have been reported to possess antibacterial, antifungal and antiprotozoal activities. Reported the bacteriostatic activity of berberine against streptococci, and that the sub-minimum inhibitory concentrations (MICs) of the compound blocked the adherence of streptococci to host cells, immobilized fibronectin, and hexadecane in epithelial cells [21]. Furthermore, blood glucose and lipid regulatory properties of *Berberis* have been demonstrated [3,22–24]; and this was due to berberine-induced improvement in insulin sensitivity through regulation of adipokine secretion [25–27]. Effectiveness of *Berberis* species in the maintenance of heart health has been demonstrated in their ability to improve hypertension, ischemic heart disease, cardiac arrhythmias and cardiomyopathy [2,28].

The health-promoting effect of *Berberis* spp. cannot be overemphasized, as well as its popularity; however, this is restricted to central and southern Europe, western Asia, as well as northwest Africa. Hence, efforts should be geared towards making the *Berberis* plant also available to other regions of the world. Furthermore, most studies on *Berberis* spp. have been on berberine; therefore, efforts should be made towards researching possible therapeutic benefits of all other important phytoconstituents of the plant. Furthermore, the synergistic or additive effect of these phytoconstituents should be studied so as to elucidate the complex molecular interaction amongst various phytochemicals leading to the observed therapeutic properties. In addition, the modulatory effect of the plant/plant materials on gene expression should be prioritized.

The aim of this review is to provide a detailed overview to the cultivation of *Berberis* species, in-depth insight on the biological properties of the phytoconstituents of this genus, regarding its food preservative applications, antimicrobial, antioxidant and anticancer effects, and lastly, special emphasis to its clinical effectiveness in humans. The present work was performed by consulting the database of PubMed, Web of Science, Embase, and Google Scholar (as a search engine) to retrieve the most updated articles on the topic under investigation (phytochemicals and biological activities of *Berberis* species). The strategy of the search included the use of the following keywords: "*Berberis*" or "barberry" and "cultivation" or "essential oil" or "antimicrobial" or "food preservative" or "antioxidant" or "anticancer". Authors carefully examined articles and for the review, prioritizing the articles published from 2013 to 2018 [29]. Only English articles having full text were considered.

2. Cultivation of *Berberis* Plants

The genus *Berberis* include about 500 different species and commonly grown in Europe, United States, South Asia and some northern areas of Iran and Pakistan (Figure 1) [11,30]. In Pakistan, majority of *Berberis* species are found in high mountains (1400 m–3500 m above sea level). In Iran, five *Berberis* species are present, including two important species, i.e., *B. orthobotrys* and *B. khorassanica*, which are grown in eastern, northern and southern regions of Iran [31]. Other important local species *zereshk* is also widely cultivated in the South-Khorasan province of Iran [32]. In Iran, 11,000 hectares of area is under cultivation of common barberry (*B. vulgaris*) species, and one of the most leading producers of barberry fruit in the world. Annually, Iran produces more than 10,000 tons of dried barberry fruits, while maximum production comes from the Southern Khorasan. More than 97% of the area located near Ghaenat County and Southern Khorasan province is cultivated with common barberry that produces 95% of fruit in Iran [33]. *B. vulgaris* is gathered from the wild in Eastern Europe in countries like Poland [34,35]. In addition, it is known as *K'otsakhuri* in Georgia, where it is grown and collected from the southeast forests of the country [36]. The yield-related important traits of many *Berberis* species are significantly affected by environmental factors, biotic stresses, seasonal variations, the climatic condition of an area, planting and harvesting date or methods, irrigation source, fruit processing and storage methods, etc.

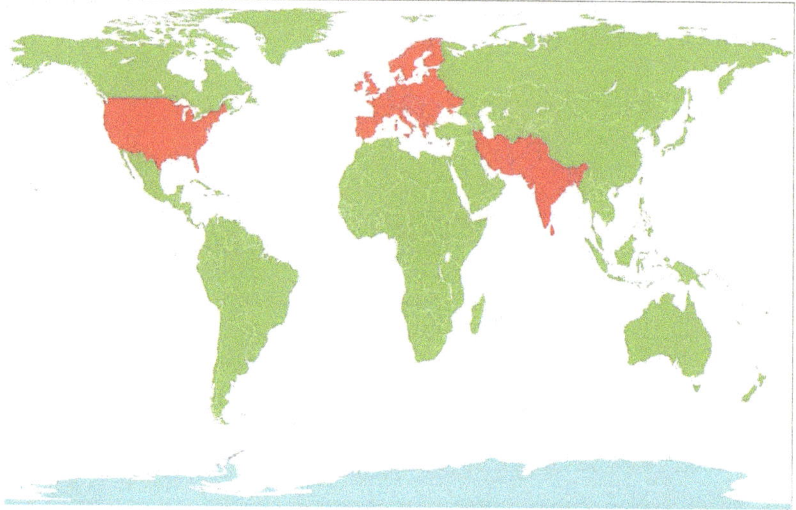

Figure 1. The areas where the *Berberis* plants are commonly grown (shown in red).

The fruit remains a vital part, producing many important secondary metabolites and used in pharmaceutical and food industries. In this section, we have highlighted the cultivation of different *Berberis* species, its status and various factors affecting its cultivation. Common barberry is a native plant in Asia's western and middle mountains and non-native to North America [37]. The hybrid species *Berberis × ottawaensis* are widely cultivated in Europe and North America [38]. The European settlers introduced common barberry to New England, and used it as a source of medicine, food, and for other aesthetic purposes [37,39]. The colonists of New England determined its cultivation spread *Puccinia graminis* fungus, causing wheat rust and important reduction of its crop [40].

The seedless type (*B. vulgaris* var. *asperma*) is commonly cultivated in the southern parts of the Khorasan province of Iran for domestic purposes [41]. *B. lyceum* represents a native species of Nepal and distributed in the temperate and subtropical regions of the world, including some part of Australia. It is distributed from Kashmir to Uttaranchal North-western Himalayas [42]. Sixteen species, and some varieties of barberry, were found in the Boaxing country, situated on the Eastern side of Hengduan Mountains in Sichuan Province, Southwest China [43]. The Japanese barberry is invasive in twenty different states of the world and five other provinces of Canada [44]. According to Reference [45] there are 21 *Berberis* species in Nepal, including two new species *Berberis pendryi* Bh. Adhikari and *Berberis karnaliensis* Bh. Adhikari. *Berberis crataegina* DC. is commonly grown in Turkey, Asia and European regions. The fruit are locally known as "karamuk" and "kadıntuzluğu" in Turkey and used as traditional medicine [46].

In Pakistan, twenty species and six subspecies of *Berberis* have been reported, and the majority of these is growing in northern parts of the country [47]. The other dominant species of *Berberis* is *B. aristata*, grown in Nepal, Pakistan and India. In Pakistan, it mainly cultivated in the Hazara division of Khyber Pakhtunkhwa (KP), and Azad Jamu and Kashmir (AJK) regions. In AJK it is locally named *sumbal* and commonly known as *daruharidra* [48]. *B. lyceum* is another key species, highly distributed in different Asian countries, including, Afghanistan, Pakistan, Nepal, India and Bangladesh. In Pakistan, it grows in different areas of KP, Punjab and Baluchistan [49]. The fruit part locally called *kashmal*, is used as a source of food, and for preparing the sauce in Himalayan regions of Jammu and Kashmir, and Himachal Pradesh [50,51].

The yield of barberry plants depends on various factors like managing operations, size and age of shrub and date and method of harvesting [52,53]. Proper harvesting method at a suitable time is one of the key steps in berry yield production because the shrubs include maximum spines in shoot part and also the fruit peel is so thin. The harvesting date plays a vital role to gain maximum yield with high quality. The local farmers set some useful sensory parameters for starting the harvest of the crop. These parameters include change of fruit color from bright red to dark red, tissue softening, the concentration of contents, and reducing sourness in fruit, etc. [11]. The optimum time for barberry harvesting in the autumn cold season, when the berries ripen. In this stage, the fruit gains dark red color due to the presence of high anthocyanin content, sweetness increases, while the berberine and sourness are reduced [53–57]. In different regions of North America and Western Europe, the common Barberry ripens in the month of August or September [58,59]. While the seeds can be mature in October [60]. The berries of common barberry remain with the stem through winter [61]. However, the delay in harvesting from 10 September to 13 November increases the anthocyanin content about 2.5 times [53]. It may also increase the yield and quality, but too much delaying may lead to early autumn chilling injury to plant. For Iranian seedless barberry 170 days after the flowering is an optimum date for harvesting [56]. Fruit maturation and development vary in different geographic regions. So, it is important to optimize a suitable harvesting date for each region.

3. *Berberis* Plants Essential Oils and Phytochemical Composition

Essential oils (EO) are volatile, complex natural compounds, which formed in aromatic plants as secondary metabolites. They are used in pharmaceutical, agricultural, and food industries, as well as are associated with antibacterial, anti-inflammatory, antioxidant, and insecticidal potential [62–64].

The gas chromatography coupled to mass spectrometry (GC-MS) analysis of various parts of *B. vulgaris* revealed that benzaldehyde, benzyl alcohol, 1-hexanol and I-2-hexenal [65] were major compounds of the EOs from fruit, while *p*-cymene, limonene and ocimene were identified as major compounds of the EOs (Figure 2) from leaves and flowers [66].

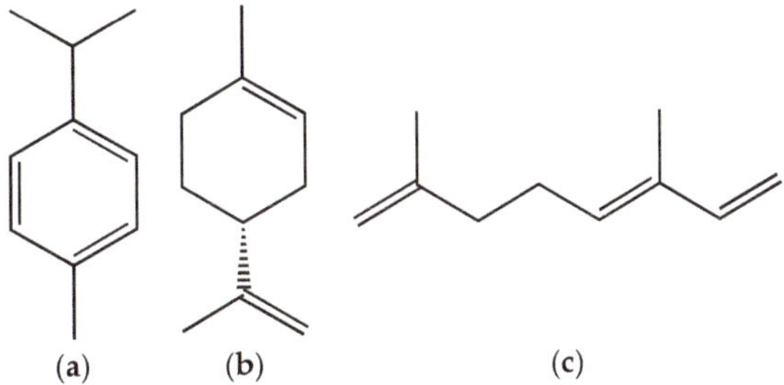

(a)   (b)   (c)

Figure 2. Major compounds of the essential oils (EOs) of *Berberis vulgaris* leaves and flowers. (a) *p*-cymene; (b) limonene; (c) ocimene.

Turkish *B. crataegina* fruit berry has 22 volatile compounds which are aldehydes had the highest concentration (5382 µg/kg), followed by alcohols (2487 µg/kg) and lactone (2422 µg/kg).

Major volatile compounds of the *B. crataegina* fruit are γ-butyrolactone, 3-hexanal and 2,6-dimethylphenol. Moreover, the olfactometric analysis of dry *B. crataegina* resulted eight aroma active compounds [67].

EOs of the roots of *B. integerrima* were analyzed by using modified microwave-assisted hydrodistillation (MAHD). Chemical diversity of 10 and 18 compounds were obtained from MAHD, MAHD with modified anyl, and with modified phenyl magnetic nanoparticles, the yields of the EOs were 0.16, 0.61 and 0.71 *w/w* %, respectively. Hexadecanoic acid was identified as a major compound for MAHD and modified MAHD methods [68].

Moreover, the GC/MS study on hexane extracts of the *B. aetnensis* and *B. libanotica* roots was showed that *B. aetnensis* have twenty-six and *B. libanotica* have thirty-seven non-polar compounds. Stigmasterol (Figure 3) is the major compound of both species [69].

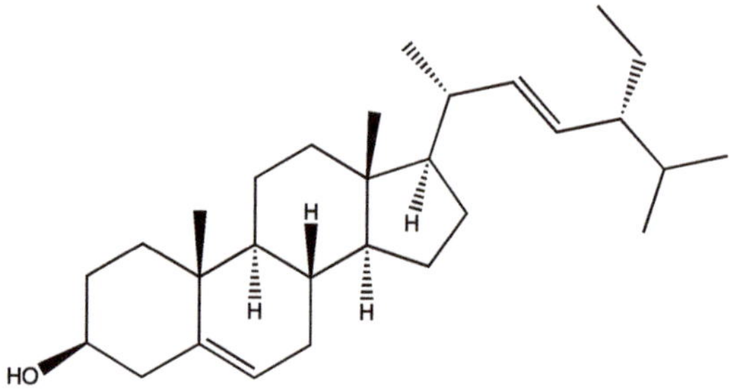

Figure 3. Stigmasterol.

On the other hand, alkaloids (Table 1) represent the main compounds in *Berberis* species, and many of them have been identified by different spectroscopic techniques previously mentioned. The most known are berberine, berbamine, palmitine, jatrorrhizine, and isotetrandrine. They are located mainly in the cortical tissues of the roots and stems and have important biological activities. In fact, in vitro and in vivo anti-proliferative and anti-metastatic effects on various types of cancers have been reported for different alkaloids. These compounds, such as vinblastine, have already used as anticancer drugs [3].

Table 1. Alkaloids from *Berberis* species.

Chemical Structure	Name	Plant
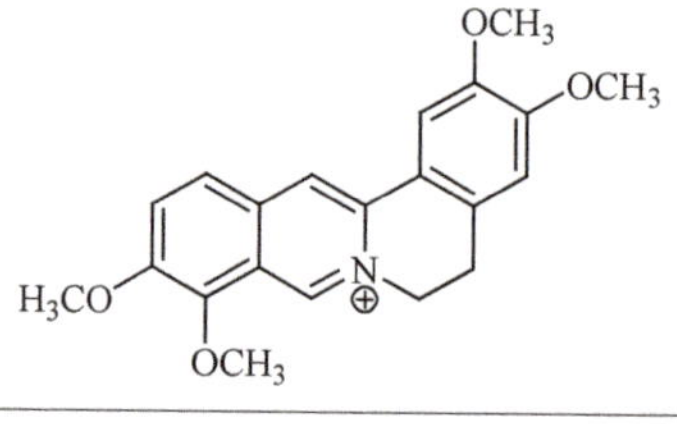	palmatine	*B. vulgaris*
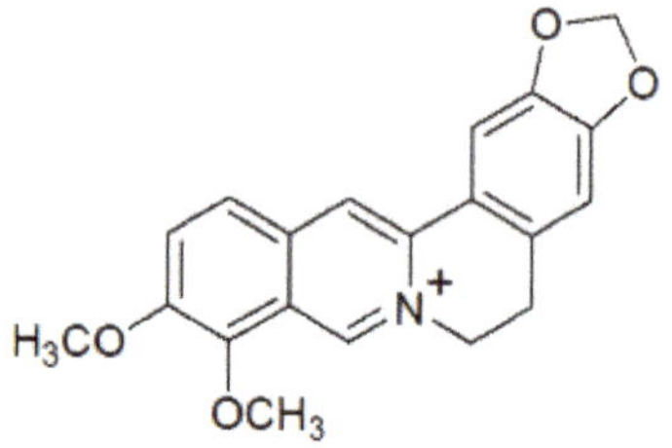	berberine	*B. vulgaris*

Table 1. *Cont.*

Chemical Structure	Name	Plant
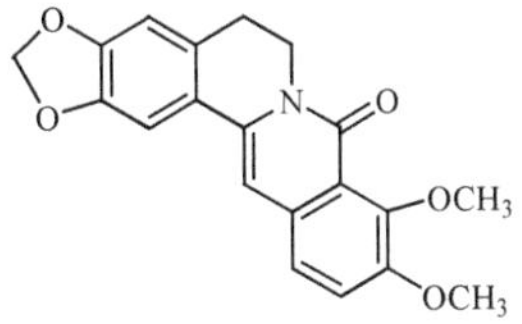	oxyberberine	*B. vulgaris*
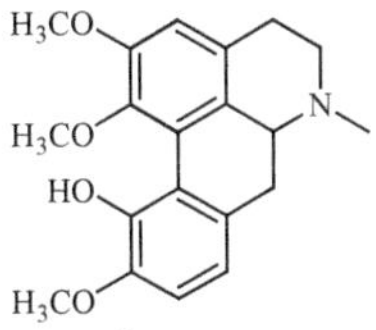	isocoridine	*B. vulgaris*
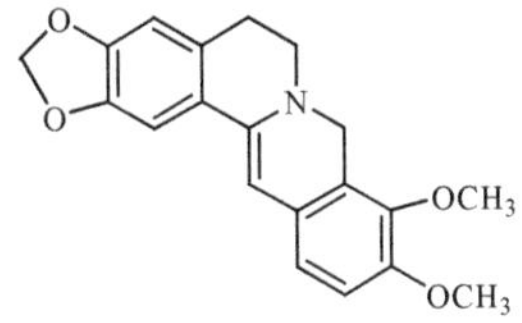	lambertine	*B. vulgaris*
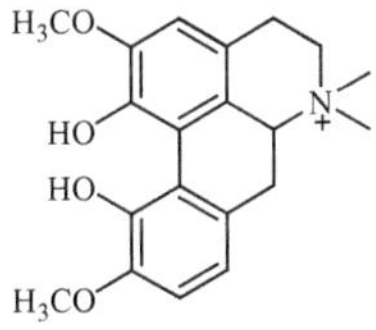	magniflorine	*B. vulgaris*
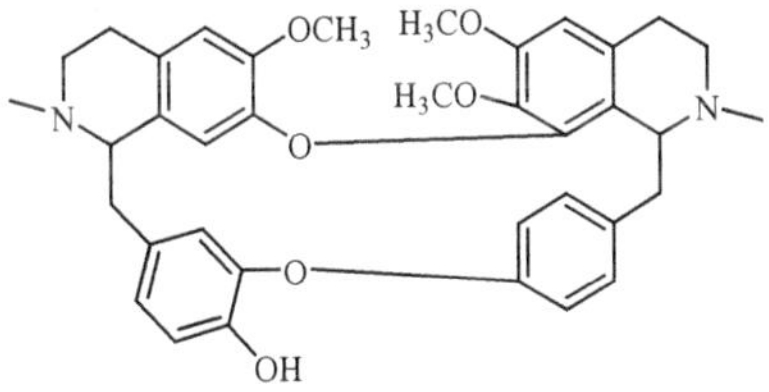	oxycanthine	*B. vulgaris*
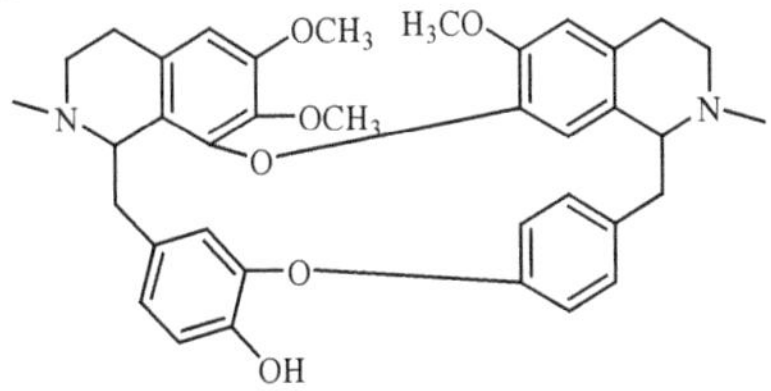	berbamine	*B. aristata*
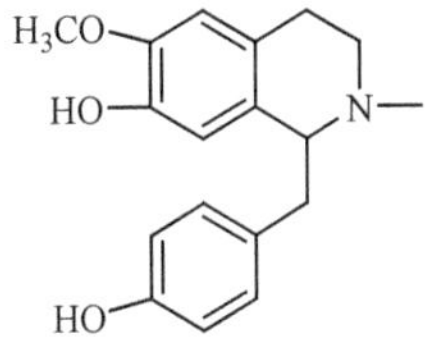	(+)-N-methylcoclaurine	*B. montana*

Table 1. *Cont.*

Chemical Structure	Name	Plant
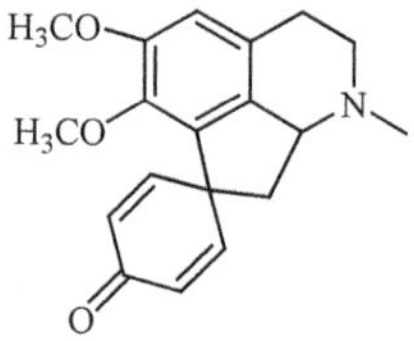	(−)-pronuciferine	*B. montana*
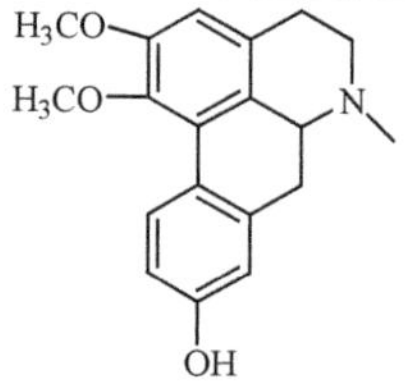	(+)-9-hydroxynuciferine	*B. montana*
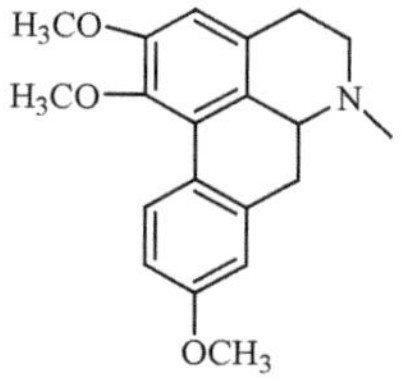	(+)-orientine	*B. montana*
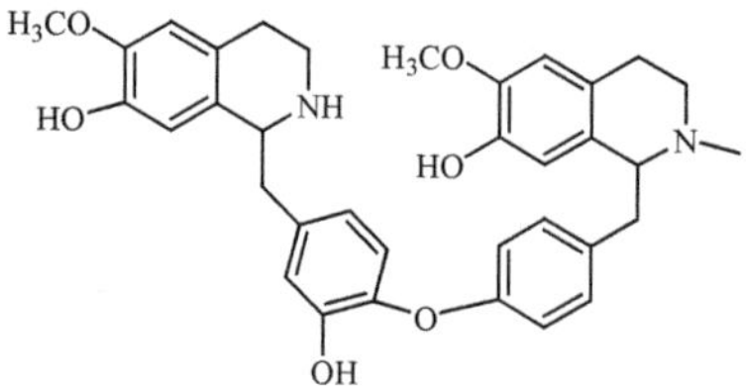	2-norberbamunine	*B. stoloniferais*
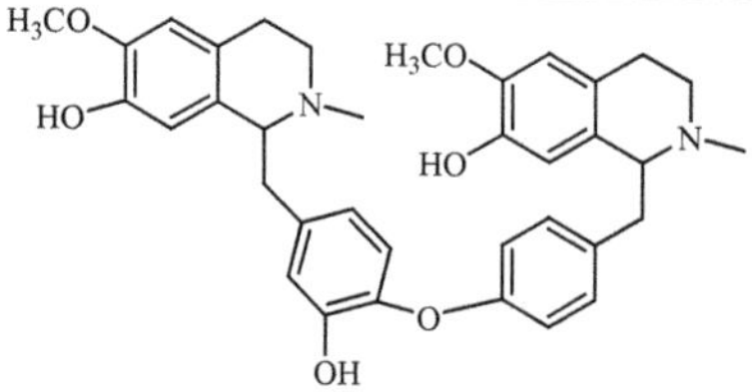	berbamunine	*B. stoloniferais*
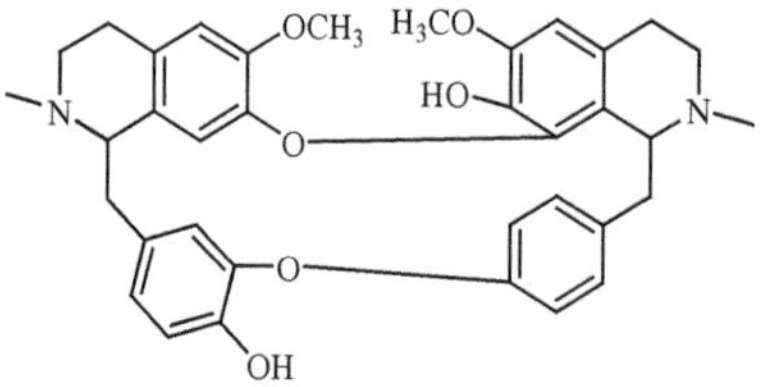	aromoline	*B. stoloniferais*

Table 1. *Cont.*

Chemical Structure	Name	Plant
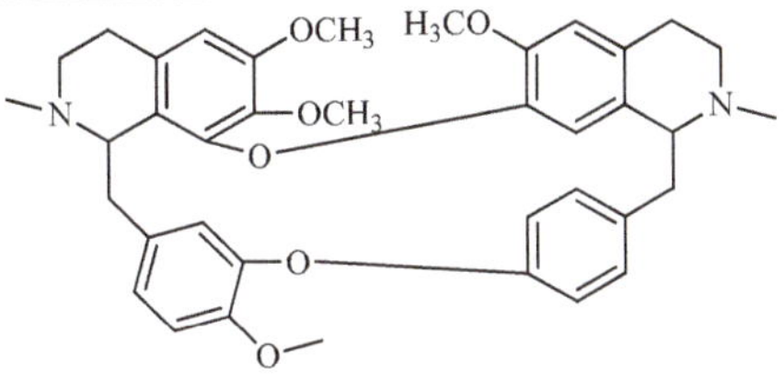	isotetrandrine	*B. stoloniferais*
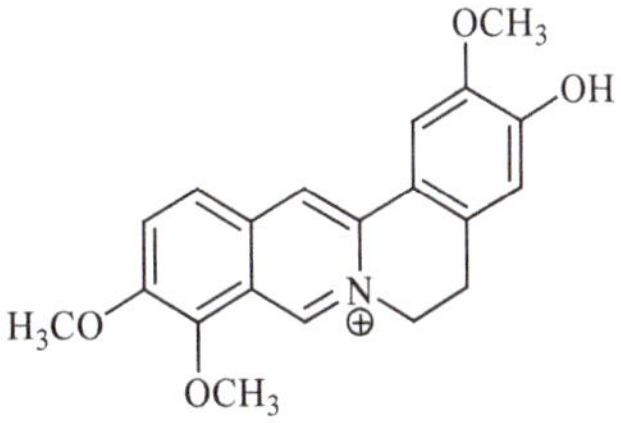	jatrorrhizine	*B. umbellate*

4. Food Preservative Applications of *Berberis* Plants

Food preservation is the most vital issue in food industries to ensure food safety for a longer period. Basically, the process of food preservation depends on the growth inhibition of undesirable microorganisms. Use of chemical agents with antimicrobial activity is commonly used a traditional method for food preservation [70]. However, antimicrobial agents also gain momentum, due to their fewer side effects and compatibility with the human body. Further, synthetic antimicrobials and their toxicological safety as food additives needed to be ensured by regulatory authorities. Moreover, processed foods with natural preservatives have great demand and considered safer and beneficial for public health [71]. The naturally occurring compounds demonstrated antimicrobial activity in foods as natural ingredients and can be used as additives to other foods.

Berberis is an important plants having enormous potential in the food industry. However, only a few reports are available on the direct application of these plants in food products. For example, seed oil and fruit extracts of *B. crataegina* were supplementing into chitosan matrix for preparation of a chitosan-based edible film. The films produced have been analyzed for the physiochemical and biological activities. Results showed that chitosan-fruit extract film exhibited higher thermal stability, antimicrobial, antioxidant, and anti-quorum sensing activity as compared to other films. Furthermore, the addition of *B. crataegina* seed oil and fruit extract into the chitosan film create a mark reduction in the UV-vis transmittance but improve the tensile strength. Likewise, hydrophobicity of the chitosan-seed oil film was found to be higher than chitosan-control film, while chitosan-fruit extract film displayed slightly lower hydrophobicity than chitosan film. These results indicated that chitosan-fruit extract film of *B. crataegina* fruit extract could be used as an effective ingredient for the production of the edible film with increased physicochemical and biological properties [72].

A list of the antimicrobial potential of the *Berberis* species evaluated across the globe is provided which support the use of *Berberis* species in food preservation (Table 2).

Table 2. A list of the antimicrobial potential of the *Berberis* species evaluated across the globe is provided which support the use of *Berberis* species in food preservation.

S. No.	Species	Part	Country	Extract/Model/Compound	Tested Micro-Organism	Results	Reference
1	*B. aristata*	Stem and leaves	Nepal	Hexane, Ethyl acetate, Methanol	*Staphylococcus aureus, Kleibsella pneumoniae, Salmonella typhimurium*	Against *S. aureus*: methanol significant zone of inhibition (21 mm), ethyl acetate extracts moderate activity, hexane extract of stem slightly active.	[73]
2	*B. aristata,* and *B. ligulata*	Bark stem Leaves	Nepal	Ethanol	*Bacillus subtilis, Escherichia coli, Pseudomona aeruginosa, Salmonella. typhi, Salmonella dyjenteriae, Salmonella cholerae*	Ethanol extract of *B. aristata*: largest zone of inhibition (21 mm) against *B. subtilis* and the smallest MBC value (90 mg/mL) for *S. aureus*. Gram positive bacteria more susceptible to the ethanol extract. *B. aristata* relatively broad-spectrum antibacterial activity.	[74]
3	B. vulgaris	Stem	Iran	Ethanol	*P. aeruginosa, Acinetobacter baumannii, E. coli* and *Salmonella enteritidis*	MIC determination: stem extracts inhibit the growth of all the studied bacteria (3900 to 37,500 µg/mL) by synergistic effects with ciprofloxacin.	[75]
4	*B. asiatica*	Leaves	Uttarakhand, India	Methanol	*E. coli, Enterobacter aerogenes, Proteus vulgaris, P. aeruginosa, K. pneumoniae, B. subtilis, S. aureus*	Methanol extracts of leaves: high inhibitory potential on *S. aureus, K. pneumoniae, E. coli, B. subtilis* and *P. vulgaris* in all concentration.	[76]
5	*B. aristata, B. asiatica, B. lycium*	Stem	Bangalore, India	Methanol	*Nocardia sp., S. aureus, S. pneumonia, P. aeruginosa, Streptococcus viridians, E. coli*	Sensitivity to *Nocardia* sp., *S. pneumonia* and *E. coli*.	[77]
6	*B. glaucocarpa*	Root wood	Pakistan	Ethanol	SMRSA, EMRSA, *Mycobacterium marinum, E. coli, Trypanosoma brucei*	Berberine (MIC = 12.5 and 25 µg/mL), berberine chloroform (MIC = 25 and 12.5 µg/mL) and syringaresinol (12.5 µg/mL): very active against SMRSA, *M. marinum* and *T. brucei*.	[78]
7	*B. vulgaris*	Stem bark	Romania	Ethanol	*Botrytis cinerea*	*B. vulgaris* bark extract, berberine, and fluconazole significantly inhibited growth of *B. cinerea*.	[79]
8	*B. vulgaris*			Ethanol	*S. aureus, Staphylococcus epidermidis, K. pneumoniae, B. subtilis, E. coli, Aspergillus niger, Trichoderma, Alternaria solanai*	20 mm zone of inhibition against *E. coli*. Good activity against *B. Subtilis*, moderate against *Trichoderma*, insignificant against other stains.	[80]

Table 2. *Cont.*

S. No.	Species	Part	Country	Extract/Model/Compound	Tested Micro-Organism	Results	Reference
9	*B. vulgaris* and its active constituent, berberine	Root	Egypt	Ethanolic extract	*Candida albicans, E. coli*	*Berberis* ethanolic extract and berberine standard can inhibit *C. albicans* and *E. coli* growth.	[81]
10	*B. vulgaris*	Fruit	Pakistan	Distilled water	*S. aureus, Proteus, S. typhi, Salmonella paratyphi A, Salmonella paratyphi B, K. pneumoniae, E. coli, P. aeruginosa*	Antibacterial activity against all tested pathogens.	[82]
11	*B. thunbergii*	Fruit	Hungary	Juice; water extract and -methanol extract	*B. subtilis, Bacillus cereus* var. *mycoides, E. coli, Serratia marcescens*	Juice, water extract and methanol extract showed activity against all bacteria.	[83]
12	*B. calliobotrys*	Stems and branches	Pakistan	Methanol	*B. subtilis, P. aeruginosa, S. aureus* fungal strains namely *C. albicans, Penicillium notatum*	The methanol extract, ethyl acetate and n-butanol fractions: maximum zone of inhibition against all bacterial strains especially *S. aureus* and antifungal effects.	[84]
13	*B. lycium*	Roots	Libya	Distilled water, ethanol, isopropanol and methanol	*Pseudomonas* sp., *E. coli, Streptococcus* sp., *Staphylococcus* sp.	Methanolic displayed maximum inhibitory zone (16 mm), isopropanol extract (13 mm) and ethanol extract (12 mm). The aqueous extract exhibited the least inhibitory zone (10 mm). The methanolic extract: maximum inhibitory zone (12 mm), *Pseudomonas* (11 mm) and *Staphylococcus* (10 mm).	[85]
14	*B. hispanica*	Root Bark	Marocco	Ethanolic extract	*Mycobactérium smegmatis, Mycobacterium aurum*	The ethanolic extract from root bark displayed an important antimycobacterial activity. The inhibition zones for *M. aurum* A+ were significantly larger than those for *M. smegmatis* MC2.	[86]
15	*B. ruscifolia*	-	Argentina	Acetone, chloroform-methanol (1:1) and methanol	*E. coli, P. aeruginosa, Listeria monocytogenes, S. aureus*	All extracts exhibited antibacterial activity with MIC varying from 16 to 2 mg/mL. The highest inhibition with acetonic and chloroform-methanolic extracts of species against *S. aureus* (MIC = 2 mg/mL). Methanolic extracts *B. ruscifolia* showed no antibacterial activity against all tested bacteria.	[87]

Table 2. *Cont.*

S. No.	Species	Part	Country	Extract/Model/Compound	Tested Micro-Organism	Results	Reference
16	*B. aristata*	Stem bark	India	Ethanol and aqueous extracts	*Shigella flexneri, Shigella sonnei, Shigella dysenteriae, Shigella boydii*	Extracts of *B. aristata*: antibacterial activity against four strains of *Shigella* (8 and 23 mm).	[88]
17	*B. aristata, B. asiatica, B. chitria* and *B. lycium*	Root and stem	India	Ethanol	*Micrococcus luteus, B. subtilis, B. cereus, Enterobacter aerogenus, E. coli, K. pneumoniae, Proteus mirabilis, P. aeruginosa, S. aureus, S. typhimurium, Streptococcus pneumonia*, Fungal strains *Aspergillus nidulans, C. albicans, Aspergillus terreus, Trichophyton rubrum, Cistus albidus, Aspergillus flavus, A. niger*	*B. lycium, B. aristata* and *B. asiatica* root extract showed significant antifungal activity against *A. terreus* and *A. flavus. B. aristata* root and *B. lycium* (stem) extracts gave very low MIC values (0.31 μg/mL) as compared to other tested species.	[89]
18	*B. Lycium*	Root	Pakistan	Ethanol, petroleum ether	*S. aureus, S. epidermidis, B. subtilis, S. typhi, E. coli, C. albicans*	The ethanolic and aqueous crud root extract: most effective antifungal and antibacterial agents.	[90]
19	*B. integerrima* Syn: *B. densiflora*	Roots	Iran	Methanol	*Brucella abortus*	MIC and MBC results, jatrorhizine exhibited higher antibacterial activity with MIC (0.78 μg/mL) and MBC (1.56 μg/mL) compared with the standard (streptomycin, 10 μg/mL).	[91]
20	*B. lycium*	Roots	Pakistan	Hydric extract	*E. coli, Pseudomonas, Staphylococcus, Proteus*	Significant activity against *E. coli* and Proteus (80 to 100%), while it demonstrated a good activity against *Pseudomonas* and *Staphylococcus* (60 to 70%).	[92]
21	*B. aristata*	Bark and leaves	India	Methanol, ethanol and hexane	*B. subtilis, Agrobacterium tumefaciens, E. coli, Xanthomonas. Phaseoli, Erwinia chrysanthemi*	All the extracts of tested plants showed variable activity against all the tested bacterial strains. Methanol extract revealed highest antibacterial activity (11 mm) recorded against *E. chrysanthemi*. Hexane extract: totally inactive against all the tested strains.	[93]

Table 2. *Cont.*

S. No.	Species	Part	Country	Extract/Model/Compound	Tested Micro-Organism	Results	Reference
22	*B. aristata*	Roots	India	Aqueous and alcohol extracts	*S. aureus, B. subtilis, E. coli, S. typhimurium*	Alcoholic and aqueous extract showed antimicrobial activity against four tested bacteria. *B. aristata* exhibited highest zone of inhibition for *B. subtilis* followed by *S. aureus, E. coli* and *S. typhimurium.*	[94]
23	*B. microphylla*	Leaves, stems and roots	Chile	Methanol	*E. coli, S. typhimurium, L. monocytogenes, E. aerogenes, S. aureus, B. cereus, S. epidermidis* and *B. subtilis*	All extract possesses significant antibacterial activity against Gram-positive bacteria but not against Gram-negative bacteria.	[95]
24	*B. lycium*	Root bark	Pakistan		*E. coli, K. pneumoniae, P. aeruginosa, S. aureus, B. subtilis*	Silver nanoparticles were very active against Gram-negative and Gram-positive bacteria Aqueous bark extract (10 µg/mL) possess highest activity against *E. coli* and *P. aeruginosa.*	[96]
25	*B. vulgaris*	Fruit	Iran		*L. monocytogenes*	Average diagonal of growing area in disk diffusion test for species: 12 mm and MIC was 125 µg/mL and MBC of *B. vulgaris* was 500 µg/mL.	[97]
26	*B. aristata*	Stem bark	Alcohol	In vivo in an animal model using Sprague Dawley rats	Carbapenem-resistant *E. coli*	An aquo-alcoholic extract of the species: effectively manage peritonitis induced by Carbapenem-resistant *E. coli* in a rat model at a single post-exposure prophylactic dose of 0.5 mg/kg body weight.	[98]
27	*B. aristata*	Roots	India	Aqueous and alcoholic extract of fresh roots, as well as aqueous extract of dried roots	*S. aureus, S. epidermidis, Streptococcus pyogenes, Streptococcus viridans, Enterococcus faecalis, B. subtilis, B. cereus, E. coli, K. pneumoniae, P. aeruginosa, P. vulgaris, P. mirabilis, S. typhi, S. paratyphi A, S. typhimurium, S. dysenteriae* type 1, *Vibrio cholerae*	All three extracts displayed wide antibacterial activity against Gram-positive bacteria. Among the Gram-negative bacteria tested, the antibacterial activity was limited to *E. coli, S. typhimurium, S. dysenteriae* type 1 and *V. cholerae.* All extracts also possess antifungal activity against the fungal species tested, except *Candida krusei.*	[99]

Table 2. *Cont.*

S. No.	Species	Part	Country	Extract/Model/Compound	Tested Micro-Organism	Results	Reference
28	*B. aristata*	Root Stem Leaf	Pakistan		*E. coli, S. typhi, S. aureus, Shigella, Citrobacter, P. vulgaris, Enterobacter, Streptococcus pyrogenes, V. cholera, Klebsiella* spp., *A. niger, Cladosporium, Rhizoctonia, Alternaria, Trichoderma, Penicillium, Curvularia, Paecilomyces* and *Rhizopus*	The extracts significantly inhibited the growth of the studied microbes, except *A. niger, Curvularia, Paecilomyces* and *Rhizopus*.	[100]
29	*B. aristata*		India		*V. cholerae, S. aureus*	All the strains of *V. cholerae* are susceptible. All the *Salmonella* sp., *Pseudomonas* sp., and some of the E. *coli* strains are highly resistant, except some strains of *E. coli* as AL26, and *Shigella* sp. are susceptible. All *Xanthomonas* sp. were highly susceptible. Berberine sulfate showed antifungal action against *C. albicans, Candida tropicalis, Trichophyton mentagrophytes, Microsporum gypseum, Cryptococcus neoformans* and *Sporothrix schenkii, Mycobacterium tuberculosis* var. *hominis* H$_{37}$RV and *Entamoeba histolytica*.	[101]
30	*B. heterophylla*	Leaves, stems and roots berberine	Argentina		*S. aureus, E. faecali, P.aeruginosa, E. coli, C. albicans, Candida glabrata, Candida haemulonii, Candida lusitaniae, C. krusei, Candida parapsilosis*	The aqueous extracts of *B. heterophylla* do not possess significant antimicrobial activity. Berberine displayed a significant antibacterial and antifungal activity against *S. aureus* and different *Candida* spp., some of them obtained from the clinical isolated.	[102]
31	*B. amurensis*	Branches and leaves	Korea		*Bacillus atrophaeus, Kocuria rhizophila, M. luteus, S. epidermidis, B. subtilis* subsp. *Spizizenii, K. pneumoniae, Enterobacter cloacae, Salmonella enterica* subsp. *enterica, P. aeruginosa*	No significant activity against gram-negative bacteria.	[103]

185

Table 2. *Cont.*

S. No.	Species	Part	Country	Extract/Model/Compound	Tested Micro-Organism	Results	Reference
32	*B. croatica* and *B. vulgaris*	Roots, leaves, and twigs	Croatia	Ethanol	*B. subtilis, S. aureus, E. coli, P. aeruginosa, C. albicans*	Extracts of both species: significant antibacterial activity against the Gram-positive bacteria. Root extracts of *B. croatica*: activity against *P. aeruginosa*, and leaf extracts against *B. subtilis*. Neither species possessed antifungal activity. Leaf extracts of *B. croatica*: antibacterial activity against *B. subtilis*. Likewise, neither of the species extracts showed activity against *E. coli* and *C. albicans*, except when were diluted. Ethanolic extracts of twigs of both species: inactive against *B. subtilis* and against *S. aureus*, with the exception of *B. croatica* twig from Kiza locality.	[104]
33	*B. lycium*	Roots	India	Hexane extract, Methanolic extract, aqueous extract and berberine	*K. pneumonia, E. coli, P. aeuroginosa, S. aureus, B. subtilis, C. albicans, A. niger, Aspergillus fumigates*	Methanolic extract of species was highly effective against *E. coli, S. aureus, B. subtilis, C. albicans, A. fumigates*. Pure berberine was effective against *E. coli* and *C. albicans*.	[105]
34	*B. aetnensis*	Roots	Italy	Ethanol ether and chloroform	*S. aureus, B. subtilis, E. faecalis, E. coli, P. aeruginosa, Stenotrophomonas maltophilia*, against 14 strains of nosocomial origin: two strains of *S. aureus* (1 Met-S, 1 Met-R); four strains of *S. epidermidis* (2 Met-S, 2 Met-R); three strains of *E. coli*; four strains of *P. aeruginosa, Hafnia alvei* and *C. albicans, C. parapsilosis, C. krusei*	The root and leaf extracts showed a greater activity against Gram-positive bacteria and yeasts than against Gram-negative bacteria, except for *P. aeruginosa*. The chloroform extract of leaves was more active than the ethanol.	[106]
35	*B. thunbergii, B. vulgaris*	Roots	USA		*E. coli, P. aeruginosa, S. aureus, S. mutans*, and *S. pyogenes*	Ethanolic extracts more active against studied bacteria, strongest activity against *S. pyogenes* and *S. aureus*.	[107]
36	*B. vulgaris*	Root bark	Algeria	Methanol and water	*S. aureus, E. faecalis, E. coli, E. cloacae, K. pneumoniae, P. aeruginosa*	The extracts of species root barks presented a strong activity against *S. aureus* (23.0 mm), a weak activity against *E. faecalis* (13.0 mm) and no activity toward other strains.	[108]

5. Antioxidant Activities of *Berberis* Plants (In Vitro and In Vivo)

Free radicals ubiquitous in the environment affect human health by oxidative stress-induced damage. Finding exogenous sources with antioxidant activity is necessary in order to support the organism against the actions of free radicals. The fruits of most plants from Berberidaceae family have a sour taste which is due mainly to the presence of ascorbic acid or vitamin C. The vitamins and antioxidant compounds in barberry plant might be useful for treating diseases [109].

The antioxidant effect of *B. vulgaris* on oxidative systems, such as liver cells oxidation, red blood cells haemolysis, and haemoglobin non-enzymatic glycosylation was demonstrated, and the highest inhibitory effect was exerted on glycosylation. The extracts of *B. vulgaris* was the most promising as antioxidants, as well as anti-inflammatory and acetylcholinesterase (AChE) inhibitors. The capacity of *B. vulgaris* for scavenging 1,1-diphenyl-2-picrylhydrazyl (DPPH) and 2,2′-azino-bis(3-ethylbenzothiazoline-6-sulphonic acid) (ABTS), the inhibitions of lipoxygenase and AChE are mainly due to the phenol and flavonoid contents [110].

B. vulgaris root extract was evaluated for the alleviation of oxidative stress by using female Japanese quails. Moreover, *B. vulgaris* root extract exerted antioxidant effects through inhibiting NF-kB, which was activated and suppressed in the heat stress environment [111]. The antioxidant potential of 50% aqueous ethanolic root extract of *B. aristata* was examined on antioxidant enzymes of the liver in diabetic rats, along with its safety parameters. The root extract of *B. aristata* has strong potential to decrease oxidative stress [112].

Significant antioxidant effects, mainly on ABTS, hydroxyl radicals and DPPH, have been reported to berberine hydrochloride. The relationship among diabetes mellitus and the increase of formation of free radicals and a decrease in antioxidant potential is well known. Since berberine hydrochloride has significant radicals scavenging and protective effects against β-cell damage and antioxidant of the pancreas in diabetes mellitus, it seems reasonable that antioxidants can play an important role in the improvement of diabetes and in screening the novel treatment drug of diabetes mellitus [113].

The extracts from the inner stem bark of *B. vulgaris* exhibited high antioxidant activity, and most of the identified compounds were isoquinoline alkaloids. The values were higher than the standard antioxidant compounds (vitamin C and butylated hydroxytoluene (BHT)) [114].

It was widely investigated the composition of the major anthocyanins and the antioxidant activities of the fruit of *B. heteropoda*. The high anthocyanins content indicated that this fruit could be considered as an excellent source of natural colorants and a functional food that benefits human health [115]. Regarding the alkaloid extract of *B. aetnensis* roots, it is also possessed antioxidant properties, and all results are in agreement with other reports on the alkaloids from the roots of *Berberis* species [116].

The findings suggested that *B. vulgaris* fruits have an important potential for their antioxidant activities depends on the content of phenolic compounds and organic acids [117].

6. Anticancer Activities of *Berberis* Plants (In Vitro and In Vivo)

Cancer is the third leading cause of death worldwide, preceded by cardiovascular and infectious diseases. In medical science, there is need for effective and acceptable cancer therapeutics agents that are non-toxic, highly efficacious against multiple cancers, cost effective, and acceptable by human population [118]. The interest in natural products has increased because they are less toxic to normal cells, and they reduce side effects and drug resistance observed in synthetic drugs.

In India, medicinal plants have been used for treating disease since ancient times. Many of these belong to the genus *Berberis*. The fruit of *B. vulgaris* is rich in polyphenols, vitamins, proteins, ascorbic acid, and anthocyanin, which are important for human health. Moreover, they are rich in alkaloids which promote anticancer activity. In particular in hepatic cancer cell (Hepg2) ethanol extract of the fruit of *B. vulgaris* reduces cell vitality and promotes the selective increase of protein expression like alkaline phosphatase (ALP), a hepatic enzyme important in the diagnosis of disease [119]. Anticancer activity of fruit extract of *B. vulgaris* was also demonstrated on human breast cancer cells (MCF-7). The extract reduces cell proliferation in time and dose-dependent manner. It has also been evidenced the importance of the solvent for the extraction processes, because as it is reported in several studies, ethanol extract is more active than water extract, probably due to its capacity to extract more compounds responsible for anticancer activity like alkaloids [120]. *B. vulgaris* ethanol extract reduces viability in the breast, colon, hepatic and cervix cancer cell lines in a dose-dependent manner after incubation at 24, 48 and 72 h. The ethanol extract has a similar activity of berberine chloride [121]. This was also confirmed by El Khalki et al. [122], who studied cytotoxicity on human breast adenocarcinoma cell (MCF-7) of *B. vulgaris* and berberine.

The antioxidant activity might play a major role in increasing efficiency of such extracts to kill cancer cells and protect normal cells, besides the inhibition of cell growth. Choi et al. [123] demonstrated that treatment with berberine reduces p53 expression in the human prostate cancer cell. In fact, berberine promotes translocation of p53 in nuclei and arrest of the cell cycle in G0/G1. This was also confirmed in in vivo studies. In this sense, the intraperitoneal administration of berberine at 10 mg/kg caused a substantial decline in tumor volume and weight of prostate cancer. The effect is more evident in cancer expressing p53 (LNCaP) both in vivo and in vitro [123].

Berberine and curcumin have been tested on different types of cancer cell line models as A549 (lung cancer cell line), Hep-G2 (liver cancer cell line), MCF-7 (breast cancer cell line), Jurkat (leukemia cancer cell line) and K562 (kidney cancer cell line) by Balakrishna et al. [124]. This work can reveal the synergetic activity of these compounds. The anticancer effects in these cells are mediated by inducing apoptosis [124]. The synergistic effect of two different compounds was also evidenced by Ren et al. [125]. They showed as galangine and berberine together demonstrated an anticancer activity stronger than that showed when used singularly, on esophageal carcinoma cells. In fact, they induce apoptosis, promote cell cycle arrest in the G2/M phase and increase reactive oxygen species (ROS) in cancer cells [125]. The anticancer activity of *B. aristata* roots was also evaluated on human osteosarcoma cells [126].

B. libanotica root extract showed potential anti-inflammatory and anticancer activity, mostly due to alkaloids and other compounds. This effect was demonstrated on human colon cancer cells [127] in which berberine inhibits COX-2 transcriptional activity. *B. libanotica* extract reduced the viability of CD4 T-cells infected by the retrovirus HTLV1, a kind of cell characteristic of an aggressive form of leukaemia [128]. Moreover, root extract showed anticancer activity on different cell lines of prostate cancer; it reduces cell viability and promotes cell cycle arrest in G0/G1 [129]. Subsequent studies on human erythroleukemia cell lines investigated molecular pathways responsible for the anticancer activity. The extract induced apoptosis of cells through the modulation of Akt/NF-Kb/COX-2 signal transduction pathways [130]. *B. libanotica* extract showed a dominant effect on K562 cells by the activation of the late markers of apoptosis with caspase-3 activation, Poly (ADP-ribose) polymerase (PARP) cleavage and DNA fragmentation. The study demonstrated that treatment with the extract induces apoptosis in erythroleukemia cell line expressing COX-2 (HEL cells) or not (K562), especially at a dose of 300 µg/mL after 48h of treatment. In particular, *B. libanotica* extract induced activation of caspase-3 and -9, correlated with PARP cleavage and DNA fragmentation. In this process, the extract is more effective than berberine tested at a dose of 40 µg/mL. Moreover, the extract reduced significantly the expression of COX-2, which prevent apoptosis in cancer cells through the activation of Akt and NF-kB. A similar mechanism was shown for 4-chlorobenzoyl berbamine; and a synthetic compound derived from berbamine. It induces apoptosis in lymphoma cell lines and G2/M cell cycle arrest through PI3K/Akt and NF-kB signaling pathways [131]. Changes as esterification, etherification or sulfonylation on the structure of berbamine allowed to obtaining new molecules capable of solving the problem of resistant in many types of tumor [132]. Many synthetic derivatives of berbamine also demonstrated antineoplastic activity. Among these, BBMD3 was shown to be the most potent as an anticancer agent on human melanoma cells. It inhibits JAK/STA3 pathways reducing pro-apoptotic gene expression [133]. This was also confirmed in osteosarcoma and glioblastoma cell lines where BBMD3 induced inhibition of Jak2/STAT3 signaling pathway and activation of the stress response

JNK pathway. Moreover, it increases the expression of miR-4284 involved in tumorigenesis and apoptosis [134,135].

Berbamine, which is contained in *B. amurensis*, has activity also tested on solid tumor. In fact, it arrests growth and migration in vitro and ex vivo of human lung cancer A549 cell line at low concentration trough down-regulation of anti-apoptotic protein Bcl-2 and up-regulation of the pro-apoptotic protein Bax [136]. *B. amurensis* extract arrests proliferation of Hepg2 and MCF-7 cells; extraction technique influences the presence of the active compound and consequently the extract activity [137].

B. orthobotrys is another species of this genus mainly grow in Iran. Roots bark leaves are used in traditional medicine in easy problem like menstrual pains, kidney stones, but the species have also demonstrated anticancer activity in HeLa cell line. As reported by Bavand et al. [138], ethanol root extract induces morphological change and apoptosis in HeLa cells then 72 h of treatment. In particular treatment of cells with 1.25 mg/mL of extract, reduced the cell viability, inhibited the cell growth, changed cell adhesion to the substrate, pigmented the cells and formed apoptotic bodies [138]. Treatment with a low concentration of the extract also presents anticancer activity on other types of cancer cell line. Engel et al. [139] reported that the reduction of cell vitality of 60% caused by treatment with different doses ranged from 100 to 1 µg/mL of root extract of *B. orthobotrys*. They also studied the possible molecular mechanism responsible for cell death. Microscopic analysis of cells showed the accumulation of lysosome, which starts programmed cell death trough liberation of ROS and hydrolytic enzymes, granularization and formation of Golgi vesicles, as well as the diffuse distribution of neutral lipids. This is pronounced at 100 µg/mL, but also lower doses causes a slight formation of lysosome vesicles [139].

In addition to alkaloids there are other secondary metabolites with anticancer activity, for example, the triterpenoids the main active constituent of the trunk of *B. koreana*. They have been identified through mass spectrometry and nuclear magnetic resonance and tested in different cancer cell lines (A549, SK-OV-3, SK-MEL-2 and HCT-15) where they reduce cell proliferation [140,141].

7. Clinical Studies of *Berberis* Plants in Human

Currently available clinical trials regarding this group of plants point on their effects in various conditions related to cardiovascular diseases and associated risk factors, neurodegenerative diseases and inflammation.

One group of clinical trials conducting by Guiseppe Derosa et al. [142–145] had a specific interest in a fixed combination that included *B. aristata* and Silybum marianum (Berberol®). The reason for this combination lies in low bioavailability of *B. aristata*, while S. marianum is there to improve its intestinal absorption. A 52-week double-blind placebo-controlled study in 136 obese patients with type-2 diabetes mellitus (T2DM) and metabolic syndrome analyzed various parameters, including: Fasting blood glucose, insulin, total cholesterol, HDL, LDL, triglycerides, and body mass index (BMI). [146]. All of these parameters have been significantly improved in the treatment group compering with the baseline and in order to control group. Previously, comparing same fixed combination vs. *B. aristata* monotherapy in clinical trial conducting by Di Pierro et al. [147] with T2DM subjects, shown that combination is more effective in decreasing of HbA1c indicated that positive effects are partly due to S. marianum. In another study with the same combination of extracts in 102 dyslipidemia subjects after three months, it has been shown reducing of total cholesterol, triglycerides and LDL, with increasing of HDL from randomization and compared to the placebo group. The same result on lipid profile has been observed in a double-blind, randomized placebo-controlled trial that included 106 patients with metabolic syndrome treated with B. vulgaris [148]. Another two double-blind, randomized, placebo-controlled, 6-months clinical studies with Berberol® conducting by Derosa et al. [142,143], followed dyslipidaemic subjects intolerant to statins at high dosages. In both studies were included patients tolerant to a half dose of statins. The lipid profile of included patients did not significantly change in the active treatment group after reduction of statins dosage and the introduction of

Barberol®. Meanwhile, in placebo group lipid profile was worsened compared to baseline and with active treatment.

The clinical trial with type 1 diabetes mellitus (T1DM) subjects treated with the same fixed combination (Berberol®) showed decreasing of insulin dose necessary to reach adequate glycemic control [144]. The clinical trial with subjects at low cardiovascular risk also confirmed hypocholesterolemia effects of a fixed combination of Berberol® [145].

Berberin has been shown to inhibit CYP3A4 in in vitro and animal models, as well as in humans, and that inhibition should increase blood levels of statins, cyclosporine, and calcium channel blockers, similar to the action of grapefruit [149]. The inhibition of enzyme CYP3A4 activity in humans has been observed in a two-phase randomized-crossover clinical study in healthy male subjects after two weeks of berberine administration (300 mg, p.o.) [150]. In a randomized double-blind placebo-controlled clinical trial, patients suffering from irritable bowel syndrome received berberine hydrochloride twice daily for two months [151]. The benefits from the treatment were observed as better IBS symptom and depression/anxiety scores.

One of the clinical trials examined the effect of aqueous extract of dried barberry taken orally as an anti-acne agent [152]. The results obtained from teenagers in this placebo-containing trial show the effectiveness of using barberry in the treatment of acne vulgaris, despite the treatment and control groups were small.

In all of these trials, no patients had serious adverse events. The limitations were the relatively small size of the sample, and relatively short follow-up period. However, the side effects of berberine have been reported in some in vitro and in vivo animal models, and observed effects were related to its neurotoxicity [153,154]. Despite that, in animal models of Alzheimer's disease neuroprotective effects of berberine have been illustrated [155]. The published results in a review article from 2015 indicated that on web page www.clinicaltrials.gov there was 17 clinical trials on the efficacy of berberine [156], and currently there are 51 clinical trials which showing us increasing interest in beneficial effects of this compound [data obtained searching web page dated 24 May 2018].

8. Conclusions

Berberis is an important genus of wild plants with a multitude of uses in pharmacology and food industry. These species are the abundant source of important natural compounds, i.e., vitamins, minerals, alkaloids and antioxidants, which can be used in a wide array of pharmaceutical and nutraceutical products. Some of the species of the genus like *B. vulgaris* are also cultivated in Iran and other countries, but information regarding its cultivation, diseases and production technology is sparse. The present study has been carried out to report on the adaptation of different *Berberis* species, suitable agro-climatic conditions for the higher yield, its production technology, diseases and harvesting methods. However, further studies should be conducted to evaluate genetic diversity in the cultivated species for selection of high yielding genotypes, development of new varieties, yield enhancement through appropriate cultivation practices and integrated pest management (IPM) techniques.

Regarding food and pharmacological applications, *Berberis* is an important group of the plants having enormous potential in the food industry, and several reports of their antimicrobial activity have been found in the literature. Several phytochemicals found in fruits, leaves, stems and root have demonstrated biological activities. Phytochemicals present in EOs confer antimicrobial and antioxidant properties that make them useful as food preservatives. On the other hand, alkaloids present pharmacological properties, such as anticancer activities reported. However, not much information is available on the direct application of these plants in food products. On the other hand, the extracts of *B. vulgaris* were the most promising as antioxidants as well as inflammatory and neurological disorders protective. In addition, positive effects related to cancer targets have been reported to reduce cell proliferation without affecting a normal human cell. For this reason, *Berberis* spp. maybe considered an alternative for cancer treatment, but it is necessary to confirm their efficacy in vivo, especially investigating the toxicity during drug therapy.

Author Contributions: All authors contributed to the manuscript. Conceptualization, B.S. and J.S.-R.; Validation investigation, resources, data curation, writing–all authors; Review and editing, J.S.-R., M.d.l.L.C.-G., and W.C.C. All the authors read and approved the final manuscript.

Funding: This research received no external funding.

Acknowledgments: This work was supported by CONICYT PIA/APOYO CCTE AFB170007.

Conflicts of Interest: The authors declare no conflict of interest.

References

1. Minaiyan, M.; Ghannadi, A.; Mahzouni, P.; Jaffari-Shirazi, E. Comparative study of berberis vulgaris fruit extract and berberine chloride effects on acetic acid-induced colitis in rats. *Iran. J. Pharm. Res.* **2011**, *10*, 97–104. [PubMed]
2. Mokhber-Dezfuli, N.; Saeidnia, S.; Gohari, A.; Kurepaz-Mahmoodabadi, M. Phytochemistry and Pharmacology of Berberis Species. *Pharmacogn. Rev.* **2014**, *8*, 8. [PubMed]
3. Rahimi-Madiseh, M.; Lorigoini, Z.; Zamani-Gharaghoshi, H.; Rafieian-Kopaei, M. Berberis vulgaris: Specifications and traditional uses. *Iran. J. Basic Med. Sci.* **2017**, *20*, 569–587. [PubMed]
4. Farhadi Chitgar, M.; Aalami, M.; Maghsoudlou, Y.; Milani, E. Comparative Study on the Effect of Heat Treatment and Sonication on the Quality of Barberry (*Berberis Vulgaris*) Juice. *J. Food Process. Preserv.* **2017**, *41*, e12956. [CrossRef]
5. Aghbashlo, M.; Kianmehr, M.H.; Hassan-Beygi, S.R. Specific heat and thermal conductivity of berberis fruit (*Berberis vulgaris*). *Am. J. Agric. Biol. Sci.* **2008**, *3*, 330–336. [CrossRef]
6. Birdsall, T.C.; Kelly, G.S. Berberine: Therapeutic potential of an alkaloid found in several medicinal plants. *Altern. Med. Rev.* **1997**, *2*, 94–103.
7. Kang, J.; Kang, Y.; Ji, X.; Guo, Q.; Jacques, G.; Pietras, M.; Łuczaj, N.; Li, D.; Łuczaj, Ł. Wild food plants and fungi used in the mycophilous Tibetan community of Zhagana (Tewo County, Gansu, China). *J. Ethnobiol. Ethnomed.* **2016**, *12*, 21. [CrossRef]
8. Tavakoli, A.; Sahari, M.A.; Barzegar, M. Antioxidant activity of Berberis integerrima seed oil as a natural antioxidant on the oxidative stability of soybean oil. *Int. J. Food Prop.* **2018**, *20*, S2914–S2925. [CrossRef]
9. Aliakbarlu, J.; Mohammadi, S.; Khalili, S. A Study on Antioxidant Potency and Antibacterial Activity of Water Extracts of Some Spices Widely Consumed in Iranian Diet. *J. Food Biochem.* **2013**, *38*, 159–166. [CrossRef]
10. Srivastava, S.; Srivastava, M.; Misra, A.; Pandey, G.; Rawat, A. A review on biological and chemical diversity in Berberis (Berberidaceae). *EXCLI J.* **2015**, *14*, 247–267.
11. Kafi, M.; Balandary, A.; Rashed-Mohasel, M.H.; Koochaki, A.; Molafilabi, A. *Berberis: Production and Processing*; Zaban va adab Press: City, Iran, 2002; ISBN 9789290814993.
12. Saied, S.; Begum, S. Phytochemical studies of Berberis vulgaris. *Chem. Nat. Compd.* **2004**, *40*, 137–140. [CrossRef]
13. Yazdani, A.; Poorbaghi, S.L.; Habibi, H.; Nazifi, S.; Rahmani Far, F.; Sepehrimanesh, M. Dietary *Berberis vulgaris* extract enhances intestinal mucosa morphology in the broiler chicken (*Gallus gallus*). *Comp. Clin. Path.* **2013**, *22*, 611–615. [CrossRef]
14. Bashir, S.; Gilani, A.H.; Siddiqui, A.A.; Pervez, S.; Khan, S.R.; Sarfaraz, N.J.; Shah, A.J. *Berberis vulgaris* root bark extract prevents hyperoxaluria induced urolithiasis in rats. *Phyther. Res.* **2010**, *24*, 1250–1255. [CrossRef] [PubMed]
15. Hermenean, A.; Popescu, C.; Ardelean, A.; Stan, M.; Hadaruga, N.; Mihali, C.V.; Costache, M.; Dinischiotu, A. Hepatoprotective effects of *Berberis vulgaris* L. extract/β cyclodextrin on carbon tetrachloride-induced acute toxicity in mice. *Int. J. Mol. Sci.* **2012**, *13*, 9014–9034. [CrossRef] [PubMed]
16. Amjad, M.S.; Arshad, M.; Qureshi, R. Ethnobotanical inventory and folk uses of indigenous plants from Pir Nasoora National Park, Azad Jammu and Kashmir. *Asian Pac. J. Trop. Biomed.* **2015**, *5*, 234–241. [CrossRef]
17. Altundag, E.; Ozturk, M. Ethnomedicinal studies on the plant resources of east Anatolia, Turkey. *Procedia Soc. Behav. Sci.* **2011**, *19*, 756–777. [CrossRef]
18. Javadzadeh, S.; Fallah, S. Therapeutic application of different parts of *Berberis vulgaris*. *Int. J. Agric. Crop Sci.* **2012**, *4*, 404–408.
19. Phillips, R.; Foy, N. *Herbs*; Pan Books Ltd.: London, UK, 2002; ISBN 0-330-30725-8.

20. Kuo, C.-L.; Chi, C.-W.; Liu, T.-Y. The anti-inflammatory potential of berberine in vitro and in vivo. *Cancer Lett.* **2004**, *203*, 127–137. [CrossRef]

21. Sun, D.; Courtney, H.S.; Beachey, E.H. Berberine sulfate blocks adherence of Streptococcus pyogenes to epithelial cells, fibronectin, and hexadecane. *Antimicrob. Agents Chemother.* **1988**, *32*, 1370–1374. [CrossRef]

22. Hajzadeh, M.A.R.; Rajaei, Z.; Shafiee, S.; Alavinejhad, A.; Samarghandian, S.; Ahmadi, M. Effect of barberry fruit (*Berberis Vulgaris*) on serum glucose and lipids in streptozotocin-diabetic rats. *Pharmacologyonline* **2011**, *1*, 809–817.

23. Meliani, N.; Dib, M.E.A.; Allali, H.; Tabti, B. Hypoglycaemic effect of *Berberis vulgaris* L. in normal and streptozotocin-induced diabetic rats. *Asian Pac. J. Trop. Biomed.* **2011**, *1*, 468–471. [CrossRef]

24. Zhou, X.; Chan, S.W.; Tseng, H.L.; Deng, Y.; Hoi, P.M.; Choi, P.S.; Or, P.M.Y.; Yang, J.M.; Lam, F.F.Y.; Lee, S.M.Y.; et al. Danshensu is the major marker for the antioxidant and vasorelaxation effects of Danshen (*Salvia miltiorrhiza*) water-extracts produced by different heat water-extractions. *Phytomedicine* **2012**, *19*, 1263–1269. [CrossRef]

25. Zhu, X.; Bian, H.; Gao, X. The Potential Mechanisms of Berberine in the Treatment of Nonalcoholic Fatty Liver Disease. *Molecules* **2016**, *21*, 1336. [CrossRef] [PubMed]

26. Yang, J.; Yin, J.; Gao, H.; Xu, L.; Wang, Y.; Xu, L.; Li, M. Berberine Improves Insulin Sensitivity by Inhibiting Fat Store and Adjusting Adipokines Profile in Human Preadipocytes and Metabolic Syndrome Patients. *Evid.-Based Complement. Altern. Med.* **2012**, *2012*, 363845. [CrossRef] [PubMed]

27. Zhang, Y.; Ye, J. Mitochondrial inhibitor as a new class of insulin sensitizer. *Acta Pharm. Sin. B* **2012**, *2*, 341–349. [CrossRef]

28. Imenshahidi, M.; Hosseinzadeh, H. *Berberis Vulgaris* and Berberine: An Update Review. *Phyther. Res.* **2016**, *30*, 1745–1764. [CrossRef]

29. West, S.; King, V.; Carey, T.S.; Lohr, K.N.; McKoy, N.; Sutton, S.F.; Lux, L. Systems to rate the strength of scientific evidence. *Evid. Rep. Technol. Assess.* **2002**, *47*, 1–11.

30. Rounsaville, T.J.; Ranney, T.G. Ploidy levels and genome sizes of berberis l. and mahonia nutt. species, hybrids, and cultivars. *HortScience* **2010**, *45*, 1029–1033. [CrossRef]

31. Mozaffarian, V. *A Dictionary of Iranian Plant Names*; Farhang Mo'aser: Tehran, Iran, 2008; ISBN 9645545196.

32. Kafi, M.; Balandri, A. Effects of gibberellic acid and ethephon on fruit characteristics and ease of harvest seed less barberry. *Iran. Res. Organ. Sci. Technol. Cent. Khorasan* **1995**, *volume*, page.

33. Peterson, P.; Leonard, K.; Miller, J.; Laudon, R.; Sutton, T. Prevalence and distribution of common barberry, the alternate host of Puccinia graminis, in Minnesota. *Plant Dis.* **2005**, *89*, 159–163. [CrossRef]

34. Łuczaj, L. Archival data on wild food plants used in Poland in 1948. *J. Ethnobiol. Ethnomed.* **2008**, *4*, 4. [CrossRef] [PubMed]

35. Łuczaj, Ł. Wild food plants used in Poland from the mid-19th century to the present. [Dziko rosnące rośliny jadalne użytkowane w Polsce od połowy XIX w. do czasów współczesnych]. *Etnobiologia Pol.* **2011**, *1*, 57–125.

36. Bussmann, R.; Zambrana, P.; Narel, Y.; Sikharulidze, S.; Kikvidze, Z.; Kikodze, D.; Tchelidze, D.; Batsatsashvili, K.; Robbie, E. Ethnobotany of Samtskhe-Javakheti, Sakartvelo (Republic of Georgia), Caucasus. *Indian J. Tradit. Knowl.* **2017**, *12*, 7–24.

37. Kern, F.D. Observations of the Dissemination of the Barberry. *Ecology* **1921**, *2*, 211–214. [CrossRef]

38. Dirr, M. *Manual of Woody Landscape Plants: Their Identification, Ornamental Characteristics, Culture, Propagation, and Uses*; 5th ed.; Stipes Publishing: Champaign, IL, USA, 1998; ISBN 9781588748706.

39. Mack, R.N.; Erneberg, M. The United States naturalized flora: Largely the product of deliberate introductions. *Ann. Missouri Bot. Gard.* **2002**, *89*, 176–189. [CrossRef]

40. Fulling, E.H. Plant life and the law of man IV barberry, currant and gooseberry, and cedar control. *Bot. Rev.* **1943**, *9*, 483–592. [CrossRef]

41. Javadzadeh, S. Effect of different methods of harvesting, drying and time on losses seedless barberry (*Berberis vulgaris* L). *Int. J. Agron. Plant* **2013**, *4*, 254–260.

42. Sharma, R. Medicinal plants of India: An Encyclopaedia. *Indian Counc. Med. Res. New Delhi* **2003**, *1*, 33.

43. Li, X.; Zhang, L.; Li, W.; Yin, X.; Yuan, S. New taxa of Berberis (Berberidaceae) with greenish flowers from a biodiversity hotspot in Sichuan Province, China. *Plant Divers.* **2017**, *39*, 94–103. [CrossRef]

44. Ward, J.S.; Worthley, T.E.; Williams, S.C. Controlling Japanese barberry (Berberis thunbergii DC) in southern New England, USA. *For. Ecol. Manag.* **2009**, *257*, 561–566. [CrossRef]

45. Adhikari, B.; Pendry, C.A.; Pennington, R.T.; Milne, R.I. A revision of berberis S.S. (Berberidaceae) in Nepal. *Edinburgh J. Bot.* **2012**, *69*, 447–522. [CrossRef]

46. Baytop, T. *Turkish Plant Names Dictionary. Atatürk Culture, Language and History High Foundation*; Turkish Language Foundation: Ankara, Turkey, 1994.

47. Khan, T.; Khan, I.A.; Rehman, A. A review on Berberis species reported from Gilgit- Baltistan and Central Karakoram National Park, Pakistan. *J. Med. Plants Stud.* **2014**, *2*, 16–20.

48. Ali, M.; Malik, A.R.; Sharma, K.R. Vegetative propagation of Berberis aristata DC. An endangered Himalayan shrub. *J. Med. Plants* **2008**, *2*, 374–377.

49. Ali, M.N.; Khan, A.A. Pharmacognostic studies on Berberis lycium Royle, and its importance as a source of raw material for the manufacture of berberine in Pakistan [angiosperm trees]. *Pakistan J. For.* **1978**, *28*, 25–27.

50. Kaur, C.; Miani, S. Fruits and vegetables healthy foods for new millennium. *Indian Hort* **2001**, *45*, 29–32.

51. Tewary, D.K.; Bhardwaj, A.; Shanker, A. Pesticidal activities in five medicinal plants collected from mid hills of western Himalayas. *Ind. Crops Prod.* **2005**, *22*, 241–247. [CrossRef]

52. Fallahi, J.; Moghaddam, R.P.; Nasiri-Mahallati, M. Effect of harvest date on quantitative and qualitative indices of seedless barberry. *Iran. J. F. Crop. Res.* **2010**, *8*, 225–234.

53. Moghaddam, P.R.; Fallahi, J.; Shajari, M.A.; Mahallati, M.N. Effects of harvest date, harvest time, and post-harvest management on quantitative and qualitative traits in seedless barberry (*Berberis vulgaris* L.). *Ind. Crops Prod.* **2013**, *42*, 30–36. [CrossRef]

54. Arena, M.E.; Curvetto, N. Berberis buxifolia fruiting: Kinetic growth behavior and evolution of chemical properties during the fruiting period and different growing seasons. *Sci. Hortic. (Amsterdam)* **2008**, *118*, 120–127. [CrossRef]

55. Chandra, P.; Todaria, N.P. Maturation and ripening of three Berberis species from different altitudes. *Sci. Hortic. (Amsterdam)* **1983**, *19*, 91–95. [CrossRef]

56. Mahmoodi, H.R.; Zamani, G.H.; Balandary, A. The study of qualitative characteristics of seedless barberry (*Berberis vulgaris* L.) as influenced by different fruit harvesting dates and two different climates. In Proceedings of the 6th Congress of Iranian Horticultural Sciences, Isfahan, Iran, 2009; pp. 1486–1489.

57. Minore, D.; Rudolf, P.O.; Berberis, L. *The Woody Plant Seed Manual, Agriculture Handbook*; Bonner, F.T., Karrfalt, R.P., Eds.; U.S. Department of Agriculture Forest Service: Washington, DC, USA, 2008; Volume 727, pp. 298–302.

58. Obesco, J.R. Fruit removal and potential seed dispersal in a southern Spanish population of *Berberis vulgaris* subsp. australis (Berberidaceae). *Acta Oecologica/Oecologia Pantarum* **1989**, *10*, 321–328.

59. Royer, F.; Dickinson, R. *Weeds of the Northern U.S. and Canada: A Guide for Identification*; University of Alberta: Edmonton, AB, Canada, 1999; ISBN 1551052210.

60. Chapman, W.K.; Bessette, A.E. *Trees and Shrubs of the Adirondacks*; North Country Books, Inc.: Utica, NY, USA, 1990.

61. Eriksson, O.; Ehrlén, J. Phenological variation in fruit characteristics in vertebrate-dispersed plants. *Oecologia* **1991**, *86*, 463–470. [CrossRef]

62. Imanshahidi, M.; Hosseinzadeh, H. Pharmacological and therapeutic effects of *Berberis vulgaris* and its active constituent, berberine. *Phyther. Res.* **2008**, *22*, 999–1012. [CrossRef]

63. Alemardan, A.; Asadi, W.; Rezaei, M.; Tabrizi, L.; Mohammadi, S. Cultivation of Iranian seedless barberry (Berberis integerrima 'Bidaneh'): A medicinal shrub. *Ind. Crops Prod.* **2013**, *50*, 276–287. [CrossRef]

64. Potdar, D.; Hirwani, R.R.; Dhulap, S. Phyto-chemical and pharmacological applications of Berberis aristata. *Fitoterapia* **2012**, *83*, 817–830. [CrossRef]

65. Dolezal, M.; Velisek, J.; Famfulikova, P. Chemical composition of less-known wild fruits. In Proceedings of the EUROFOODCHEM XI Meeting, Norwich, UK, 26–28 September 2001; Volume 269, pp. 241–244.

66. Hamedi, A.; Moheimani, S.M.; Sakhteman, A.; Etemadfard, H.; Moein, M. An Overview on Indications and Chemical Composition of Aromatic Waters (Hydrosols) as Functional Beverages in Persian Nutrition Culture and Folk Medicine for Hyperlipidemia and Cardiovascular Conditions. *J. Evid.-Based Complement. Altern. Med.* **2017**, *22*, 544–561. [CrossRef]

67. Sonmezdag, A.S.; Kelebek, H.; Selli, S. Volatile and key odourant compounds of Turkish Berberis crataegina fruit using GC-MS-Olfactometry. *Nat. Prod. Res.* **2018**, *32*, 777–781. [CrossRef]

68. Hashemi-Moghaddam, H.; Mohammadhosseini, M.; Azizi, Z. Impact of amine- and phenyl-functionalized magnetic nanoparticles impacts on microwave-assisted extraction of essential oils from root of Berberis integerrima Bunge. *J. Appl. Res. Med. Aromat. Plants* **2018**, *10*, 1–8. [CrossRef]

69. Bonesi, M.; Loizzo, M.R.; Conforti, F.; Passalacqua, N.G.; Saab, A.; Menichini, F.; Tundis, R. *Berberis aetnensis* and *B. libanotica*: A comparative study on the chemical composition, inhibitory effect on key enzymes linked to Alzheimer's disease and antioxidant activity. *J. Pharm. Pharmacol.* **2013**, *65*, 1726–1735. [CrossRef]

70. Jay, J. *Modern Food Microbiology*; Aspen Publishers Inc.: Gaithersburg, MD, USA, 1998; ISBN 978-0-387-23180-8.

71. Pszczola, D.E. Emerging ingredients: Believe it or not! *Food Technol.* **1999**, *53*, 98–100.

72. Kaya, M.; Ravikumar, P.; Ilk, S.; Mujtaba, M.; Akyuz, L.; Labidi, J.; Salaberria, A.M.; Cakmak, Y.S.; Erkul, S.K. Production and characterization of chitosan based edible films from Berberis crataegina's fruit extract and seed oil. *Innov. Food Sci. Emerg. Technol.* **2018**, *45*, 287–297. [CrossRef]

73. Thusa, R.; Mulmi, S. Analysis of Phytoconstituents and Biological Activities of Different Parts of Mahonia nepalensis and Berberis aristata. *Nepal J. Biotechnol.* **2017**, *5*, 5–13. [CrossRef]

74. Pokhrel, N.R.; Adhikari, R.P.; Baral, M.P. In Vitro screening and evaluation of antimicrobial activities of some medicinal plants of Nepal. *Nepal J. Sci. Technol.* **2003**, *5*, 1.

75. Ebrahimi, A.; Chavoushpour, M.; Mahzoonieh, M.R.; Lotfalian, S. Antibacterial activity and ciprofloxacin-potentiation property of Berberis vulgaris asperma stem extracts on pathogenic bacteria. *J. HerbMed Pharmacol.* **2016**, *5*, 112–115.

76. Singh, S.K.; Vishnoi, R.; Dhingra, G.K.; Kishor, K. Antibacterial activity of leaf extracts of some selected traditional medicinal plants of Uttarakhand, North East India. *J. Appl. Nat. Sci.* **2012**, *4*, 47–50. [CrossRef]

77. Saravanakumar, T.; Manonmani, E.; Venkatasubramanian, P.; Vasanthi, N.S. Antimicrobial potential of Daruharidra (Berberis aristata DC) against the pathogens causing eye infection. *Int. J. Green Pharm.* **2014**, *8*, 153. [CrossRef]

78. Alamzeb, M. Bioassay guided isolation and characterization of anti-microbial and anti-trypanosomal agents from Berberis glaucocarpa Stapf. *African J. Pharm. Pharmacol.* **2013**, *7*, 2564–2570. [CrossRef]

79. Parvu, M.; Parvu, A.E.; Craciun, C.; Barbu-Tudoran, L.; Vlase, L.; Tamas, M.; Rosca-Casian, O.; PersecA, O.; Molnar, A.M. Changes in Botrytis cinerea conidia caused by Berberis vulgaris extract. *Not. Bot. Horti Agrobot. Cluj-Napoca* **2010**, *38*, 15–20.

80. Shah, Z.; Ilyas, M.; Khan, M.; Ahmad, A.; Khan, M.; Khan, N. Antimicrobial activities of selected medicinal plants collected from Northern districts of Khyber Pakhtunkhwa, Pakistan. *J. Pharm. Res.* **2012**, *5*, 1729–1733.

81. Ghareeb, D.A.; El-Wahab, A.E.A. Biological assessment of *Berberis vulgaris* and its active constituent, berberine: Antibacterial, antifungal and anti-hepatitis C virus (HCV) effect. *J. Med. Plants Res.* **2013**, *7*, 1529–1536.

82. Shahid, T.; Memon, M.; Malik, R.A.; Ikram, N.; Malik, W.; Ali, A. A study of Antimicrobial Activity of *Berberis vulgaris* (Zirishk) Aqueous Plant Extract using Pathogenic Isolates from Patients of Islamabad and Rawalpindi. *Imp. J. Interdiscip. Res.* **2017**, *3*, 1365–1371.

83. Krisch, J.; Galgóczy, L.; Tölgyesi, M.; Papp, T.; Vágvölgyi, C. Effect of fruit juices and pomace extracts on the growth of Gram-positive and Gram-negative bacteria. *Acta Biol. Szeged.* **2008**, *52*, 267–270.

84. Rasool, S.; Khan, F.Z.; Hassan, S.U.; Ahmed, M.; Ahmed, M.; Tareen, R.B. Anticonvulsant, antimicrobial and cytotoxic activities of berberis calliobotrys aitch ex koehne (Berberidaceae). *Trop. J. Pharm. Res.* **2015**, *14*, 2031–2039. [CrossRef]

85. Irshad, A.H.; Pervaiz, A.H.; Abrar, Y.B.; Fahelboum, I.; Awen, B.Z.S. Antibacterial activity of Berberis lycium root extract. *Trakia J. Sci.* **2013**, *11*, 88–90.

86. Haouat, A.C.; Haggoud, A.; David, S.; Ibnsouda, S.; Iraqui, M. In vitro evaluation of the antimycobacterial activity and fractionation of Berberis hispanica root bark. *J. Pure Appl. Microbiol.* **2014**, *8*, 917–925.

87. Mattana, C.M.; Satorres, S.E.; Juan, V.; Cifuente, D.; Tonn, C.; Laciar, A.L. Antibacterial activity study of single and combined extracts of Berberis ruscifolia, Baccharis sagittalis, Euphorbia dentata and Euphorbia schikendanzii, native plants from Argentina. *BLACPMA* **2012**, *11*, 428–434.

88. Joshi, P.V.; Shirkhedkar, A.A.; Prakash, K.; Maheshwari, V.L. Antidiarrheal activity, chemical and toxicity profile of Berberis aristata. *Pharm. Biol.* **2011**, *49*, 94–100. [CrossRef]

89. Singh, M.; Srivastava, S.; Rawat, A.K.S. Antimicrobial activities of Indian Berberis species. *Fitoterapia* **2007**, *78*, 574–576. [CrossRef]

90. Hussain, M.A.; Khan, M.Q.; Habib, T.; Hussain, N. Antimicronbial activity of the crude root extract of berberis lycium royle. *Adv. Environ. Biol.* **2011**, *5*, 585–588.

91. Azimi, G.; Hakakian, A.; Ghanadian, M.; Joumaa, A.; Alamian, S. Bioassay-directed isolation of quaternary benzylisoquinolines from Berberis integerrima with bactericidal activity against Brucella abortus. *Res. Pharm. Sci.* **2018**, *13*, 149–158.

92. Bukhari, I.; Hassan, M.; Abbasi, F.; Mujtaba, G.; Mahmood, N.; Fatima, A.; Afzal, M.; Rehman, M.; Perveen, P.; Khan, T. A study on comparative pharmacological efficacy of Berberis lycium and penicillin G. *African J. Microbiol. Res.* **2011**, *5*, 725–727.

93. Sati, S.C.; Takuli, P.; Kumar, P.; Khulbe, K. Antibacterial activity of three medicinal plants of Kumaun Himalaya against some pathogenic bacteria. *Int. J. Pharma Sci. Res.* **2015**, *6*, 1361–1368.

94. Malik, Z.; Jain, K.; Ravindran, K.; Sathiyaraj, G. In vitro antimicrobial activity and preliminary phytochemical analysis of Berberis aristata. *Int. J. Ethnobiol. Ethnomed.* **2017**, *4*, 1–6.

95. Manosalva, L.; Mutis, A.; Urzúa, A.; Fajardo, V.; Quiroz, A. Antibacterial activity of alkaloid fractions from berberis microphylla G. Forst and study of synergism with ampicillin and cephalothin. *Molecules* **2016**, *21*, 76. [CrossRef]

96. Mehmood, A.; Murtaza, G.; Bhatti, T.M.; Kausar, R.; Ahmed, M.J. Biosynthesis, characterization and antimicrobial action of silver nanoparticles from root bark extract of Berberis lycium Royle. *Pak. J. Pharm. Sci.* **2016**, *29*, 131–137.

97. Anzabi, Y. In vitro study of Berberis vulgaris, Actinidia deliciosa and Allium cepa L. antibacterial effects on Listeria monocytogenes. *Crescent J. Med. Biol. Sci.* **2015**, *2*, 111–115.

98. Thakur, P.; Chawla, R.; Narula, A.; Sharma, R.K. Protective effect of *Berberis aristata* against peritonitis induced by carbapenem-resistant *Escherichia coli* in a mammalian model. *J. Glob. Antimicrob. Resist.* **2017**, *9*, 21–29. [CrossRef]

99. Shahid, M.; Rahim, T.; Shahzad, A.; Latif, T.A.; Fatma, T.; Rashid, M.; Raza, A.; Mustafa, S. Ethnobotanical studies on Berberis aristata DC. root extracts. *African J. Biotechnol.* **2009**, *8*, 556–563.

100. Rizwan, M.; Nasir, H.; Shah, S.Z. Phytochemical and biological screening of Berberis aristata. *Adv. Life Sci.* **2017**, *57*, 1–7.

101. Amin, A.H.; Subbaiah, T.V.; Abbasi, K.M. Berberine sulfate: Antimicrobial activity, bioassay, and mode of action. *Can. J. Microbiol.* **1969**, *15*, 1067–1076. [CrossRef]

102. Freile, M.L.; Giannini, F.; Pucci, G.; Sturniolo, A.; Rodero, L.; Pucci, O.; Balzareti, V.; Enriz, R.D. Antimicrobial activity of aqueous extracts and of berberine isolated from Berberis heterophylla. *Fitoterapia* **2003**, *74*, 702–705. [CrossRef]

103. Hyun, T.K.; Kim, H.C.; Kim, J.S. In vitro Screening for Antioxidant, Antimicrobial, and Antidiabetic Properties of Some Korean Native Plants on Mt. Halla, Jeju Island. *Indian J. Pharm. Sci.* **2015**, *77*, 668–674. [PubMed]

104. Kosalec, I.; Gregurek, B.; Kremer, D.; Zovko, M.; Sanković, K.; Karlović, K. Croatian barberry (*Berberis croatica* Horvat): A new source of berberine—Analysis and antimicrobial activity. *World J. Microbiol. Biotechnol.* **2009**, *25*, 145–150. [CrossRef]

105. Malik, T.A.; Kamili, A.N.; Chishti, M.Z.; Ahad, S.; Tantry, M.A.; Hussain, P.R.; Johri, R.K. Breaking the resistance of Escherichia coli: Antimicrobial activity of Berberis lycium Royle. *Microb. Pathog.* **2017**, *102*, 12–20. [CrossRef] [PubMed]

106. Musumeci, R.; Speciale, A.; Costanzo, R.; Annino, A.; Ragusa, S.; Rapisarda, A.; Pappalardo, M.S.S.; Iauk, L. *Berberis aetnensis* C. Presl. extracts: Antimicrobial properties and interaction with ciprofloxacin. *Int. J. Antimicrob. Agents* **2003**, *22*, 48–53. [CrossRef]

107. Villinski, J.R.; Dumas, E.R.; Chai, H.B.; Pezzuto, J.M.; Angerhofer, C.K.; Gafner, S. Antibacterial activity and alkaloid content of Berberis thunbergii, *Berberis vulgaris* and *Hydrastis canadensis. Pharm. Biol.* **2003**, *41*, 551–557. [CrossRef]

108. Bereksi, M.S.; Hassaïne, H.; Bekhechi, C.; Abdelouahid, D.E. Evaluation of Antibacterial Activity of some Medicinal Plants Extracts Commonly Used in Algerian Traditional Medicine against some Pathogenic Bacteria. *Pharmacogn. J.* **2018**, *10*, 507–512. [CrossRef]

109. Maznah, I.; Teoh, S.L.; Loh, P. Determination of total antioxidant activity of selected local medicinal plants. In Proceedings of the Herbs an International Conference and Exhibitions, Mines, Seri Kembangan, Malaysia, 9–11 November 1999; pp. 124–128.

110. Eddouks, M.; Maghrani, M.; Lemhadri, A.; Ouahidi, M.-L.; Jouad, H. Ethnopharmacological survey of medicinal plants used for the treatment of diabetes mellitus, hypertension and cardiac diseases in the south-east region of Morocco (Tafilalet). *J. Ethnopharmacol.* **2002**, *82*, 97–103. [CrossRef]

111. Sahin, K.; Orhan, C.; Tuzcu, M.; Borawska, M.H.; Jabłonski, J.; Guler, O.; Sahin, N.; Hayirli, A. *Berberis vulgaris* root extract alleviates the adverse effects of heat stress via modulating hepatic nuclear transcription factors in quails. *Br. J. Nutr.* **2013**, *110*, 609–616. [CrossRef]

112. Komal, S.; Ranjan, B.; Neelam, C.; Birendra, S.; Kumar, S.N. Berberis Aristata: A Review. *Int. J. Res. Ayurveda Pharm.* **2011**, *2*, 383–388.

113. Luo, A.; Fan, Y. Antioxidant activities of berberine hydrochloride. *J. Med. Plants Res.* **2011**, *5*, 3702–3707.

114. Pyrkosz-Biardzka, K.; Kucharska, A.Z.; Sokół-Łętowska, A.; Strugała, P.; Gabrielska, J. A comprehensive study on antioxidant properties of crude extracts from fruits of *Berberis vulgaris* L., *Cornus mas* L. and *Mahonia aquifolium* nutt. *Polish J. Food Nutr. Sci.* **2014**, *64*, 91–99. [CrossRef]

115. Sun, L.L.; Gao, W.; Zhang, M.M.; Li, C.; Wang, A.G.; Su, Y.L.; Ji, T.F. Composition and antioxidant activit y of the anthocyanins of the fruit of berberis heteropoda schrenk. *Molecules* **2014**, *19*, 19078–19096. [CrossRef] [PubMed]

116. Campisi, A.; Acquaviva, R.; Bonfanti, R.; Raciti, G.; Amodeo, A.; Mastrojeni, S.; Ragusa, S.; Iauk, L. Antioxidant Properties of *Berberis aetnensis* C. Presl (Berberidaceae) Roots Extract and Protective Effects on Astroglial Cell Cultures. *Sci. World J.* **2014**, *2014*, 315473. [CrossRef] [PubMed]

117. Gundogdu, M. Determination of antioxidant capacities and biochemical compounds of *Berberis vulgaris* L. Fruits. *Adv. Environ. Biol.* **2013**, *7*, 344–348.

118. Zaorsky, N.G.; Churilla, T.M.; Egleston, B.L.; Fisher, S.G.; Ridge, J.A.; Horwitz, E.M.; Meyer, J.E. Causes of death among cancer patients. *Ann. Oncol.* **2017**, *28*, 400–407. [CrossRef]

119. Hanachi, P.; Kua, S.H.; Asmah, R.; Motalleb, G.; Fauziah, O. Cytotoxic Effect of *Berberis vulgaris* Fruit Extract on the Proliferation of Human Liver Cancer Cell line (HepG2) and Its Antioxidant Properties. *Int. J. Cancer Res.* **2006**, *2*, 1–9.

120. Hoshyar, R.; Mahboob, Z.; Zarban, A. The antioxidant and chemical properties of *Berberis vulgaris* and its cytotoxic effect on human breast carcinoma cells. *Cytotechnology* **2016**, *68*, 1207–1213. [CrossRef]

121. Abd El-Wahab, A.E.; Ghareeb, D.A.; Sarhan, E.E.M.; Abu-Serie, M.M.; El Demellawy, M.A. In vitro biological assessment of berberis vulgaris and its active constituent, berberine: Antioxidants, anti-acetylcholinesterase, anti-diabetic and anticancer effects. *BMC Complement. Altern. Med.* **2013**, *13*, 218. [CrossRef]

122. El Khalki, L.; Tilaoui, M.; Jaafari, A.; Ait Mouse, H.; Zyad, A. Studies on the Dual Cytotoxicity and Antioxidant Properties of *Berberis vulgaris* Extracts and Its Main Constituent Berberine. *Adv. Pharmacol. Sci.* **2018**, *2018*, 3018498. [CrossRef]

123. Choi, M.S.; Oh, J.H.; Kim, S.M.; Jung, H.Y.; Yoo, H.S.; Lee, Y.M.; Moon, D.C.; Han, S.B.; Hong, J.T. Berberine inhibits p53-dependent cell growth through induction of apoptosis of prostate cancer cells. *Int. J. Oncol.* **2009**, *34*, 1221–1230.

124. Balakrishna, A.; Kumar, M.H. Evaluation of synergetic anticancer activity of berberine and curcumin on different models of A549, Hep-G2, MCF-7, Jurkat, and K562 cell lines. *Biomed Res. Int.* **2015**, *2015*, 354614. [CrossRef]

125. Ren, K.; Zhang, W.; Wu, G.; Ren, J.; Lu, H.; Li, Z.; Han, X. Synergistic anti-cancer effects of galangin and berberine through apoptosis induction and proliferation inhibition in oesophageal carcinoma cells. *Biomed. Pharmacother.* **2016**, *84*, 1748–1759. [CrossRef]

126. Sengupta, P.; Raman, S.; Chowdhury, R.; Lohitesh, K.; Saini, H.; Mukherjee, S.; Paul, A. Evaluation of Apoptosis and Autophagy Inducing Potential of Berberis aristata, Azadirachta indica, and Their Synergistic Combinations in Parental and Resistant Human Osteosarcoma Cells. *Front. Oncol.* **2017**, *7*, 296. [CrossRef]

127. Fukuda, K.; Hibiya, Y.; Mutoh, M.; Koshiji, M.; Akao, S.; Fujiwara, H. Inhibition by berberine of cyclooxygenase-2 transcriptional activity in human colon cancer cells. *J. Ethnopharmacol.* **1999**, *66*, 227–233. [CrossRef]

128. Safi, S.; Esseily, F.; El Ezzy, M.; Gali-Muhtasib, H.; Esseily, J.; Diab-Assaf, M.; Lampronti, I.; Saab, A. The ethanol fraction from the stem of Berberis libanotica inhibits the viability of adult T cell leukemia. *Minerva Biotecnol.* **2012**, *24*, 129–133.

129. El-Merahbi, R.; Liu, Y.N.; Eid, A.; Daoud, G.; Hosry, L.; Monzer, A.; Mouhieddine, T.H.; Hamade, A.; Najjar, F.; Abou-Kheir, W. Berberis libanotica ehrenb extract shows anti-neoplastic effects on prostate cancer stem/progenitor cells. *PLoS ONE* **2014**, *9*, e112453. [CrossRef]

130. Diab, S.; Ftdanzi, C.; Léger, D.Y.; Ghezali, L.; Millot, M.; Martin, F.; Azar, R.; Esseily, F.; Saab, A.; Sol, V.; et al. Berberis libanotica extract targets NF-κB/COX-2, PI3K/Akt and mitochondrial/caspase signalling to induce human erythroleukemia cell apoptosis. *Int. J. Oncol.* **2015**, *47*, 220–230. [CrossRef]

131. Du, H.P.; Shen, J.K.; Yang, M.; Wang, Y.G.; Yuan, X.G.; Ma, Q.L.; Jin, J. 4-Chlorobenzoyl berbamine induces apoptosis and G2/M cell cycle arrest through the PI3K/Akt and NF-κB signal pathway in lymphoma cells. *Oncol. Rep.* **2010**, *23*, 709–716.

132. Xie, J.; Ma, T.; Gu, Y.; Zhang, X.; Qiu, X.; Zhang, L.; Xu, R.; Yu, Y. Berbamine derivatives: A novel class of compounds for anti-leukemia activity. *Eur. J. Med. Chem.* **2009**, *44*, 3293–3298. [CrossRef]

133. Nam, S.; Xie, J.; Perkins, A.; Ma, Y.; Yang, F.; Wu, J.; Wang, Y.; Zhen Xu, R.; Huang, W.; Horne, D.A.; et al. Novel synthetic derivatives of the natural product berbamine inhibit Jak2/Stat3 signaling and induce apoptosis of human melanoma cells. *Mol. Oncol.* **2012**, *6*, 484–493. [CrossRef] [PubMed]

134. Yang, F.; Nam, S.; Brown, C.E.; Zhao, R.; Starr, R.; Horne, D.A.; Malkas, L.H.; Jove, R.; Hickey, R.J. A novel berbamine derivative inhibits cell viability and induces apoptosis in cancer stem-like cells of human glioblastoma, via up-regulation of miRNA-4284 and JNK/AP-1 signaling. *PLoS ONE* **2014**, *9*, e94443. [CrossRef] [PubMed]

135. Yang, F.; Nam, S.; Zhao, R.; Tian, Y.; Liu, L.; Horne, D.A.; Jove, R. A novel synthetic derivative of the natural product berbamine inhibits cell viability and induces apoptosis of human osteosarcoma cells, associated with activation of JNK/AP-1 signaling. *Cancer Biol. Ther.* **2013**, *14*, 1024–1031. [CrossRef] [PubMed]

136. Duan, H.; Luan, J.; Liu, Q.; Yagasaki, K.; Zhang, G. Suppression of human lung cancer cell growth and migration by berbamine. *Cytotechnology* **2010**, *62*, 341–348. [CrossRef]

137. Wu, J.; Yu, D.; Sun, H.; Zhang, Y.; Zhang, W.; Meng, F.; Du, X. Optimizing the extraction of anti-tumor alkaloids from the stem of *Berberis amurensis* by response surface methodology. *Ind. Crops Prod.* **2015**, *69*, 68–75. [CrossRef]

138. Bavand, R.; Nemati, F. Cytotoxic effect of the root extract of Berberis orthobotrys on hela cell line. *IIOAB J.* **2016**, *7*, 204–208.

139. Engel, N.; Ali, I.; Adamus, A.; Frank, M.; Dad, A.; Ali, S.; Nebe, B.; Atif, M.; Ismail, M.; Langer, P.; et al. Antitumor evaluation of two selected Pakistani plant extracts on human bone and breast cancer cell lines. *BMC Complement. Altern. Med.* **2016**, *16*, 1. [CrossRef]

140. Kim, K.H.; Choi, S.U.; Lee, K.R. Bioactivity-guided isolation of cytotoxic triterpenoids from the trunk of *Berberis koreana*. *Bioorganic Med. Chem. Lett.* **2010**, *20*, 1944–1947. [CrossRef]

141. Kim, K.H.; Choi, S.U.; Lee, K.R. Cytotoxic triterpenoids from Berberis koreana. *Planta Med.* **2012**, *78*, 86–89. [CrossRef]

142. Derosa, G.; Romano, D.; D'Angelo, A.; Maffioli, P. *Berberis aristata* combined with *Silybum marianum* on lipid profile in patients not tolerating statins at high doses. *Atherosclerosis* **2015**, *239*, 87–92. [CrossRef]

143. Derosa, G.; Romano, D.; D'Angelo, A.; Maffioli, P. Berberis aristata/Silybum marianum fixed combination (Berberol®) effects on lipid profile in dyslipidemic patients intolerant to statins at high dosages: A randomized, placebo-controlled, clinical trial. *Phytomedicine* **2015**, *22*, 231–237. [CrossRef]

144. Derosa, G.; D'Angelo, A.; Maffioli, P. The role of a fixed Berberis aristata/Silybum marianum combination in the treatment of type 1 diabetes mellitus. *Clin. Nutr.* **2016**, *35*, 1091–1095. [CrossRef]

145. Derosa, G.; D'Angelo, A.; Romano, D.; Maffioli, P. Effects of a Combination of *Berberis aristata*, *Silybum marianum* and Monacolin on Lipid Profile in Subjects at Low Cardiovascular Risk; A Double-Blind, Randomized, Placebo-Controlled Trial. *Int. J. Mol. Sci.* **2017**, *18*, 343.

146. Guarino, G.; Strollo, F.; Carbone, L.; Della Corte, T.; Letizia, M.; Marino, G.; Gentile, S. Bioimpedance analysis, metabolic effects and safety of the association Berberis aristata/Bilybum marianum: A 52-week double-blind, placebo-controlled study in obese patients with type 2 diabetes. *J. Biol. Regul. Homeost. Agents* **2017**, *31*, 495–502.

147. Di Pierro, F.; Putignano, P.; Villanova, N.; Montesi, L.; Moscatiello, S.; Marchesini, G. Preliminary study about the possible glycemic clinical advantage in using a fixed combination of Berberis aristata and Silybum marianum standardized extracts versus only Berberis aristata in patients with type 2 diabetes. *Clin. Pharmacol. Adv. Appl.* **2013**, *5*, 167–174. [CrossRef]

148. Zilaee, M.; Kermany, T.; Tavalaee, S.; Salehi, M.; Ghayour-Mobarhan, M.; Ferns, G.A. Barberry treatment reduces serum anti-heat shock protein 27 and 60 antibody titres and high-sensitivity c-reactive protein in patients with metabolic syndrome: A double-blind, randomized placebo-controlled trial. *Phytother. Res.* **2014**, *28*, 1211–1215. [CrossRef]

149. Brenyo, A.; Aktas, M.K. Review of complementary and alternative medical treatment of arrhythmias. *Am. J. Cardiol.* **2014**, *113*, 897–903. [CrossRef]

150. Guo, Y.; Chen, Y.; Tan, Z.R.; Klaassen, C.D.; Zhou, H.H. Repeated administration of berberine inhibits cytochromes P450 in humans. *Eur. J. Clin. Pharmacol.* **2012**, *68*, 213–217. [CrossRef]

151. Chen, C.; Tao, C.; Liu, Z.; Lu, M.; Pan, Q.; Zheng, L.; Li, Q.; Song, Z.; Fichna, J. A Randomized Clinical Trial of Berberine Hydrochloride in Patients with Diarrhea-Predominant Irritable Bowel Syndrome. *Phyther. Res.* **2015**, *29*, 1822–1827. [CrossRef]

152. Fouladi, R.F. Aqueous extract of dried fruit of berberis vulgaris L. in acne vulgaris, a clinical trial. *J. Diet. Suppl.* **2012**, *9*, 253–261. [CrossRef]

153. Shin, K.S.; Choi, H.S.; Zhao, T.T.; Suh, K.H.; Kwon, I.H.; Choi, S.O.; Lee, M.K. Neurotoxic effects of berberine on long-term l-DOPA administration in 6-hydroxydopamine-lesioned rat model of Parkinson's disease. *Arch. Pharm. Res.* **2013**, *36*, 759–767. [CrossRef]

154. Kwon, I.H.; Choi, H.S.; Shin, K.S.; Lee, B.K.; Lee, C.K.; Hwang, B.Y.; Lim, S.C.; Lee, M.K. Effects of berberine on 6-hydroxydopamine-induced neurotoxicity in PC12 cells and a rat model of Parkinson's disease. *Neurosci. Lett.* **2010**, *486*, 29–33. [CrossRef]

155. Qian, C.; Zhu, F. Berberine chloride can ameliorate the spatial memory impairment and increase the expression of interleukin-1beta and inducible nitric oxide synthase in the rat model of Alzheimer's disease. *BioMed* **2006**, *3*, 1–9.

156. Ahmed, T.; Gilani, A.U.H.; Abdollahi, M.; Daglia, M.; Nabavi, S.F.; Nabavi, S.M. Berberine and neurodegeneration: A review of literature. *Pharmacol. Rep.* **2015**, *67*, 970–979. [CrossRef]

 *foods*

Review

Targeting Gut Microbiota for the Prevention and Management of Diabetes Mellitus by Dietary Natural Products

Bang-Yan Li [1], Xiao-Yu Xu [1], Ren-You Gan [2,3,*], Quan-Cai Sun [4], Jin-Ming Meng [1], Ao Shang [1], Qian-Qian Mao [1] and Hua-Bin Li [1,*]

[1]   Guangdong Provincial Key Laboratory of Food, Nutrition and Health, Department of Nutrition, School of Public Health, Sun Yat-sen University, Guangzhou 510080, China; liby35@mail2.sysu.edu.cn (B.-Y.L.); xuxy53@mail2.sysu.edu.cn (X.-Y.X.); mengjm@mail2.sysu.edu.cn (J.-M.M.); shangao@mail2.sysu.edu.cn (A.S.); maoqq@mail2.sysu.edu.cn (Q.-Q.M.)
[2]   Institute of Urban Agriculture, Chinese Academy of Agricultural Sciences, Chengdu 610213, China
[3]   Department of Food Science & Technology, School of Agriculture and Biology, Shanghai Jiao Tong University, Shanghai 200240, China
[4]   School of Food and Biological Engineering, Jiangsu University, Zhenjiang 212013, China; sqctp8@ujs.edu.cn
*   Correspondence: ganrenyou@caas.cn (R.-Y.G.); lihuabin@mail.sysu.edu.cn (H.-B.L.); Tel.: +86-28-8020-3191 (R.-Y.G.); +86-20-873-323-91 (H.-B.L.)

Received: 17 August 2019; Accepted: 23 September 2019; Published: 25 September 2019

Abstract: Diabetes mellitus is one of the biggest public health concerns worldwide, which includes type 1 diabetes mellitus, type 2 diabetes mellitus, gestational diabetes mellitus, and other rare forms of diabetes mellitus. Accumulating evidence has revealed that intestinal microbiota is closely associated with the initiation and progression of diabetes mellitus. In addition, various dietary natural products and their bioactive components have exhibited anti-diabetic activity by modulating intestinal microbiota. This review addresses the relationship between gut microbiota and diabetes mellitus, and discusses the effects of natural products on diabetes mellitus and its complications by modulating gut microbiota, with special attention paid to the mechanisms of action. It is hoped that this review paper can be helpful for better understanding of the relationships among natural products, gut microbiota, and diabetes mellitus.

Keywords: gut microbiota; natural products; diabetes mellitus; complications; mechanisms

1. Introduction

Over the past 20 years, the prevalence of diabetes mellitus (DM) and its complications has rapidly increased across the world, which poses a serious threat to global health [1–3]. The international diabetes federation (IDF) estimated that one in eleven adults aged 20 to 79 (415 million) have DM globally in 2015, and predicts that the number can rise to 642 million by 2040 [4]. DM is a complex metabolic disorder characterized with the functional impairment or a lack of insulin-producing β-cells, alone or in combination with insulin resistance [5]. DM has several types, including type 1 diabetes mellitus (T1DM), type 2 diabetes mellitus (T2DM), gestational diabetes mellitus (GDM), and other rare forms of diabetes mellitus [6,7]. T1DM is an autoimmune disease that is associated with the aberrant immune responses to specific β-cell autoantigens, which causes the reduction of insulin [8–11]. T2DM is due to insulin resistance in the target tissue and a relative lack of insulin secreted by islet β-cells, and is mainly affected by the combination of genetic susceptibility and lifestyle factors [12–15]. GDM is a kind of diabetes during pregnancy due to the increased severity of insulin resistance and an impairment of the compensatory increase in insulin secretion [16–18]. Other rare forms of diabetes mellitus, such as maturity-onset of diabetes of the young (MODY) resulted from mutations in a single gene. In addition, maternally inherited diabetes with deafness (MIDD) is caused by mitochondrial mutations, and rare forms resulted from insulin gene mutations [19,20]. In this review paper, only T2DM and T1DM will be discussed, because T2DM and T1DM are the main types of diabetes mellitus, accounting for 90%–95% and 5%–10%, respectively [4,21,22]. In addition, T2DM can induce various complications, which contributes to high morbidity and mortality, such as diabetic cardiovascular diseases, nephropathy, retinopathy, neuropathy, and erectile dysfunctions [23,24]. Furthermore, many studies have shown that some natural products and their bioactive components could prevent and manage diabetes mellitus and its complications by several possible mechanisms, such as enhancing insulin action, ameliorating insulin resistance, activating insulin signaling pathway, protecting islet β-cells, scavenging free radicals, decreasing inflammation, and modulating gut microbiota [25–29]. In recent years, the role of gut microbiota in the prevention and management of diabetes mellitus and its complications has been a hot topic in research [14,30–32]. Special attention will be paid to the role of gut microbiota in this paper.

Gut microbiota has been recently reported to be involved in the pathogenesis of several diseases, such as obesity [33–37], liver diseases [38–41], cardiovascular diseases [42–44], cancer [45,46], and DM [47–50]. The alteration of intestinal microbiota and the production of metabolites have been demonstrated to play a critical role in the initiation and development of DM [51–54]. Some bacteria like *Saccharomyces boulardii* Biocodex, *Bifidobacterium animalis* subsp. lactis 420, *Lactobacillus casei* CCFM419, and *Clostridium butyricum* CGMCC0313.1 have been shown to ameliorate DM [55–59]. However, it was reported that some bacteria, like *Bacteroides* and *Candida albicans*, promoted the incidence and progression of T2DM in animal studies [60,61]. On the other hand, natural products have attracted wide attention in recent years due to their diverse bioactivities, such as cardiovascular protective, hepatoprotective, anti-cancer, anti-obesity, and anti-diabetic effects [25,36,62–71]. Furthermore, natural products are also important factors in regulating gut microbiota [72–74]. A variety of dietary natural products and their bioactive components have shown anti-diabetic effects by regulating gut microbiota composition and abundance, the change of gut permeability, the production of short-chain fatty acids (SCFAs), the decrease of lipopolysaccharides (LPS), and the inhibition of inflammation [58,75–80]. In addition, natural products can be effective adjuvant agents in the therapy of T2DM with lower cost and less side effects [81,82]. In this review paper, we first address the relationship between gut microbiota and T1DM as well as T2DM based on animal and epidemiological studies, and then discuss the effects of natural products on T2DM and its complications via modulation of gut microbiota, based on the literature from animal and clinical studies in the last five years (2015–2019).

2. The Association Between Gut Microbiota and DM Based on Animal and Epidemiological Studies

Many animal studies have revealed that the gut microbiota plays an important role in the incidence and development of T1DM as well as T2DM [54,75,83,84]. In addition, increasing investigations have found that gut microbiota is closely associated with T1DM and T2DM patients [85–87], and they will be discussed in detail below.

2.1. The Association between Gut Microbiota and T1DM

The animal studies have shown that T1DM has been associated with the composition of the gut microbiota [88]. For example, the abundance of *Proteobacteria* showed a more marked change than that of the predominant phyla, such as *Firmicutes* and *Bacteroidetes* compared with the healthy control in the rat model of T1DM induced by streptozotocin [89]. In another study, the gut microbiota from non-obese diabetic (NOD) mice without myeloid differentiation primary response gene 88 (MyD88) was transferred to wild-type NOD mice, which reduced the intensity of insulitis and delayed the onset of autoimmune diabetes in recipients, by decreasing the amount of *Lactobacillaceae* and increasing the number of *Lachnospiraceae* and *Clostridiaceae* [59]. Moreover, the deficiency of TIR-domain-containing adapter-inducing interferon-β in NOD mice could prevent diabetes via the change in the gut microbiota, such as the proportion of *Sutterella*, *Rikenella*, and *Turicibacter* species [90]. In addition, the mouse β-defensin 14 (mBD14) in pancreatic endocrine cells in NOD mice could prevent autoimmune diabetes, while the microbiota dysbiosis could induce the deficiency of the pancreatic expression of mBD14, which results in the higher incidence of T1DM [91].

The relationship between gut microbiota and T1DM has been summarized in a review paper published in 2015 [54]. In addition, the epidemiological studies have reported that some intestinal microbiota like phylum *Bacteroidetes*, genus *Clostridium*, *Bacteroides*, and *Veillonella* as well as *Candida albicans* increased in T1DM patients, while some other gut microbiota like phylum *Actinobacteria* and *Firmicutes*, genus *Lactobacillus*, *Bifidobacterium*, and *Faecalibacterium* as well as the ratio of *Firmicutes* to *Bacteroidetes* decreased in T1DM patients [92–96].

2.2. The Association between Gut Microbiota and T2DM

The animal studies have found that there have been prominent differences in the composition of gut microbiota between T2DM animal models and control subjects. For example, an *in vivo* study found that the diabetic *db/db* mice had higher abundance of phyla *Firmicutes*, *Proteobacteria*, and *Fibrobacteres* [97]. Additionally, the streptozotocin-induced T2DM mice showed an increase in *Brevibacterium*, *Corynebacterium*, and *Facklamia* compared with wild type mice [98]. A decrease in the diversity of gut microbiota and a lack of butyrate-producing bacteria were observed in T2DM cats [99]. The level of serum fructosamine was inversely associated with the abundance of *Prevotellaceae*, and was positively associated with the abundance of *Enterobacteriaceae*. In addition, the transfer of *Lachnospiraceae* (strain AJ110941) from the feces of hyperglycemic obese mice to germ-free *ob/ob* mice contributed to the development of T2DM in *ob/ob* mice. The colonization of *Lachnospiraceae* could induce an increase in the level of fasting blood glucose (FBG) and a decrease in the level of plasma insulin as well as homeostasis model assessment-β (HOMA-β). The decrease in HOMA-β values indicated the dysfunction of pancreatic β-cells [100].

The epidemiological studies have also investigated the relationship between gut microbiota and T2DM patients [101–104]. For example, a case-control study has shown that the proportions of phylum *Firmicutes* and class *Clostridia* as well as the ratio of *Firmicutes* to *Bacteroidetes* decreased, whereas the abundance of class *Betaproteobacteria* increased in T2DM patients [101]. In addition, the concentration of *Faecalibacterium prausnitzii* and the abundance of genus *Blautia* was reduced in T2DM patients [102,103]. Moreover, a case-control study demonstrated that T2DM and obese patients had a lower quantity of *Lactobacillus*, especially its subgroups (*L. acidophilus*, *L. plantarum*, and *L. reuteri*) in comparison with controls [104].

Overall, several animal studies have revealed that the intestinal microbiota play an important role in T1DM and T2DM. In addition, many epidemiological investigations have demonstrated the closed association between gut microbiota and T2DM (Table 1). Furthermore, there are few prospective studies on the relationship between gut microbiota and DM, which should be further investigated in the future.

Table 1. The epidemiological studies on the association between gut microbiota and T2DM.

Study Types	Participants	Alterations in Gut Microbiota Composition	Reference
Case-control study	Adults with T2DM ($n = 18$) Healthy male adults ($n = 18$)	↓ The proportions of phylum *Firmicutes* and class *Clostridia*, and the ratios of *Firmicutes* to *Bacteroidetes* ↑ The proportion of *betaproteobacteria*	[101]
Case-control study	Patients with T2DM ($n = 18$) Healthy individuals ($n = 18$)	↓ The concentration of *Faecalibacterium prausnitzii*	[102]
Case-control study	Patients with T2DM ($n = 10$) Healthy individuals ($n = 12$)	↓ The abundance of genus *Blautia*	[103]
Case-control study	Patients with T2DM ($n = 100$) Healthy individuals ($n = 100$)	↓ The counts of *Lactobacillus* sp. (*L. acidophilus*, *L. plantarum*, and *L. reuteri*) and *Bifidobacterium*	[104]

Abbreviations: T2DM, Type 2 diabetes mellitus. ↓, Decrease. ↑, Increase.

3. The Relationships among Natural Products, Gut Microbiota, and T2DM as well as T1DM Based on Animal Studies

Some natural products, such as vegetables, fruits, dietary fibers, and medicinal plants, have shown potential preventive effects against T2DM and T1DM as well as diabetic complications with mechanisms of action, at least partly, via the modulation of gut microbiota [80,105–108].

3.1. Diabetes Mellitus

The extracts of vegetables and their bioactive components have been demonstrated to alleviate T2DM by modulating the gut microbiota [75,79,80]. Capsaicin could improve the glucose homeostasis and insulin tolerance in obese diabetic *ob/ob* mice by increasing the production of SCFAs and modulating gut microbiota. It could increase the ratio of *Firmicutes* to *Bacteroidetes* and the quantity of *Roseburia*, and reduce the quantities of *Bacteroides* and *Parabacteroides* at the genus level, which could decrease the levels of pro-inflammatory cytokines, such as TNF-α and IL-6 [79]. Additionally, pumpkin polysaccharide alleviated the T2DM in mice fed with a high-fat diet (HFD). Its antidiabetic effects were also related to the increase in SCFAs production and selective enhancement of some bacteria, such as *Bacteroidetes*, *Prevotella*, and *Deltaproteobacteria* [75]. Moreover, the carrot juice fermented with *Lactobacillus rhamnosus* GG was demonstrated to ameliorate T2DM in rats via the increase in SCFAs in the cecum and the change in the composition and abundance of gut microbiota, such as *Oscillibacter* and *Akkermansia* [80]. In addition, the anti-diabetic activity of *Momordica charantia* (bitter melon) was enhanced by the fermentation of *Lactobacillus*, which could induce an increase in the abundance of *Bacteroides caecigallinarwn*, *Bacteroides thetaiotaomicron*, *Prevotella loescheii*, *Prevotella oralis*, and *Prevotella melaninogenica* as well as the level of SCFAs [109].

It has also been reported that the extracts of several fruits and their bioactive components could improve T2DM via gut microbiota [105,110,111]. The phlorizin in many fruits could competitively inhibit the sodium-glucose symporters and regulate the level of blood glucose by decreasing the level of serum LPS and insulin resistance, and increasing the level of SCFAs, especially butyric acid. It could also increase the quantity of *Akkermansia muciniphila* and *Prevotella*, which modulates the gut microbial community structure [110]. In addition, polysaccharides from the mulberry fruit have shown protective effects against T2DM via gut microbiota in *db/db* mice, and the treatment with polysaccharides could enrich the functional bacteria, such as *Bacteroidales*, *Lactobacillus*, *Allobaculum*, *Bacteroides*, and *Akkermansia* [105]. The extracts from cinnamon bark were reported to improve glucose tolerance and insulin resistance by reducing the abundance of genus *Peptococcus*, and the extracts from grape pomace could reduce *Desulfovibrio* and *Lactococcus*, and increase *Allobaculum* and *Roseburia* in diabetic mice [111].

The inulin was also found to alleviate different stages of T2DM in diabetic mice by modulating gut microbiota. It increased the relative abundance of *Cyanobacteria* and *Bacteroides*, and reduced the relative abundance of *Deferribacteres* and *Tenericutes* [112]. In another study, the oral administration with inulin-type fructan could reduce the FBG level, increase the glucagon-like peptide-1 (GLP-1) level, and alleviate glucose intolerance as well as blood lipid in T2DM rats induced by HFD and streptozotocin [108]. The treatment with inulin also increased the level of GLP-1 and improved the gut microbiota dysbiosis by enriching probiotic bacteria *Lactobacillus* and SCFAs-producing bacteria *Lachnospiraceae*, *Phascolarctobacterium*, and *Bacteroides*. Furthermore, the supplementation of the long-chain inulin-type fructan fibers from chicory root to female NOD mice could delay the progression of T1DM via the modulation of gut microbiota. This fiber could increase the ratio of *Firmicutes* to *Bacteroidetes* and the abundance of *Ruminococcaceae* and *Lactobacilli*, which might contribute to the increase in the expression of tight junction proteins occludin and claudin-2 and an antidiabetogenic effect [106].

Some traditional Chinese medicine have been found to be effective in preventing and treating T2DM. The administration of ethanol extract of *Atractylodis macrocephalae* rhizoma improved the glucose metabolism by regulating the gut microbiota in diabetic *db/db* mice, and it could increase the abundance of *Bacteroides thetaiotaomicron* and *Methanobrevibacter smithii* [113]. In addition, the extract of *Alpinia oxyphylla* Miq. had protective effects against diabetes in T2DM *db/db* mice. It decreased the blood glucose levels by modulating the intestinal microbiota composition, which increased the abundance of *Akkermansia* and decreased *Helicobacter* [114]. The water extract of *Potentilla discolor* Bunge could alleviate diabetes, which might be related to the change in the gut microbiota. The treatment reduced the relative abundance of *Proteobacteria* in T2DM mice induced by HFD and streptozotocin [77]. It was related to an increase in levels of fecal acetic acid, butyric acid, and their specific receptors including the expression of G-protein-coupled receptor 41 (GPR41) and G-protein-coupled receptor 43 (GPR43), which increased the insulin sensitivity and decreased the fat accumulation. Additionally, the expressions of toll-like receptor-4 (TLR4), MyD88, nuclear factor-kappa B (NF-κB), and gut mucosal tight junction proteins (caudin-3 and occludin) were decreased, which could reduce the gut permeability and inflammation. In addition, the oral administration of total saponins and polysaccharides in *Polygonatum kingianum* could prevent T2DM by regulating the gut microbiota [115]. The treatment improved micro-ecology in the gut by reducing the abundance of *Bacteroidetes* and *Proteobacteria*, and increasing *Firmicutes*. Furthermore, the baicalein showed anti-diabetic effects on diabetic rats, which reduced the levels of blood glucose, LPS, and insulin resistance. The treatment of baicalein could regulate the gut microbiota by increasing the relative abundance of *Bacteroides* and *Bacteroidales* S24-7 [78]. It could also increase the production of SCFAs as well as the thickness of the gut mucus layer.

Overall, many studies have focused on the effects of different natural products, such as vegetables, fruits, dietary fibers, and medicinal plants, on T2DM by regulating gut microbiota and improving microecology in the gut. On the other hand, there are few studies about the effect of natural products on T1DM by modulating gut microbiota. In the future, more studies are needed to investigate the effects of different natural products and bioactive components on T1DM via gut microbiota-related mechanisms.

3.2. Diabetic Complications

Long-term diabetes increases the likelihood of its complications like diabetic erectile dysfunctions, nephropathy, cardiovascular diseases, retinopathy, and neuropathy [116,117]. The gut microbiota has also been found to have a close relationship with the development of the diabetic complications. In the rat models with T2DM erectile dysfunction, there was a reduction in the relative abundance of beneficial bacteria like *Allobaculum*, *Bifidobacterium*, *Eubacterium*, and *Anaerotruncus*. In addition, the relative abundance of opportunistic pathogens increased, such as *Enterococcus*, *Corynebacterium*, *Aerococcus*, and *Facklamia* [118]. Some natural products have been reported to have protective effects on diabetic nephropathy or kidney injury in diabetic mice by modulating the gut microbiota balance [107,119]. For example, the polyphenolic extract of *Dendrobium loddigesii* could improve the symptoms of diabetes and complications in diabetic *db/db* mice. These effects were likely related to the improvement of the intestinal flora balance, with an increase in the relative abundance of *Prevotella* and *Akkermansia*, and the reduction in the relative abundance of S24-7/*Rikenella*/*Escherichia coli* [119]. Additionally, the treatment with the water-ethanolic extract of green macroalgae *Enteromorpha prolifera* reduced the inflammation in the liver and kidney by significantly increasing the abundance of *Lachnospiraceae* and *Alisties* as well as regulating the insulin signaling pathway in T2DM mice induced by HFD and high sucrose diet (HSD) and streptozotocin [107].

To sum up, several studies have demonstrated that the gut microbiota has a tight association with diabetic complications. The increase or decrease in different types of gut microbiota could exert different effects on the progression of diabetic complications. Moreover, several natural products have showed potential efficacy on the prevention and management against diabetic complications, and these effects were mainly associated with the modulation of gut microbiota composition and abundance, regulation of the insulin signaling pathway, and inhibition of inflammation.

Lastly, the relationship among gut microbiota, T2DM, and its complications is shown in Figure 1, and the association among natural products, gut microbiota, T2DM, and its complications is given in Table 2 and Figure 2.

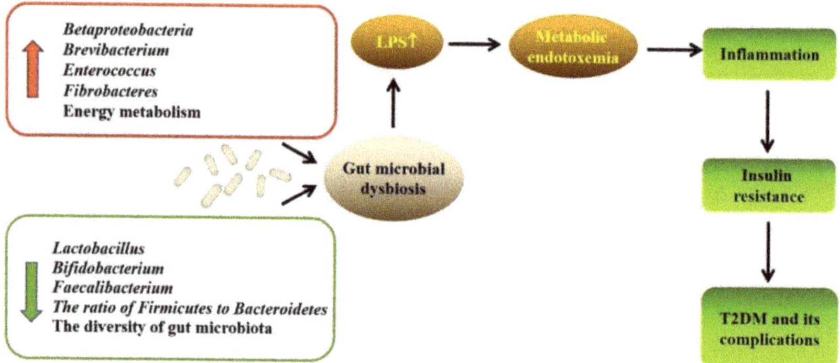

Figure 1. The association among gut microbiota, T2DM, and its complications. The changes of gut microbiota caused the increase in LPS, which could cause inflammation and insulin resistance. It indicates that gut microbiota would play an important role in the initiation and development of T2DM and its complications. Abbreviations: LPS, lipopolysaccharides; T2DM, type 2 diabetes mellitus.

Table 2. The animal studies of natural products on DM and its complications by modulating gut microbiota.

Natural Products	Disease	Study Types	Models	Effects	Mechanisms	Reference
Long-chain inulin-type fructans fibers	T1DM	*In vivo*	NOD diabetic mice	Promoting modulatory T-cell responses. Delaying the development of T1DM.	↑ The ratio of *Firmicutes* to *Bacteroidetes* ↑ The abundance of *Ruminococcaceae* and *Lactobacilli* ↑ The production of SCFAs ↑ The expression of tight junction proteins occludin and claudin-2	[106]
Capsaicin	T2DM	*In vivo*	Obese T2DM *ob/ob* mice	Inhibiting the levels of FBG and insulin. Improving the glucose homeostasis and insulin tolerance.	↑ The ratio of *Firmicutes* to *Bacteroidetes* ↑ The abundance of *Roseburia* ↑ The production of butyrate ↓ The quantities of *Bacteroides* and *Parabacteroides* ↓ The levels of TNF-α and IL-6	[79]
Pumpkin polysaccharide	T2DM	*In vivo*	T2DM rats	Improving insulin tolerance. Decreasing the levels of glucose, TC, LDL-C. Increasing the level of HDL-C.	↑ The quantities of some bacteria, such as *Bacteroidetes*, *Prevotella*, *Deltaproteobacteria*, *Oscillospira*, *Veillonellaceae*, *Phascolarctobacterium*, *Sutterella*, and *Bilophila* ↑ The production of SCFAs	[75]
Fermented carrot juice	T2DM	*In vivo*	T2DM rats	Regulating the levels of blood glucose and insulin as well as the morphology of the pancreas and kidney.	↑ The quantities of *Christensenellaceae*, *Oscillibacter*, *Ruminococcaceae*, *Lachnospiraceae*, and *Akkermansia* ↑ The level of SCFAs	[80]
Fermented *Momordica charantia* juice	T2DM	*In vivo*	T2DM rats induced by HFD and STZ	Relieving the hyperglycemia, hyperinsulinemia, and hyperlipidemia.	↑ The abundance of *Bacteroides caecigallinarum*, *Bacteroides thetaiotaomicron*, *Prevotella loescheii*, *Prevotella oralis*, and *Prevotella melaninogenica* ↑ The concentrations of acetic acid, propionic acid, and butyric acid	[109]
Phlorizin	T2DM	*In vivo*	T2DM *db/db* mice	Reducing insulin resistance. Regulating the level of blood glucose.	↑ The abundance of *Akkermansia muciniphila* and *Prevotella* ↑ The gut microbial diversity ↑ The production of butyric acid ↓ The level of LPS	[110]

Table 2. *Cont.*

Natural Products	Disease	Study Types	Models	Effects	Mechanisms	Reference
Mulberry fruit polysaccharide	T2DM	*In vivo*	T2DM *db/db* mice	Improving glucose tolerance. Increasing the level of HDL-C. Decreasing the levels of TC TG, LDL-C, and FFA. Inhibiting body weight gain.	↑ The abundance of *Bacteroidales*, *Lactobacillus*, *Allobaculum*, *Bacteroides*, and *Akkermansia* ↓ The gut microbial diversity	[105]
Cinnamon bark and grape pomace extracts	T2DM	*In vivo*	T2DM C57BL/6J mice induced by HFD	Improving glucose tolerance and insulin resistance. Decreasing fat mass gain and adipose tissue inflammation.	↑ The abundances of *Allobaculum* and *Roseburia*, ↑ The expression of tight junction proteins. ↓ The abundance of genus *Peptococcus*, *Desulfovibrio*, and *Lactococcus*	[111]
Inulin	T2DM	*In vivo*	Mice	Decreasing the levels of FBG and glycated hemoglobin, and body weight.	↑ The relative abundance of *Cyanobacteria* and *Bacteroides* ↓ The relative abundance of *Deferribacteres* and *Tenericutes*	[112]
Inulin-type fructan	T2DM	*In vivo*	T2DM rats induced by HFD/STZ	Reducing the levels of FBG, IL-6, and alleviated glucose intolerance.	↑ The abundance of *Lactobacillus* and SCFAs-producing bacteria, such as *Lachnospiraceae*, *Phascolarctobacterium*, and *Bacteroides* ↑ The level of GLP-1	[108]
Atractylodis macrocephalae Rhizoma ethanol extract	T2DM	*In vivo*	T2DM *db/db* mice	Decreasing the levels of blood glucose, TG, TC, endotoxin, and IL-10	↑ The abundance of *Bacteroides thetaiotaomicron* and *Methanobrevibacter smithii*	[113]
Alpinia oxyphylla Miq. extract	T2DM	*In vivo*	T2DM *db/db* mice	Improving glycemic control and renal function	↑ The ratio of *Bacteroidetes* to *Firmicutes* and ↑ The abundance of *Akkermansia* ↓ The abundance of *Helicobacter*	[114]
Potentilla discolor Bunge water extract	T2DM	*In vivo*	T2DM C57BL/6J mice induced by HFD and STZ	Decreasing the level of pro-inflammatory cytokines. Improving inflammation.	↑ The concentrations of fecal acetic acid butyric acid and their specific receptors including the expression of GPR41 and GPR43 ↓ The ratio of *Firmicutes* to *Bacteroidetes* and ↓ The abundance of *Proteobacteria* ↓ The expression of gut mucosal tight junction proteins (Claudin3, ZO-1, and Occludin), TLR4, MyD88, and NF-κB	[77]

Table 2. *Cont.*

Natural Products	Disease	Study Types	Models	Effects	Mechanisms	Reference
Polygonatum kingianu	T2DM	*In vivo*	T2DM SD rats induced by HFD and STZ	Increasing the level of fasting insulin. Preventing the increase of FBG. Improving the intestinal microecology.	↑ The abundance of *Firmicutes* ↓The abundances of *Bacteroidetes* and *Proteobacteria*	[115]
Baicalein	T2DM	*In vivo*	Diabetic rats induced by HFD, HSD, and STZ	Decreasing the level of blood glucose. Improved inflammation and lipid metabolism.	↑ The relative abundances of *Bacteroides* and *Bacteroidales* S24-7 ↑ The production of SCFAs ↑ The thickness of the gut mucus layer	[78]
Dendrobium loddigesii polyphenols extract	Diabetic nephropathy	*In vivo*	Diabetic nephropathy *db/db* mice	Improving diabetic nephropathy. Decreasing blood glucose level. Increasing the level of insulin.	↑ The relative abundance of *Prevotella/Akkermansia* ↓ The relative abundance of S24-7/*Rikenella/Escherichia coli*	[119]
Green seaweed *Enteromorpha prolifera* flavonoids	T2DM	*In vivo*	T2DM mice induced by HFD and HSD and STZ	Decreasing the level of FBG. Improving glucose tolerance. Reducing inflammation. Preventing liver and kidney injury.	↑ The abundances of *Lachnospiraceae* and *Alisties* ↑ The IRS1/PI3K/AKT pathway ↓ The JNK1/2 insulin pathway in liver	[107]

Abbreviations: DM, diabetes mellitus. T1DM, type 1 diabetes mellitus. T2DM, type 2 diabetes mellitus. NOD mice, non-obese diabetic mice. HFD, high-fat diet. HSD, high sucrose diet. STZ, streptozotocin. LPS, lipopolysaccharides. FBG, fasting blood glucose. GLP-1, glucagon-like peptide-1. TG, triglyceride. HOMA-IR, homeostasis model assessment of insulin resistance. TC, total cholesterol. FFA, free fatty acid. SCFAs, short-chain fatty acids. IL-6, interleukin 6.IL-10, interleukin 10. TLR4, toll-like receptor-4. TNF-α, tumor necrosis factor-α. MyD88, myeloid differentiation primary response gene 88. HDL-C, high-density lipoprotein cholesterol. LDL-C, low-density lipoprotein cholesterol. NF-κB, nuclear factor-kappa B. GPR41/43, G-protein-coupled receptor 41/43. IRS1, insulin receptor substrate 1. PI3K, phosphatidylinositol-3-kinase. AKT, protein kinase B. JNK, Jun N-terminal kinase.

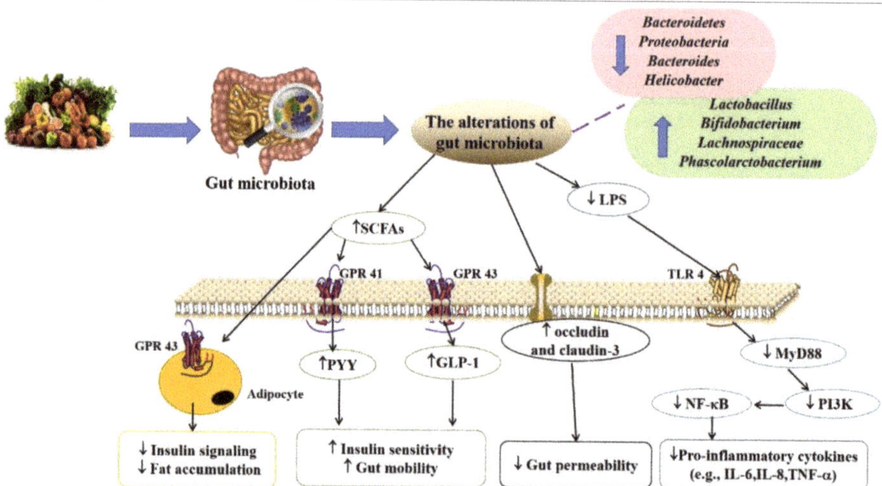

Figure 2. The relationship among natural products, gut microbiota, T2DM, and its complications. Some natural products and their bioactive components alleviate T2DM by changing the composition and abundance of gut microbiota. It decreases the permeability of gut and the level of LPS, increases the production of SCFAs, and inhibits the inflammation. The arrows out of the box indicates the direction of action, and the up arrows in the box mean the upregulation or activation, and the down arrows in the box mean the downregulation or inhibition. Abbreviations: LPS, lipopolysaccharides. SCFAs, short-chain fatty acids. T2DM, type 2 diabetes mellitus. GPR41/43, G-protein-coupled receptor41/43. TLR4, toll-like receptor-4. GLP-1, glucagon-like peptide-1. PYY, peptide YY. MyD88, myeloid differentiation primary response gene 88. PI3K, phosphatidylinositol-3-kinase. NF-κB, nuclear factor-kappa B. IL-6, interleukin 6. IL-8, interleukin 8. TNF-α, tumor necrosis factor-α.

4. The Relationships among Natural Products, Gut Microbiota, and T2DM Based on Clinical Studies

Several clinical trials have evaluated the protective effects of natural products on T2DM via gut microbiota [120,121]. A double-blind, randomized, controlled clinical trial (RCT) involving 60 patients with T2DM found that the supplementation of 10 g/day inulin powder promoted the gut health by increasing the proportion of *Akkermansia muciniphila* [120]. Moreover, an RCT study found that the fiber-rich macrobiotic Ma-Pi 2 diet could regulate gut microbiome dysbiosis in obese T2DM patients by increasing the gut microbiota ecosystem diversity, recovering the SCFAs-producing bacteria community, such as *Roseburia*, *Lachnospira*, *Faecalibacterium*, *Bacteroides*, and *Akkermansia*, and inhibiting the increase of the pro-inflammatory group, such as *Collinsella* and *Streptococcus* [122]. Furthermore, an RCT study including 100 patients with T2DM reported that a Chinese herbal formula could ameliorate T2DM and improved the glucose and lipid homeostasis by increasing the abundance of *Faecalibacterium* spp. [116]. Another RCT study observed that the treatment with Gegen Qinlian decoction, which is a type of Chinese herbal formula, at the moderate and high doses, significantly reduced the mean alterations in adjusted FBG and glycated hemoglobin A1c (HbA1c) levels by enriching the beneficial bacteria, such as *Faecalibacterium prausnitzii* [121].

To sum up, several clinical trials demonstrate that natural products can ameliorate T2DM by regulating gut microbiota (Table 3). Some natural products, such as inulin and medicine plants, are able to enrich certain beneficial bacteria in T2DM patients, such as *Akkermansia muciniphila*, *Faecalibacterium*, and *Lachnospira*.

Table 3. The clinical trials of natural products on T2DM by modulating gut microbiota.

Natural Products	Study Types	Participants	Dose	Duration	Results	Reference
Inulin	RCT	T2DM patients (*n* = 60)	10 g/day	N/A	↑ The proportion of *Akkermansia muciniphila*	[120]
Macrobiotic Ma-Pi 2 diet enriched fiber	RCT	Obese T2DM patients (*n* = 56)	N/A	N/A	↑ Gut microbiota ecosystem diversity ↑ SCFA-producing bacteria, such as *Faecalibacterium, Roseburia, Lachnospira, Bacteroides* and *Akkermansia* ↓ Pro-inflammatory bacteria, such as *Collinsella* and *Streptococcus*	[122]
Chinese herbal formula	RCT	T2DM patients (*n* = 100)	N/A	12 weeks	↑ The abundance of *Faecalibacterium spp* ↓ Hyperglycemia and hyperlipidemia	[123]
Ge gen Qin lian decoction	RCT	T2DM patients (*n* = 187)	N/A	12 weeks	↑ The number of *Faecalibacterium prausnitzii* ↓ The mean changes of FBG and HbA1c levels	[121]

Abbreviations: N/A, not available. RCT, randomized control clinical trial. FBG, fasting blood glucose. T2DM, type 2 diabetes mellitus. HbA1c, hemoglobin A1c. SCFA, short-chain fatty acids.

5. Conclusions

In conclusion, the relationship among diabetes mellitus (including T1DM and T2DM), gut microbiota, and natural products has been summarized and discussed. Animal and epidemiological studies found significant differences in intestinal microbiota composition and abundance between diabetic patients and healthy controls. The abundance of *Bifidobacterium* and *Lactobacillus* as well as the ratio of *Firmicutes* to *Bacteroidetes* decreased, while the abundance of *Bacteroidetes* and *Proteobacteria* increased in T1DM and T2DM patients. However, the consequences were not always consistent due to other factors like types of diets and progression of diseases. In addition, animal studies have demonstrated that gut microbiota is one of the most important factors for diabetes initiation and development. Several natural products possessing prebiotic effects like fruits, vegetables, and medicinal plants, have been found to ameliorate T2DM by modulating gut microbiota composition and abundance, reducing the gut permeability, increasing the production of SCFAs, decreasing the level of LPS, and inhibiting the inflammation. Moreover, clinical trials have further confirmed that several natural products are effective in preventing and treating T2DM, with regulation of gut microbiota as one of the potential mechanisms. In the future, it is necessary to explore the effects of more natural products and their bioactive components on DM via gut microbiota-related mechanisms. In addition, more well-designed clinical trials on natural products and their various bioactive compounds should be carried out to verify their effects on T2DM and its complications by regulating gut microbiota. Furthermore, functional foods based on natural products can be researched and developed by targeting gut microbiota for the prevention and management of T2DM. On the other hand, present studies mainly focus on modulating the action of natural products and their bioactive components on gut microbiota for preventing and managing T2DM. However, seldom studies have been carried out about the metabolism of natural products and their bioactive components by gut microbiota, even though the metabolites could directly play a role in the prevention and management of DM or, in return, could modulate gut microbiota. Thus, metabolisms of natural products and their bioactive components by gut microbiota should be widely studied in the future for the prevention and management of DM.

Author Contributions: Conceptualization, B.-Y.L., R.-Y.G., and H.-B.L. Writing—original draft preparation, B.-Y.L., X.-Y.X., J.-M.M., A.S., and Q.-Q.M. Writing—review and editing, R.-Y.G., Q.-C.S., and H.-B.L. Supervision, R.-Y.G. and H.-B.L. Funding acquisition, R.-Y.G. and H.-B.L.

Funding: The Shanghai Basic and Key Program (No. 18JC1410800), Technology Innovation Program of Chinese Academy of Agricultural Sciences (ASTIP), the Shanghai Pujiang Talent Plan (No. 18PJ1404600), the Agri-X Interdisciplinary Fund of Shanghai Jiao Tong University (No. Agri-X2017004), and the Key Project of Guangdong Provincial Science and Technology Program (No. 2014B020205002) supported this study.

Acknowledgments: Thanks Shi-Yu Cao for her assistance.

Conflicts of Interest: The authors declare no conflicts of interest.

References

1. Zimmet, P.; Alberti, K.G.; Magliano, D.J.; Bennett, P.H. Diabetes mellitus statistics on prevalence and mortality: Facts and fallacies. *Nat. Rev. Endocrinol.* **2016**, *12*, 616–622. [CrossRef] [PubMed]

2. Chen, Y.; Wang, T.; Liu, X.; Shankar, R.R. Prevalence of type 1 and type 2 diabetes among US pediatric population in the marketscan multi-state database, 2002 to 2016. *Pediatr. Diabetes* **2019**, *20*, 523–529. [CrossRef] [PubMed]

3. Rosenbauer, J.; Icks, A.; Giani, G. Incidence and prevalence of childhood type 1 diabetes mellitus in germany–model-based national estimates. *J. Pediatr. Endocrinol. Metab.* **2002**, *15*, 1497–1504. [CrossRef]

4. Zheng, Y.; Ley, S.H.; Hu, F.B. Global aetiology and epidemiology of type 2 diabetes mellitus and its complications. *Nat. Rev. Endocrinol.* **2018**, *14*, 88–98. [CrossRef] [PubMed]

5. King, G.L. The role of inflammatory cytokines in diabetes and its complications. *J. Periodontol.* **2008**, *79*, 1527–1534. [CrossRef] [PubMed]

6. Peltier, A.; Goutman, S.A.; Callaghan, B.C. Painful diabetic neuropathy. *BMJ* **2014**, *348*, g1799. [CrossRef] [PubMed]

7. Wang, P.; Fiaschi-Taesch, N.M.; Vasavada, R.C.; Scott, D.K.; Garcia-Ocana, A.; Stewart, A.F. Diabetes mellitus–advances and challenges in human β-cell proliferation. *Nat. Rev. Endocrinol.* **2015**, *11*, 201–212. [CrossRef] [PubMed]

8. Kleinberger, J.W.; Maloney, K.A.; Pollin, T.I. The genetic architecture of diabetes in pregnancy: Implications for clinical practice. *Am. J. Perinatol.* **2016**, *33*, 1319–1326. [CrossRef] [PubMed]

9. Vana, D.R.; Adapa, D.; Prasad, V.S.S.; Choudhury, A. Diabetes mellitus types: Key genetic determinants and risk assessment. *Genet. Mol. Res.* **2019**, *18*. [CrossRef]

10. Hober, D.; Sauter, P. Pathogenesis of type 1 diabetes mellitus: Interplay between enterovirus and host. *Nat. Rev. Endocrinol.* **2010**, *6*, 279–289. [CrossRef]

11. Roep, B.O.; Tree, T.I. Immune modulation in humans: Implications for type 1 diabetes mellitus. *Nat. Rev. Endocrinol.* **2014**, *10*, 229–242. [CrossRef] [PubMed]

12. Van Belle, T.L.; Coppieters, K.T.; von Herrath, M.G. Type 1 diabetes: Etiology, immunology, and therapeutic strategies. *Physiol. Rev.* **2011**, *91*, 79–118. [CrossRef] [PubMed]

13. DiMeglio, L.A.; Evans-Molina, C.; Oram, R.A. Type 1 diabetes. *Lancet* **2018**, *391*, 2449–2462. [CrossRef]

14. Canfora, E.E.; Meex, R.C.R.; Venema, K.; Blaak, E.E. Gut microbial metabolites in obesity, NAFLD and T2DM. *Nat. Rev. Endocrinol.* **2019**, *15*, 261–273. [CrossRef] [PubMed]

15. Gross, B.; Pawlak, M.; Lefebvre, P.; Staels, B. PPARs in obesity-induced T2DM, dyslipidaemia and NAFLD. *Nat. Rev. Endocrinol.* **2017**, *13*, 36–49. [CrossRef] [PubMed]

16. Kootte, R.S.; Vrieze, A.; Holleman, F.; Dallinga-Thie, G.M.; Zoetendal, E.G.; de Vos, W.M.; Groen, A.K.; Hoekstra, J.B.L.; Stroes, E.S.; Nieuwdorp, M. The therapeutic potential of manipulating gut microbiota in obesity and type 2 diabetes mellitus. *Diabetes Obes. Metab.* **2012**, *14*, 112–120. [CrossRef]

17. Mao, Q.Q.; Xu, X.Y.; Cao, S.Y.; Gan, R.Y.; Corke, H.; Beta, T.; Li, H.B. Bioactive compounds and bioactivities of ginger (*Zingiber officinale* roscoe). *Foods* **2019**, *8*, 185. [CrossRef]

18. Buchanan, T.A.; Xiang, A.H.; Page, K.A. Gestational diabetes mellitus: Risks and management during and after pregnancy. *Nat. Rev. Endocrinol.* **2012**, *8*, 639–649. [CrossRef]

19. Coustan, D.R. Gestational diabetes mellitus. *Clin. Chem.* **2013**, *59*, 1310–1321. [CrossRef]

20. Johns, E.C.; Denison, F.C.; Norman, J.E.; Reynolds, R.M. Gestational diabetes mellitus: Mechanisms, treatment, and complications. *Trends Endocrinol. Metab.* **2018**, *29*, 743–754. [CrossRef]

21. Chakraborty, C.; Bandyopadhyay, S.; Doss, C.G.; Agoramoorthy, G. Exploring the genomic roadmap and molecular phylogenetics associated with mody cascades using computational biology. *Cell Biochem. Biophys.* **2015**, *71*, 1491–1502. [CrossRef] [PubMed]

22. Malecki, M.T. Genetics of type 2 diabetes mellitus. *Diabetes Res. Clin. Pract.* **2005**, *68* (Suppl. 1), S10–S21. [CrossRef]

23. American Diabetes, A. Diagnosis and classification of diabetes mellitus. *Diabetes Care* **2010**, *33* (Suppl. 1), S62–S69. [CrossRef]

24. Tauschmann, M.; Hovorka, R. Technology in the management of type 1 diabetes mellitus - current status and future prospects. *Nat. Rev. Endocrinol.* **2018**, *14*, 464–475. [CrossRef] [PubMed]

25. Meng, J.M.; Cao, S.Y.; Wei, X.L.; Gan, R.Y.; Wang, Y.F.; Cai, S.X.; Xu, X.Y.; Zhang, P.Z.; Li, H.B. Effects and mechanisms of tea for the prevention and management of diabetes mellitus and diabetic complications: An updated review. *Antioxidants* **2019**, *8*, 170. [CrossRef] [PubMed]

26. Xu, L.; Li, Y.; Dai, Y.; Peng, J. Natural products for the treatment of type 2 diabetes mellitus: Pharmacology and mechanisms. *Pharmacol. Res.* **2018**, *130*, 451–465. [CrossRef]

27. Prabhakar, P.K.; Doble, M. Mechanism of action of natural products used in the treatment of diabetes mellitus. *Chin. J. Integr. Med.* **2011**, *17*, 563–574. [CrossRef]

28. Patel, S.S.; Udayabanu, M. Effect of natural products on diabetes associated neurological disorders. *Rev. Neurosci.* **2017**, *28*, 271–293. [CrossRef]

29. Dragan, S.; Andrica, F.; Serban, M.C.; Timar, R. Polyphenols-rich natural products for treatment of diabetes. *Curr. Med. Chem.* **2015**, *22*, 14–22. [CrossRef]

30. He, C.X.; Shan, Y.J.; Song, W. Targeting gut microbiota as a possible therapy for diabetes. *Nutr. Res.* **2015**, *35*, 361–367. [CrossRef]

31. Hu, Y.J.; Wong, F.S.; Wen, L. Antibiotics, gut microbiota, environment in early life and type 1 diabetes. *Pharmacol. Res.* **2017**, *119*, 219–226. [CrossRef] [PubMed]

32. Hu, Y.J.; Peng, J.; Li, F.Y.; Wong, F.S.; Wen, L. Evaluation of different mucosal microbiota leads to gut microbiota-based prediction of type 1 diabetes in NOD mice. *Sci. Rep.* **2018**, *8*, 15451. [CrossRef] [PubMed]

33. Seganfredo, F.B.; Blume, C.A.; Moehlecke, M.; Giongo, A.; Casagrande, D.S.; Spolidoro, J.V.N.; Padoin, A.V.; Schaan, B.D.; Mottin, C.C. Weight-loss interventions and gut microbiota changes in overweight and obese patients: A systematic review. *Obes. Rev.* **2017**, *18*, 832–851. [CrossRef] [PubMed]

34. Bouter, K.E.; van Raalte, D.H.; Groen, A.K.; Nieuwdorp, M. Role of the gut microbiome in the pathogenesis of obesity and obesity-related metabolic dysfunction. *Gastroenterology* **2017**, *152*, 1671–1678. [CrossRef] [PubMed]

35. Sun, L.; Ma, L.; Ma, Y.; Zhang, F.; Zhao, C.; Nie, Y. Insights into the role of gut microbiota in obesity: Pathogenesis, mechanisms, and therapeutic perspectives. *Protein Cell* **2018**, *9*, 397–403. [CrossRef] [PubMed]

36. Cao, S.Y.; Zhao, C.N.; Yu Xu, X.Y.; Tang, G.Y.; Corke, H.; Gan, R.Y.; Li, H.B. Dietary plants, gut microbiota and obesity: Effects and mechanisms. *Trends Food Sci. Technol.* **2019**, *92*, 194–204. [CrossRef]

37. Zhang, Y.J.; Li, S.; Gan, R.Y.; Zhou, T.; Xu, D.P.; Li, H.B. Impacts of gut bacteria on human health and diseases. *Int. J. Mol. Sci.* **2015**, *16*, 7493–7519. [CrossRef] [PubMed]

38. Meng, X.; Li, S.; Li, Y.; Gan, R.Y.; Li, H.B. Gut microbiota's relationship with liver disease and role in hepatoprotection by dietary natural products and probiotics. *Nutrients* **2018**, *10*, 1457. [CrossRef]

39. Hrncir, T.; Hrncirova, L.; Kverka, M.; Tlaskalova-Hogenova, H. The role of gut microbiota in intestinal and liver diseases. *Lab. Anim.* **2019**, *53*, 271–280. [CrossRef]

40. Petrov, P.D.; Garcia-Mediavilla, M.V.; Guzman, C.; Porras, D.; Nistal, E.; Martinez-Florez, S.; Castell, J.V.; Gonzalez-Gallego, J.; Sanchez-Campos, S.; Jover, R. A network involving gut microbiota, circulating bile acids, and hepatic metabolism genes that protects against non-alcoholic fatty liver disease. *Mol. Nutr. Food Res.* **2019**, 1900487. [CrossRef]

41. Ezzaidi, N.; Zhang, X.; Coker, O.O.; Yu, J. New insights and therapeutic implication of gut microbiota in non-alcoholic fatty liver disease and its associated liver cancer. *Cancer Lett.* **2019**, *459*, 186–191. [CrossRef] [PubMed]

42. Li, J.; Zhao, F.; Wang, Y.; Chen, J.; Tao, J.; Tian, G.; Wu, S.; Liu, W.; Cui, Q.; Geng, B.; et al. Gut microbiota dysbiosis contributes to the development of hypertension. *Microbiome* **2017**, *5*, 14. [CrossRef] [PubMed]

43. Tang, W.H.; Kitai, T.; Hazen, S.L. Gut microbiota in cardiovascular health and disease. *Circ. Res.* **2017**, *120*, 1183–1196. [CrossRef] [PubMed]

44. Zuo, K.; Li, J.; Li, K.; Hu, C.; Gao, Y.; Chen, M.; Hu, R.; Liu, Y.; Chi, H.; Wang, H.; et al. Disordered gut microbiota and alterations in metabolic patterns are associated with atrial fibrillation. *GigaScience* **2019**, *8*, giz058. [CrossRef] [PubMed]

45. Tao, J.; Li, Y.; Li, S.; Li, H.-B. Plant foods for the prevention and management of colon cancer. *J. Funct. Food.* **2018**, *42*, 95–110. [CrossRef]

46. Xu, X.Y.; Zhao, C.N.; Cao, S.Y.; Tang, G.Y.; Gan, R.Y.; Li, H.B. Effects and mechanisms of tea for the prevention and management of cancers: An updated review. *Crit. Rev. Food. Sci. Nutr.* **2019**, 1–13. [CrossRef]

47. Li, X.; Watanabe, K.; Kimura, I. Gut microbiota dysbiosis drives and implies novel therapeutic strategies for diabetes mellitus and related metabolic diseases. *Front. Immunol.* **2017**, *8*, 1882. [CrossRef] [PubMed]

48. Li, J.; Lin, S.; Vanhoutte, P.M.; Woo, C.W.; Xu, A. *Akkermansia muciniphila* protects against atherosclerosis by preventing metabolic endotoxemia-induced inflammation in Apoe(−/−) mice. *Circulation* **2016**, *133*, 2434–2446. [CrossRef] [PubMed]

49. Goldacre, R.R. Associations between birthweight, gestational age at birth and subsequent type 1 diabetes in children under 12: A retrospective cohort study in England, 1998–2012. *Diabetologia* **2018**, *61*, 616–625. [CrossRef]

50. Hanninen, A.; Toivonen, R.; Poysti, S.; Belzer, C.; Plovier, H.; Ouwerkerk, J.P.; Emani, R.; Cani, P.D.; De Vos, W.M. *Akkermansia muciniphila* induces gut microbiota remodelling and controls islet autoimmunity in NOD mice. *Gut* **2018**, *67*, 1445–1453. [CrossRef]

51. Moossavi, S.; Bishehsari, F. Microbes: Possible link between modern lifestyle transition and the rise of metabolic syndrome. *Obes. Rev.* **2019**, *20*, 407–419. [CrossRef] [PubMed]

52. Cao, Y.; Yao, G.W.; Sheng, Y.Y.; Yang, L.; Wang, Z.X.; Yang, Z.; Zhuang, P.W.; Zhang, Y.J. Jinqi jiangtang tablet regulates gut microbiota and improve insulin sensitivity in type 2 diabetes mice. *J. Diabetes Res.* **2019**, *2019*, 1872134. [CrossRef] [PubMed]

53. Han, H.; Li, Y.Y.; Fang, J.; Liu, G.; Yin, J.; Li, T.J.; Yin, Y.L. Gut microbiota and type 1 diabetes. *Int. J. Mol. Sci.* **2018**, *19*, 995. [CrossRef] [PubMed]

54. Gulden, E.; Wong, F.S.; Wen, L. The gut microbiota and type 1 diabetes. *Clin. Immunol.* **2015**, *159*, 143–153. [CrossRef] [PubMed]

55. Wang, G.; Li, X.F.; Zhao, J.X.; Zhang, H.; Chen, W. *Lactobacillus casei* CCFM419 attenuates type 2 diabetes via a gut microbiota dependent mechanism. *Food Funct.* **2017**, *8*, 3155–3164. [CrossRef] [PubMed]

56. Everard, A.; Matamoros, S.; Geurts, L.; Delzenne, N.M.; Cani, P.D. *Saccharomyces boulardii* administration changes gut microbiota and reduces hepatic steatosis, low-grade inflammation, and fat mass in obese and type 2 diabetic *db/db* mice. *Mbio* **2014**, *5*, e01011-14. [CrossRef] [PubMed]

57. Jia, L.L.; Li, D.Y.; Feng, N.H.; Shamoon, M.; Sun, Z.H.; Ding, L.; Zhang, H.; Chen, W.; Sun, J.; Chen, Y.Q. Anti-diabetic effects of *Clostridium butyricum* CGMCC0313.1 through promoting the growth of gut butyrate-producing bacteria in type 2 diabetic mice. *Sci. Rep.* **2017**, *7*, 7046. [CrossRef] [PubMed]

58. Amar, J.; Chabo, C.; Waget, A.; Klopp, P.; Vachoux, C.; Bermudez-Humaran, L.G.; Smirnova, N.; Berge, M.; Sulpice, T.; Lahtinen, S.; et al. Intestinal mucosal adherence and translocation of commensal bacteria at the early onset of type 2 diabetes: Molecular mechanisms and probiotic treatment. *EMBO Mol. Med.* **2011**, *3*, 559–572. [CrossRef]

59. Peng, J.; Narasimhan, S.; Marchesi, J.R.; Benson, A.; Wong, F.S.; Wen, L. Long term effect of gut microbiota transfer on diabetes development. *J. Autoimmun.* **2014**, *53*, 85–94. [CrossRef]

60. Plotkin, B.J.; Paulson, D.; Chelich, A.; Jurak, D.; Cole, J.; Kasimos, J.; Burdick, J.R.; Casteel, N. Immune responsiveness in a rat model for type ii diabetes (Zucker rat, fa/fa): Susceptibility to candida albicans infection and leucocyte function. *J. Med. Microbiol.* **1996**, *44*, 277–283. [CrossRef]

61. Raffel, L.; Pitsakis, P.; Levison, S.P.; Levison, M.E. Experimental candida albicans, staphylococcus aureus, and streptococcus faecalis pyelonephritis in diabetic rats. *Infect. Immun.* **1981**, *34*, 773–779. [PubMed]

62. Tang, G.Y.; Meng, X.; Li, Y.; Zhao, C.N.; Liu, Q.; Li, H.B. Effects of vegetables on cardiovascular diseases and related mechanisms. *Nutrients* **2017**, *9*, 857. [CrossRef]

63. Zhao, C.N.; Meng, X.; Li, Y.; Li, S.; Liu, Q.; Tang, G.Y.; Li, H.B. Fruits for prevention and treatment of cardiovascular diseases. *Nutrients* **2017**, *9*, 598. [CrossRef] [PubMed]

64. Meng, X.; Li, Y.; Li, S.; Gan, R.-Y.; Li, H.-B. Natural products for prevention and treatment of chemical-induced liver injuries. *Compr. Rev. Food. Sci. Food Saf.* **2018**, *17*, 472–495. [CrossRef]

65. Shang, A.; Cao, S.Y.; Xu, X.Y.; Gan, R.Y.; Tang, G.Y.; Corke, H.; Mavumengwana, V.; Li, H.B. Bioactive compounds and biological functions of garlic (*Allium sativum* L.). *Foods* **2019**, *8*, 246. [CrossRef] [PubMed]

66. Cao, S.Y.; Li, Y.; Meng, X.; Zhao, C.N.; Li, S.; Gan, R.Y.; Li, H.B. Dietary natural products and lung cancer: Effects and mechanisms of action. *J. Funct. Food.* **2019**, *52*, 316–331. [CrossRef]

67. Liu, Q.; Meng, X.; Li, Y.; Zhao, C.N.; Tang, G.Y.; Li, S.; Gan, R.Y.; Li, H.-B. Natural products for the prevention and management of *Helicobacter pylori* infection. *Compr. Rev. Food. Sci. Food Saf.* **2018**, *17*, 937–952. [CrossRef]

68. Li, Y.; Li, S.; Meng, X.; Gan, R.Y.; Zhang, J.J.; Li, H.B. Dietary natural products for prevention and treatment of breast cancer. *Nutrients* **2017**, *9*, 728. [CrossRef]

69. Zhou, Y.; Li, Y.; Zhou, T.; Zheng, J.; Li, S.; Li, H.B. Dietary natural products for prevention and treatment of liver cancer. *Nutrients* **2016**, *8*, 156. [CrossRef]

70. Gan, R.Y.; Li, H.B.; Sui, Z.Q.; Corke, H. Absorption, metabolism, anti-cancer effect and molecular targets of epigallocatechin gallate (EGCG): An updated review. *Crit. Rev. Food. Sci. Nutr.* **2018**, *58*, 924–941. [CrossRef]

71. Xu, X.Y.; Meng, X.; Li, S.; Gan, R.Y.; Li, Y.; Li, H.B. Bioactivity, health benefits, and related molecular mechanisms of curcumin: Current progress, challenges, and perspectives. *Nutrients* **2018**, *10*, 1553. [CrossRef] [PubMed]

72. Arora, T.; Backhed, F. The gut microbiota and metabolic disease: Current understanding and future perspectives. *J. Intern. Med.* **2016**, *280*, 339–349. [CrossRef] [PubMed]

73. Munch, N.S.; Fang, H.Y.; Ingermann, J.; Maurer, H.C.; Anand, A.; Kellner, V.; Sahm, V.; Wiethaler, M.; Baumeister, T.; Wein, F.; et al. High-fat diet accelerates carcinogenesis in a mouse model of barrett's esophagus via interleukin 8 and alterations to the gut microbiome. *Gastroenterology* **2019**, *157*, 492–495. [CrossRef] [PubMed]

74. Tadokoro, Y.; Hoshii, T.; Yamazaki, S.; Eto, K.; Ema, H.; Kobayashi, M.; Ueno, M.; Ohta, K.; Arai, Y.; Hara, E.; et al. Spred1 safeguards hematopoietic homeostasis against diet-induced systemic stress. *Cell stem cell* **2018**, *22*, 713–718. [CrossRef] [PubMed]

75. Liu, G.M.; Liang, L.; Yu, G.Y.; Li, Q.H. Pumpkin polysaccharide modifies the gut microbiota during alleviation of type 2 diabetes in rats. *Int. J. Biol. Macromol.* **2018**, *115*, 711–717. [CrossRef] [PubMed]

76. Han, J.L.; Lin, H.L.; Huang, W.P. Modulating gut microbiota as an anti-diabetic mechanism of berberine. *Med. Sci. Monit.* **2011**, *17*, RA164–RA167. [CrossRef] [PubMed]

77. Han, L.H.; Li, T.G.; Du, M.; Chang, R.; Zhan, B.Y.; Mao, X.Y. Beneficial effects of *Potentilla discolor* Bunge water extract on inflammatory cytokines release and gut microbiota in high-fat diet and streptozotocin-induced type 2 diabetic mice. *Nutrients* **2019**, *11*, 670. [CrossRef]

78. Zhang, B.W.; Sun, W.L.; Yu, N.; Sun, J.; Yu, X.X.; Li, X.; Xing, Y.; Yan, D.; Ding, Q.Z.; Xiu, Z.L.; et al. Anti-diabetic effect of baicalein is associated with the modulation of gut microbiota in streptozotocin and high-fat-diet induced diabetic rats. *J. Funct. Food.* **2018**, *46*, 256–267. [CrossRef]

79. Song, J.X.; Ren, H.; Gao, Y.F.; Lee, C.Y.; Li, S.F.; Zhang, F.; Li, L.; Chen, H. Dietary capsaicin improves glucose homeostasis and alters the gut microbiota in obese diabetic *ob/ob* mice. *Front. Physiol.* **2017**, *8*, 602. [CrossRef]

80. Hu, R.K.; Zeng, F.; Wu, L.X.; Wan, X.Z.; Chen, Y.F.; Zhang, J.C.; Liu, B. Fermented carrot juice attenuates type 2 diabetes by mediating gut microbiota in rats. *Food Funct.* **2019**, *10*, 2935–2946. [CrossRef]

81. Alam, F.; Islam, M.A.; Kamal, M.A.; Gan, S.H. Updates on managing type 2 diabetes mellitus with natural products: Towards antidiabetic drug development. *Curr. Med. Chem.* **2018**, *25*, 5395–5431. [CrossRef] [PubMed]

82. Buchholz, T.; Melzig, M.F. Polyphenolic compounds as pancreatic lipase inhibitors. *Planta Med.* **2015**, *81*, 771–783. [CrossRef]

83. Bibbo, S.; Dore, M.P.; Pes, G.M.; Delitala, G.; Delitala, A.P. Is there a role for gut microbiota in type 1 diabetes pathogenesis? *Ann. Med.* **2017**, *49*, 11–22. [CrossRef] [PubMed]

84. Sircana, A.; Framarin, L.; Leone, N.; Berrutti, M.; Castellino, F.; Parente, R.; De Michieli, F.; Paschetta, E.; Musso, G. Altered gut microbiota in type 2 diabetes: Just a coincidence? *Curr. Diabetes Rep.* **2018**, *18*, 98. [CrossRef] [PubMed]

85. Vatanen, T.; Franzosa, E.A.; Schwager, R.; Tripathi, S.; Arthur, T.D.; Vehik, K.; Lernmark, A.; Hagopian, W.A.; Rewers, M.J.; She, J.X.; et al. The human gut microbiome in early-onset type 1 diabetes from the teddy study. *Nature* **2018**, *562*, 589–594. [CrossRef] [PubMed]

86. Gulden, E.; Palm, N.; Herold, K.C. Mait cells: A link between gut integrity and type 1 diabetes. *Cell. Metab.* **2017**, *26*, 813–815. [CrossRef] [PubMed]

87. Sanna, S.; van Zuydam, N.R.; Mahajan, A.; Kurilshikov, A.; Vich Vila, A.; Vosa, U.; Mujagic, Z.; Masclee, A.A.M.; Jonkers, D.; Oosting, M.; et al. Causal relationships among the gut microbiome, short-chain fatty acids and metabolic diseases. *Nat. Genet.* **2019**, *51*, 600–605. [CrossRef] [PubMed]

88. Wen, L.; Ley, R.E.; Volchkov, P.Y.; Stranges, P.B.; Avanesyan, L.; Stonebraker, A.C.; Hu, C.; Wong, F.S.; Szot, G.L.; Bluestone, J.A.; et al. Innate immunity and intestinal microbiota in the development of type 1 diabetes. *Nature* **2008**, *455*, 1109–1113. [CrossRef] [PubMed]

89. Wirth, R.; Bodi, N.; Maroti, G.; Bagyanszki, M.; Talapka, P.; Fekete, E.; Bagi, Z.; Kovacs, K.L. Regionally distinct alterations in the composition of the gut microbiota in rats with streptozotocin-induced diabetes. *PLoS ONE* **2014**, *9*, e110440. [CrossRef]

90. Gulden, E.; Chao, C.; Tai, N.W.; Pearson, J.A.; Peng, J.; Majewska-Szczepanik, M.; Zhou, Z.G.; Wong, F.S.; Wen, L. TRIF deficiency protects non-obese diabetic mice from type 1 diabetes by modulating the gut microbiota and dendritic cells. *J. Autoimmun.* **2018**, *93*, 57–65. [CrossRef] [PubMed]

91. Miani, M.; Le Naour, J.; Waeckel-Enee, E.; Verma, S.C.; Straube, M.; Emond, P.; Ryffel, B.; Van Endert, P.; Sokol, H.; Diana, J. Gut microbiota-stimulated innate lymphoid cells support beta-defensin 14 expression in pancreatic endocrine cells, preventing autoimmune diabetes. *Cell Metab.* **2018**, *28*, 557–562. [CrossRef] [PubMed]

92. De Goffau, M.C.; Fuentes, S.; van den Bogert, B.; Honkanen, H.; de Vos, W.M.; Welling, G.W.; Hyoty, H.; Harmsen, H.J.M. Aberrant gut microbiota composition at the onset of type 1 diabetes in young children. *Diabetologia* **2014**, *57*, 1569–1577. [CrossRef] [PubMed]

93. Salamon, D.; Sroka-Oleksiak, A.; Kapusta, P.; Szopa, M.; Mrozinska, S.; Ludwig-Slomczynska, A.H.; Wolkow, P.P.; Bulanda, M.; Klupa, T.; Malecki, M.T.; et al. Characteristics of gut microbiota in adult patients with type 1 and type 2 diabetes based on next-generation sequencing of the 16s rRNA gene fragment. *Pol. Arch. Intern. Med.* **2018**, *128*, 336–343. [CrossRef] [PubMed]

94. Huang, Y.; Li, S.C.; Hu, J.; Ruan, H.B.; Guo, H.M.; Zhang, H.H.; Wang, X.; Pei, Y.F.; Pan, Y.; Fang, C. Gut microbiota profiling in Han Chinese with type 1 diabetes. *Diabetes Res. Clin. Pract.* **2018**, *141*, 256–263. [CrossRef] [PubMed]

95. Murri, M.; Leiva, I.; Gomez-Zumaquero, J.M.; Tinahones, F.J.; Cardona, F.; Soriguer, F.; Queipo-Ortuno, M.I. Gut microbiota in children with type 1 diabetes differs from that in healthy children: A case-control study. *BMC Med.* **2013**, *11*, 46. [CrossRef] [PubMed]

96. Soyucen, E.; Gulcan, A.; Aktuglu-Zeybek, A.C.; Onal, H.; Kiykim, E.; Aydin, A. Differences in the gut microbiota of healthy children and those with type 1 diabetes. *Pediatr. Int.* **2014**, *56*, 336–343. [CrossRef] [PubMed]

97. Geurts, L.; Lazarevic, V.; Derrien, M.; Everard, A.; Van Roye, M.; Knauf, C.; Valet, P.; Girard, M.; Muccioli, G.G.; Francois, P.; et al. Altered gut microbiota and endocannabinoid system tone in obese and diabetic leptin-resistant mice: Impact on apelin regulation in adipose tissue. *Front. Microbiol.* **2011**, *2*, 149. [CrossRef]

98. Ohtsu, A.; Takeuchi, Y.; Katagiri, S.; Suda, W.; Maekawa, S.; Shiba, T.; Komazaki, R.; Udagawa, S.; Sasaki, N.; Hattori, M.; et al. Influence of *Porphyromonas gingivalis* in gut microbiota of streptozotocin-induced diabetic mice. *Oral Dis.* **2019**, *25*, 868–880. [CrossRef]

99. Kieler, I.N.; Osto, M.; Hugentobler, L.; Puetz, L.; Gilbert, M.T.P.; Hansen, T.; Pedersen, O.; Reusch, C.E.; Zini, E.; Lutz, T.A.; et al. Diabetic cats have decreased gut microbial diversity and a lack of butyrate producing bacteria. *Sci. Rep.* **2019**, *9*, 4822. [CrossRef]

100. Kameyama, K.; Itoii, K. Intestinal colonization by a *Lachnospiraceae* bacterium contributes to the development of diabetes in obese mice. *Microbes Environ.* **2014**, *29*, 427–430. [CrossRef]

101. Larsen, N.; Vogensen, F.K.; van den Berg, F.W.J.; Nielsen, D.S.; Andreasen, A.S.; Pedersen, B.K.; Abu Al-Soud, W.; Sorensen, S.J.; Hansen, L.H.; Jakobsen, M. Gut microbiota in human adults with type 2 diabetes differs from NOD-diabetic adults. *PLoS ONE* **2010**, *5*, e9085. [CrossRef] [PubMed]

102. Navab-Moghadam, F.; Sedighi, M.; Khamseh, M.E.; Alaei-Shahmiri, F.; Talebi, M.; Razavi, S.; Amirmozafari, N. The association of type II diabetes with gut microbiota composition. *Microb. Pathog.* **2017**, *110*, 630–636. [CrossRef] [PubMed]

103. Inoue, R.; Ohue-Kitano, R.; Tsukahara, T.; Tanaka, M.; Masuda, S.; Inoue, T.; Yamakage, H.; Kusakabe, T.; Hasegawa, K.; Shimatsu, A.; et al. Prediction of functional profiles of gut microbiota from 165 rRNA metagenomic data provides a more robust evaluation of gut dysbiosis occurring in Japanese type 2 diabetic patients. *J. Clin. Biochem. Nutr.* **2017**, *61*, 217–221. [CrossRef] [PubMed]

104. Suceveanu, A.I.; Stoian, A.P.; Parepa, I.; Voinea, C.; Hainarosie, R.; Manuc, D.; Nitipir, C.; Mazilu, L.; Suceveanu, A.P. Gut microbiota patterns in obese and type 2 diabetes (T2D) patients from Romanian Black Sea Coast Region. *Rev. Chim.* **2018**, *69*, 2260–2267.

105. Chen, C.; You, L.J.; Huang, Q.; Fu, X.; Zhang, B.; Liu, R.H.; Li, C. Modulation of gut microbiota by mulberry fruit polysaccharide treatment of obese diabetic db/db mice. *Food Funct.* **2018**, *9*, 3732–3742. [CrossRef] [PubMed]

106. Chen, K.; Chen, H.; Faas, M.M.; de Haan, B.J.; Li, J.H.; Xiao, P.; Zhang, H.; Diana, J.; de Vos, P.; Sun, J. Specific inulin-type fructan fibers protect against autoimmune diabetes by modulating gut immunity, barrier function, and microbiota homeostasis. *Mol. Nutr. Food Res.* **2017**, *61*, 1601006. [CrossRef] [PubMed]

107. Yan, X.; Yang, C.F.; Lin, G.P.; Chen, Y.Q.; Miao, S.; Liu, B.; Zhao, C. Antidiabetic potential of green seaweed *Enteromorpha prolifera* flavonoids regulating insulin signaling pathway and gut microbiota in type 2 diabetic mice. *J. Food Sci.* **2019**, *84*, 165–173. [CrossRef] [PubMed]

108. Zhang, Q.; Yu, H.Y.; Xiao, X.H.; Hu, L.; Xin, F.J.; Yu, X.B. Inulin-type fructan improves diabetic phenotype and gut microbiota profiles in rats. *PeerJ* **2018**, *6*, e4446. [CrossRef]

109. Gao, H.; Wen, J.J.; Hu, J.L.; Nie, Q.X.; Chen, H.H.; Xiong, T.; Nie, S.P.; Xie, M.Y. Fermented *Momordica charantia* L. Juice modulates hyperglycemia, lipid profile, and gut microbiota in type 2 diabetic rats. *Food Res. Int.* **2019**, *121*, 367–378. [CrossRef]

110. Mei, X.R.; Zhang, X.Y.; Wang, Z.G.; Gao, Z.Y.; Liu, G.; Hu, H.L.; Zou, L.; Li, X.L. Insulin sensitivity-enhancing activity of phlorizin is associated with lipopolysaccharide decrease and gut microbiota changes in obese and type 2 diabetes (*db/db*) mice. *J. Agric. Food Chem.* **2016**, *64*, 7502–7511. [CrossRef]

111. Van Hul, M.; Geurts, L.; Plovier, H.; Druart, C.; Everard, A.; Stahlman, M.; Rhimi, M.; Chira, K.; Teissedre, P.L.; Delzenne, N.M.; et al. Reduced obesity, diabetes, and steatosis upon cinnamon and grape pomace are associated with changes in gut microbiota and markers of gut barrier. *Am. J. Physiol. Endocrinol. Metab.* **2018**, *314*, E334–E352. [CrossRef] [PubMed]

112. Li, K.; Zhang, L.; Xue, J.; Yang, X.L.; Dong, X.Y.; Sha, L.P.; Lei, H.; Zhang, X.X.; Zhu, L.L.; Wang, Z.; et al. Dietary inulin alleviates diverse stages of type 2 diabetes mellitus via anti-inflammation and modulating gut microbiota in *db/db* mice. *Food Funct.* **2019**, *10*, 1915–1927. [CrossRef] [PubMed]

113. Zhang, W.Y.; Zhang, H.H.; Yu, C.H.; Fang, J.; Ying, H.Z. Ethanol extract of *Atractylodis macrocephalae* rhizoma ameliorates insulin resistance and gut microbiota in type 2 diabetic *db/db* mice. *J. Funct. Food.* **2017**, *39*, 139–151. [CrossRef]

114. Xie, Y.Q.; Xiao, M.; Ni, Y.L.; Jiang, S.F.; Feng, G.Z.; Sang, S.G.; Du, G.K. *Alpinia oxyphylla* Miq. extract prevents diabetes in mice by modulating gut microbiota. *J. Diabetes Res.* **2018**, *2018*, 4230590. [CrossRef] [PubMed]

115. Yan, H.L.; Lu, J.M.; Wang, Y.F.; Gu, W.; Yang, X.X.; Yu, J. Intake of total saponins and polysaccharides from *Polygonatum kingianum* affects the gut microbiota in diabetic rats. *Phytomedicine* **2017**, *26*, 45–54. [CrossRef] [PubMed]

116. Singh, D.K.; Winocour, P.; Farrington, K. Oxidative stress in early diabetic nephropathy: Fueling the fire. *Nat. Rev. Endocrinol.* **2011**, *7*, 176–184. [CrossRef] [PubMed]

117. Thorve, V.S.; Kshirsagar, A.D.; Vyawahare, N.S.; Joshi, V.S.; Ingale, K.G.; Mohite, R.J. Diabetes-induced erectile dysfunction: Epidemiology, pathophysiology and management. *J. Diabetes Complicat.* **2011**, *25*, 129–136. [CrossRef]

118. Li, H.; Qi, T.; Huang, Z.S.; Ying, Y.; Zhang, Y.; Wang, B.; Ye, L.; Zhang, B.; Chen, D.L.; Chen, J. Relationship between gut microbiota and type 2 diabetic erectile dysfunction in Sprague-Dawley rats. *J. Huazhong Univ. Sci. Tech. Med. Sci.* **2017**, *37*, 523–530. [CrossRef]

119. Li, X.W.; Chen, H.P.; He, Y.Y.; Chen, W.L.; Chen, J.W.; Gao, L.; Hu, H.Y.; Wang, J. Effects of rich-polyphenols extract of *Dendrobium loddigesii* on anti-diabetic, anti-inflammatory, anti-oxidant, and gut microbiota modulation in *db/db* mice. *Molecules* **2018**, *23*, 3245. [CrossRef]

120. Roshanravan, N.; Mahdavi, R.; Jafarabadi, M.A.; Alizadeh, E.; Ghavami, A.; Saadat, Y.R.; Alamdari, N.M.; Dastouri, M.R.; Alipour, S.; Ostadrahimi, A. The effects of sodium butyrate and high-performance inulin supplementation on the promotion of gut bacterium *Akkermansia muciniphila* growth and alterations in miR-375 and KLF5 expression in type 2 diabetic patients: A randomized, double-blind, placebo-controlled trial. *Eur. J. Integr. Med.* **2018**, *18*, 1–7.

121. Xu, J.; Lian, F.M.; Zhao, L.H.; Zhao, Y.F.; Chen, X.Y.; Zhang, X.; Guo, Y.; Zhang, C.H.; Zhou, Q.; Xue, Z.S.; et al. Structural modulation of gut microbiota during alleviation of type 2 diabetes with a chinese herbal formula. *ISME J.* **2015**, *9*, 552–562. [CrossRef] [PubMed]

122. Candela, M.; Biagi, E.; Soverini, M.; Consolandi, C.; Quercia, S.; Severgnini, M.; Peano, C.; Turroni, S.; Rampelli, S.; Pozzilli, P.; et al. Modulation of gut microbiota dysbioses in type 2 diabetic patients by macrobiotic Ma-Pi 2 diet. *Br. J. Nutr.* **2016**, *116*, 80–93. [CrossRef] [PubMed]

123. Tong, X.L.; Xu, J.; Lian, F.M.; Yu, X.T.; Zhao, Y.F.; Xu, L.P.; Zhang, M.H.; Zhao, X.Y.; Shen, J.; Wu, S.P.; et al. Structural alteration of gut microbiota during the amelioration of human type 2 diabetes with hyperlipidemia by metformin and a traditional Chinese herbal formula: A Multicenter, Randomized, Open label Clinical Trial. *Mbio* **2018**, *9*, e02392-17. [CrossRef] [PubMed]

MDPI

St. Alban-Anlage 66

4052 Basel

Switzerland

Tel. +41 61 683 77 34

Fax +41 61 302 89 18

www.mdpi.com

Foods Editorial Office

E-mail: foods@mdpi.com

www.mdpi.com/journal/foods